Power Electronics with MATLAB

Power electronics play an important role in the functioning of AC drives, semiconductors, power supplies, converters and inverters. The aim of studying the tools and techniques of power electronics is to obtain familiarity with advanced electronic applications and systems. With this aim in mind, this textbook is designed specifically for undergraduate and graduate students of electrical and electronics engineering, and electronics and communication engineering. It presents the fundamental concepts of power electronics including semiconductor devices, rectifiers, inverters, converters, choppers and cyclo-converters. It provides a comprehensive discussion on power electronic drives and electrical circuits. The latest developments including diac, diode, light-emitting diode, thyristors, power MOSFET and static induction transistor are also discussed in detail.

The book elaborates the essential concepts with the MATLAB examples and simulations. Plenty of applications based on MATLAB models using fuzzy logic and neural networks are interspersed throughout the text. Numerous solved examples, practice questions, objective type questions and review questions are provided at the end of each chapter.

The text can also serve as a reference book for researchers who need preliminary knowledge of the design and investigation of power electronic circuits.

L. Ashok Kumar is Professor in the Department of Electrical and Electronics Engineering at PSG College of Technology, Coimbatore. He has authored six books and has completed sixteen Government of India funded projects and is currently working on five projects. His areas of interest include wearable electronics, power electronics and drives, smart grids, solar PV and wind energy systems.

A. Kalaiarasi is Assistant Professor in the Department of Electrical and Electronics Engineering, RVS College of Engineering and Technology, Coimbatore. Her areas of interest are medical instrumentation, power electronics and drives, big data analytics, embedded system technologies and industrial automation.

Y. Uma Maheswari is Technology Manager at Pramura Software Private Limited, Coimbatore. She has around fifteen years of industrial experience. Her expertise is in PCB designing and simulation software.

Power Electronics with MATLAB®

L. Ashok Kumar
A. Kalaiarasi
Y. Uma Maheswari

CAMBRIDGE
UNIVERSITY PRESS

University Printing House, Cambridge CB2 8BS, United Kingdom

One Liberty Plaza, 20th Floor, New York, NY 10006, USA

477 Williamstown Road, Port Melbourne, VIC 3207, Australia

314 to 321, 3rd Floor, Plot No.3, Splendor Forum, Jasola District Centre, New Delhi 110025, India

79 Anson Road, #06–04/06, Singapore 079906

Cambridge University Press is part of the University of Cambridge.

It furthers the University's mission by disseminating knowledge in the pursuit of education, learning and research at the highest international levels of excellence.

www.cambridge.org
Information on this title: www.cambridge.org/9781316642313

© Cambridge University Press 2018

This publication is in copyright. Subject to statutory exception and to the provisions of relevant collective licensing agreements, no reproduction of any part may take place without the written permission of Cambridge University Press.

First published 2018

Printed in India at Rajkamal Electric Press, Kundli, Haryana.

A catalogue record for this publication is available from the British Library

ISBN 978-1-316-64231-3 Paperback

Additional resources for this publication at www.cambridge.org/9781316642313

Cambridge University Press has no responsibility for the persistence or accuracy of URLs for external or third-party internet websites referred to in this publication, and does not guarantee that any content on such websites is, or will remain, accurate or appropriate.

MATLAB and Simulink are registered trademarks of The MathWorks, Inc. See www.mathworks.com/trademarks for a list of additional trademarks. The MathWorks Publisher Logo identifies books that contain "MATLAB®" and "Simulink®" content. Used with permission.
The MathWorks does not warrant the accuracy of the text or exercises in this book. This book's use or discussion of "MATLAB®" and/or "Simulink®" software or related products does not constitute endorsement or sponsorship by The MathWorks of a particular use of the "MATLAB®" and/or "Simulink®" software or related products.

To our families

Contents

Figures	xv
Tables	xxv
Acknowledgements	xxvii
Preface	xxix

1. Introduction to MATLAB

1.1	Introduction and Outlook		1
1.2	How to Start with MATLAB?		2
	1.2.1	Installing and activation	3
1.3	MATLAB: A Calculator		4
	1.3.1	Basic arithmetic operations	5
	1.3.2	Assigning values to variables	6
1.4	Basic Features of MATLAB		10
	1.4.1	Investigation of a MATLAB function	10
	1.4.2	Mathematical functions	10
	1.4.3	Vector and matrix operations	12
	1.4.4	Arrays	21
	1.4.5	Basic plotting	24
1.5	Programming with MATLAB		26
	1.5.1	Creating M-files	26
	1.5.2	M-file functions	27
	1.5.3	Control structures and operators	28
	1.5.4	Debugging M-files	31
	1.5.5	Creating plots	32

1.6	Circuit Descriptions		33
	1.6.1	Format and layout	33
	1.6.2	Electrical circuit description	34
	1.6.3	Simulink library browser	34
	1.6.4	Circuit elements	39
	1.6.5	DC analysis	41
	1.6.6	AC analysis	42
1.7	Examples of MATLAB Simulations		45
	1.7.1	Steady state analysis of a linear circuit	45
	1.7.2	Resonant switch converter using metal oxide semiconductor field effect transistor (MOSFET)	46
	1.7.3	Gate turn off (GTO) thyristor-based converter	47
	1.7.4	Regulation of zener diode	49
	1.7.5	Regulation of pulse generator using thyristor converter	50
1.8	Other Types Circuit Simulators		52
	1.8.1	PSpice	52
	1.8.2	LabVIEW	52
	1.8.3	PSIM	52
	1.8.4	Scilab	53
	1.8.5	VisSim	53
1.9	Merits and Demerits of MATLAB		54
	1.9.1	Merits	54
	1.9.2	Demerits	54
Summary			54
Review Questions			54
Practice Questions			55
Multiple Choice Questions			57

2. MATLAB Simulation of Power Semiconductor Devices

2.1	Introduction and Outlook		60
2.2	Why is Power Electronics Important?		61
2.3	Features of Power Electronics		61
2.4	Applications of Power Electronics		63
2.5	Power Semiconductor Devices in MATLAB/Simulink		64
	2.5.1	Power diode and its characteristics	66
	2.5.2	Zener diode	73
	2.5.3	Fast recovery diode	75
	2.5.4	Thyristors	76
	2.5.5	Power MOSFET	100

				Page
	2.5.6	Gate turn off thyristors		102
	2.5.7	Insulated-gate bipolar transistor (IGBT)		105
2.6	Other Semiconductor Devices			107
	2.6.1	DIAC		107
	2.6.2	TRIAC		107
	2.6.3	MOS controlled thyristor		108
	2.6.4	Integrated gate-commutated thyristors		108
2.7	MATLAB/Simulink Model of Semiconductor Devices in Electronics			109
	2.7.1	Schottky diode		109
	2.7.2	Bipolar junction transistors		111
	2.7.3	MOSFET		113
	2.7.4	IGBT		114
2.8	Gate Triggering Methods			116
	2.8.1	Resistance firing circuit		116
	2.8.2	Resistance–capacitance firing circuit		117
	2.8.3	UJT firing circuit		118
	2.8.4	Pulse transformers		119
	2.8.5	Optocoupler		119
	2.8.6	Ramp-pedestrial triggering		120
2.9	Comparison of Power Semiconductor Devices with Industry Applications			120
	2.9.1	Other devices		122
Summary				*125*
Solved Examples				*125*
Practice Questions				*141*
Review Questions				*142*
Multiple Choice Questions				*143*

3. Phase-Controlled Rectifiers Using MATLAB (AC–DC Converters)

3.1	Introduction			146
3.2	Rectification and Its Classification			147
	3.2.1	Based on control characteristics		147
	3.2.2	Based on period of conduction		152
	3.2.3	Based on number of phases		153
	3.2.4	Based on number of pulses		153
3.3	Selection of Components from the Simulink Library Browser			153
3.4	One Pulse Converters			155
	3.4.1	Single-phase half-wave-controlled rectifiers		155
3.5	Two Pulse Converters			163

		3.5.1	Single-phase full-wave bridge rectifiers	163

		3.5.1	Single-phase full-wave bridge rectifiers	163
		3.5.2	Single-phase midpoint bridge rectifiers	169
		3.5.3	Single-phase semiconverter half-controlled bridge rectifiers	173
	3.6	Three Pulse Converters		178
		3.6.1	Three-phase half-wave-controlled rectifiers	178
		3.6.2	Three-phase half-controlled bridge rectifier with *RL* load	181
	3.7	Six Pulse Converters		183
		3.7.1	Six pulse converter with *R* load	183
		3.7.2	Six pulse converter with *RL* load	186
	3.8	Dual Converter		187
	3.9	Role of Source Inductance in Rectifier Circuits (L_s)		188
	3.10	Applications of Controlled Rectifiers		189
	Summary			*189*
	Solved Examples			*189*
	Objective Type Questions			*200*
	Review Questions			*203*
	Practice Questions			*204*
4.	**DC Choppers Using MATLAB (DC–DC Converters)**			
	4.1	Introduction		206
	4.2	Choppers and their Classification		207
	4.3	Control Strategies of Chopper		208
		4.3.1	Pulse width modulation or constant frequency system	208
		4.3.2	Variable frequency control or frequency modulation	208
		4.3.3	Current limit control	209
	4.4	Selection of Components from the Simulink Library Browser		209
	4.5	Principle of Operation of a Step-down Chopper		211
	4.6	Principle of Operation of a Step-up Chopper		212
	4.7	Performance Parameters of Step-up and Step-down Choppers		213
	4.8	Chopper Configuration		215
		4.8.1	Type A chopper	216
		4.8.2	Type B chopper	217
		4.8.3	Type C chopper (regenerative chopper)	220
		4.8.4	Type D chopper	222
		4.8.5	Type E chopper	224
	4.9	Switching Mode Regulators		226
		4.9.1	Buck converter	226
		4.9.2	Boost converter	228
		4.9.3	Buck–boost converter	230

	4.9.4 Cuk converter	231
4.10	Chopper Commutation	231
	4.10.1 Voltage-commutated chopper	231
	4.10.2 Current-commutated chopper	234
	4.10.3 Load-commutated chopper	236
4.11	Jones Chopper	237
4.12	Morgan Chopper	238
4.13	AC Choppers	239
4.14	Source Filter	239
4.15	Multiphase Chopper	240
4.16	Applications of Choppers	240

Summary — 241
Solved Problems — 241
Objective Type Questions — 251
Review Questions — 253
Practice Questions — 254

5. Inverters Using MATLAB (DC–DC Converters)

5.1	Introduction	256
5.2	Inverters and their Classification	257
	5.2.1 Classification based on input source	257
	5.2.2 Classification based on output voltage	257
	5.2.3 Classification based on technique for substitution	258
	5.2.4 Classification based on associations with other devices	258
5.3	Selection of Components from Simulink Library Browser	258
5.4	Voltage Source Inverters	260
	5.4.1 Single-phase voltage source inverters	260
5.5	Performance Parameters of Inverters	265
5.6	McMurray Inverter (Auxiliary-Commutated Inverter)	266
5.7	Modified McMurray Half-Bridge and Full-Bridge Inverter	267
	5.7.1 Modified McMurray half-bridge inverter	269
	5.7.2 Modified McMurray full-bridge inverter	270
5.8	PWM Inverters	270
	5.8.1 Single pulse width modulation	271
	5.8.2 Multiple pulse width modulation	271
5.9	Three-Phase Bridge Inverter	273
	5.9.1 180° Conduction mode	273
	5.9.2 120° Conduction mode	274
5.10	Current Source Inverters	275

		5.10.1 Single-phase capacitor-commutated current source inverter with R load	275
5.11	Resonant Converters		277
	5.11.1	Series resonant converters	277
	5.11.2	Parallel resonant converters	279
	5.11.3	ZVS and ZCS PWM converters	280
5.12	Applications of Inverters		280

Summary 281
Solved Problems 281
Objective Type Questions 283
Review Questions 288
Practice Questions 288

6. Controllers Using MATLAB (AC–AC Converters)

6.1	Introduction		290
	6.1.1	ON–OFF control	291
	6.1.2	Phase control	291
6.2	Classification of AC Voltage Controllers		291
6.3	Single-Phase AC Voltage Controllers		292
	6.3.1	Single-phase half-wave AC voltage controller with R load	293
	6.3.2	Single-phase full-wave AC voltage controller with R load	296
	6.3.3	Single-phase full-wave AC voltage controller with RL load	299
6.4	Cycloconverters and Its Types		302
	6.4.1	Single-phase cycloconverters	303
	6.4.2	Three-phase cycloconverters	307
6.5	Load-Commutated Cycloconverter		317
6.6	Matrix Converter		317
6.7	Applications of Voltage Controllers		319

Summary 319
Solved Problems 319
Objective Type Questions 334
Review Questions 336
Practice Questions 337

7. Simulation and Digital Control Using MATLAB

7.1	Introduction		339
7.2	Fuzzy Logic Principles		341
	7.2.1	Fuzzy logic tool box	341
	7.2.2	Implementation	346

		7.2.3	Description and design of FLC	347
		7.2.4	Simulation and results	350
	7.3	Neural Network Principles		352
		7.3.1	Background of neural networks	353
		7.3.2	Implementation	355
		7.3.3	Algorithm for ANN	357
		7.3.4	Simulation results	359
	7.4	Converter Control Using Microprocessors and Microcontrollers		362
	Summary			*363*
	Solved Examples			*363*
	Practice Questions			*366*
	Review Questions			*367*
	Multiple Choice Questions			*367*
8.	**Power Electronics Applications**			
	8.1	Introduction		372
	8.2	Uninterruptible Power Supply (UPS)		373
		8.2.1	Static systems	373
	8.3	Switch-Mode Power Supply		377
		8.3.1	Forward-mode SMPS	378
		8.3.2	Flyback-mode SMPS	378
	8.4	High-Voltage DC Transmission		380
	8.5	VAR Compensators		380
	8.6	Battery Charger		382
	8.7	Switch-Mode Welding		383
	8.8	RF Heating		383
	8.9	Electronic Ballast		383
		8.9.1	Characteristics of fluorescent lamps	383
	8.10	Brushless DC (BLDC) Motors		384
	8.11	Thermal Management and Heat Sinks		385
	Summary			*385*
	Multiple Choice Questions			*385*
	Review Questions			*387*
	Practice Questions			*387*
9.	**Introduction to Electrical Drives**			
	9.1	Introduction		388
		9.1.1	Merits and demerits of electrical drive systems	389
	9.2	DC Drives		390

	9.2.1	Steady-state operation of a separately excited DC motor	390
	9.2.2	Four quadrant operation	394
	9.2.3	Single-phase and three-phase DC drive	396
	9.2.4	Reversal of DC motor	401
	9.2.5	DC chopper drives	402
9.3	AC Drives		404
	9.3.1	Induction motor drive	404
9.4	Synchronous Motor Drive		409
9.5	Phase-Locked Loop (PLL)		411

Summary — 411
Solved Problems — 411
Objective Type Questions — 414
Review Questions — 417
Practice Questions — 418

Appendix 1 Block Parameter Settings — 419
Appendix 2 List of MATLAB Projects — 422
Appendix 3 MATLAB Functions — 432
Appendix 4 Useful Formulae — 441
Appendix 5 Table of Laplace and Z Transforms — 474
Appendix 6 Gate Questions — 478
Resources for MATLAB — 511

Index — 513

Figures

1.1	Startup Page of MATLAB.	4
1.2	MATLAB as a Calculator.	7
1.3	Working with Variables.	8
1.4	Basic Plotting 1.	24
1.5	Basic Plotting 2.	25
1.6	Plot with Label and Annotation.	26
1.7	Editor Window.	27
1.8	Plotting.	32
1.9	Simulink Library Browser.	34
1.10	Electrical Sources.	35
1.11	Measurement.	36
1.12	Elements.	36
1.13	Commonly Used Blocks.	37
1.14	Simple Electrical Circuit.	37
1.15	Parameters.	38
1.16	Circuit Elements.	39
1.17	Passive Elements.	40
1.18	DC Analysis.	41
1.19	Simulink Model.	42
1.20	AC Analysis.	42
1.21	Simulink Model.	44
1.22	Scope Output.	44
1.23	Steady State Analysis of a Linear Circuit with RLC.	45
1.24	Scope Output for Steady State Analysis of a Linear Circuit.	46

1.25	Resonant Switch Converter using MOSFET.	46
1.26	Scope Output of Resonant Switch Converter Using MOSFET.	47
1.27	GTO-based Converter.	48
1.28	Scope Output of GTO-based Converter.	48
1.29	Regulation of Zener Diode.	49
1.30	Scope Output of Regulation of Zener Diode.	50
1.31	Regulation of Pulse Generator Using Thyristor Converter.	51
1.32	Scope Output of Regulation of Pulse Generator Using Thyristor Converter.	51
2.1	Basic Block Diagram of a Power Electronic System.	62
2.2	Power Semiconductor Devices.	65
2.3	Symbol of Diode.	66
2.4	Schematic Representation of a Diode Circuit.	66
2.5	Reverse Leakage Current.	66
2.6	Diode On and Off Characteristics.	67
2.7	Reverse Recovery Characteristics.	68
2.8	Simulink Model of a Diode (Forward Bias).	68
2.9	Parameters of a Diode.	70
2.10	Switching Waveform of a Diode (Forward Bias).	70
2.11	V–I Characteristics of a Diode (Forward Bias).	71
2.12	Simulink Model of a Diode (Reverse Bias).	72
2.13	Switching Waveform of a Diode (Reverse Bias).	72
2.14	V–I Characteristics of a Diode (Reverse Bias).	73
2.15	Symbol of a Zener Diode.	73
2.16	Simulink Model of a Zener Diode.	74
2.17	Switching Waveforms of a Zener Diode.	75
2.18	Symbol of an SCR.	76
2.19	Structure of an SCR.	77
2.20	V–I characteristics of an SCR.	78
2.21	Simulink Model of a Thyristor (Forward Bias).	79
2.22	Switching Waveform of a Thyristor (Forward Bias).	80
2.23	Simulink Model of a Thyristor (Reverse Bias).	81
2.24	Switching Waveform of a Thyristor (Reverse Bias).	81
2.25	Effect of Holding Current and Latching Current.	82
2.26	Gate Characteristics.	82
2.27	dI/dt Protection.	84
2.28	dV/dt Protection.	84
2.29	Turn on of Thyristor.	85
2.30	Turn off of Thyristor.	86
2.31	Off State Characteristics of Two Thyristors of the Same Type.	87

2.32	Series Operation of Thyristors.	88
2.33	Equal Voltage Sharing.	88
2.34	Voltage Sharing and Reverse Recovery Time.	90
2.35	Characteristics of Two Parallel Thyristors.	91
2.36	Parallel Operation of Thyristors.	91
2.37	Two-transistor Model of Thyristor.	92
2.38	Natural Commutation.	96
2.39	Class A Commutation.	97
2.40	Class B Commutation.	97
2.41	Class C Commutation.	98
2.42	Class D Commutation.	98
2.43	Class E Commutation.	99
2.44	Symbol of MOSFET.	100
2.45	Simulink Model of MOSFET.	101
2.46	Switching Waveforms of MOSFET.	102
2.47	Symbol of GTO.	103
2.48	Simulink Model of a GTO.	104
2.49	Switching Waveform of GTO.	104
2.50	Symbol of IGBT.	105
2.51	Simulink Model of an IGBT.	106
2.52	Switching Waveforms of IGBT.	107
2.53	Symbol of DIAC.	107
2.54	Symbol of TRIAC.	108
2.55	Symbol of MCT.	108
2.56	Symbol of IGCT.	108
2.57	Symbol of a Schottky Diode.	109
2.58	Model of a Schottky Diode.	110
2.59	Block Parameter of a Schottky Diode.	110
2.60	I–V Plot (Schottky Diode).	111
2.61	Symbol of BJT.	111
2.62	Simulink Model of a *PNP* Transistor.	112
2.63	Simulink Model of an *NPN* Transistor.	112
2.64	Switching Characteristics of a Transistor.	113
2.65	Simulink Model of *N* Channel MOSFET.	113
2.66	Switching Characteristics of MOSFET.	114
2.67	Simulink Model of *N* Channel IGBT.	115
2.68	Switching Characteristics of IGBT.	115
2.69	Resistance Firing Circuits.	116
2.70	Waveform of Resistance Firing Circuits.	116

2.71	Resistance–Capacitance Firing Circuit.	117
2.72	Waveform of a Resistance–Capacitance Firing Circuit.	117
2.73	UJT Firing Circuit.	118
2.74	Waveform of UJT Firing Circuit.	118
2.75	Pulse Transformer.	119
2.76	Optocoupler.	119
2.77	Ramp-Pedestrial Triggering Circuit.	120
3.1	Simulation Model of a Half-Wave Diode Rectifier with R Load.	148
3.2	Output Waveforms for a Half-Wave Diode Rectifier with R Load.	149
3.3	Simulation Model of a Center Taped and Bridge Rectifier.	150
3.4	Simulation Results of a Center Taped Rectifier.	150
3.5	Simulation Model of a Single-Phase Full-Wave Bridge Rectifier with R Load.	151
3.6	Simulation Results of a Full-Wave Bridge Rectifier.	152
3.7	Simulation Model of a Single-Phase Half-Wave-Controlled Rectifier with R Load.	156
3.8	Simulation Waveform of a Single-Phase Half-Wave-Controlled Rectifier with R Load.	156
3.9	Simulation Model of a Single-Phase Half-Wave-Controlled Rectifier with RL Load.	158
3.10	Single-Phase Half-Wave-Controlled Rectifier with RL Load.	159
3.11	Single-Phase Half-Wave-Controlled Rectifier with RL Load and Freewheeling Diode.	160
3.12	Switching Waveforms of a Single-Phase Half-Wave-Controlled Rectifier with RL Load and Freewheeling Diode.	161
3.13	Single-Phase Half-Wave-Controlled Rectifier with RLE Load.	162
3.14	Switching Waveforms of a Single-Phase Half-Wave-Controlled Rectifier with RLE Load.	163
3.15	Simulink Model of Single-Phase Full-Wave Bridge Rectifier with R Load.	164
3.16	Switching Waveforms of a Single-Phase Full-Wave Bridge Rectifier with R Load.	165
3.17	Simulink Model of a Single-Phase Full-Wave Bridge Rectifier with RL Load.	166
3.18	Switching Waveforms of a Single-Phase Full-Wave Bridge Rectifier with RL Load.	167
3.19	Simulink Model of a Single-Phase Full-Wave Bridge Rectifier with RLE Load.	168
3.20	Switching Waveforms of a Single-Phase Full-Wave Bridge Rectifier with RLE Load.	168
3.21	Simulink Model of a Single-Phase Midpoint Bridge Rectifier with R Load.	169
3.22	Switching Waveform of a Single-Phase Midpoint Bridge Rectifier with R Load.	170
3.23	Simulink Model of a Single-Phase Midpoint Bridge Rectifier with RL Load.	171
3.24	Switching Waveforms of a Single-Phase Midpoint Bridge Rectifier with RL Load.	172
3.25	Simulink Model of a Single-Phase Midpoint Bridge Rectifier with Freewheeling Diode.	172
3.26	Switching Waveforms of a Single-Phase Midpoint Bridge Rectifier with Freewheeling Diode.	173

3.27	Simulink Model of a Single-Phase Semiconverter Half-Controlled Bridge Rectifier with R Load.	174
3.28	Switching Waveforms of a Single-Phase Semiconverter Half-Controlled Bridge Rectifier with R Load.	174
3.29	Simulink Model of a Single-Phase Semiconverter Half-Controlled Bridge Rectifier with RL Load.	176
3.30	Switching Waveforms of a Single-Phase Semiconverter Half-Controlled Bridge Rectifier with RL Load.	176
3.31	Simulink Model of a Single-Phase Semiconverter Half-Controlled Bridge Rectifier with RLE Load.	177
3.32	Switching Waveforms of a Single-Phase Semiconverter Half-Controlled Bridge Rectifier with RLE Load.	177
3.33	Simulink Model of a Three-Phase Half-Wave-Controlled Rectifier with R Load.	178
3.34	Switching Waveforms of a Three-Phase Half-Wave-Controlled Rectifier with R Load.	179
3.35	Simulink Model of a Three-Phase Half-Wave-Controlled Rectifier with RL Load.	180
3.36	Switching Waveforms of Three-Phase Half-Wave-Controlled Rectifier with RL Load.	180
3.37	Simulink Model of a Three-Phase Half-Controlled Bridge Rectifier with RL Load.	181
3.38	Switching Waveforms for Continuous Conduction Mode, $\alpha = 30°$.	182
3.39	Switching Waveforms for Discontinuous Conduction Mode, $\alpha = 90°$.	183
3.40	Simulink Model of a Six Pulse Rectifier with R Load.	184
3.41	Pulse Generator Waveform.	185
3.42	Switching Waveforms for a Six Pulse Converter with R Load.	185
3.43	Simulink Model of a Six Pulse Rectifier with RL Load.	186
3.44	Pulse Generator Waveforms.	187
3.45	Switching Waveforms Model of a Six Pulse Rectifier with RL Load.	187
3.46	Dual Converter.	188
4.1	Principle of Pulse Width Modulation System.	208
4.2	Principle of Variable Frequency Control.	209
4.3	Operation of Step-down Chopper with R Load.	211
4.4	Step-down Chopper Circuit with RLE Load.	212
4.5	Step-up Chopper Circuit.	212
4.6	Type A Chopper.	216
4.7	Performance of Type A Chopper.	216
4.8	Simulation Model of a Type A Chopper.	217
4.9	Simulation Waveform of a Type A Chopper.	217
4.10	Type B Chopper.	218
4.11	Performance of a Type B Chopper.	218
4.12	Simulation Model of a Type B Chopper.	219

4.13	Simulation Waveform of a Type B Chopper.	219
4.14	Type C Chopper.	220
4.15	Performance of a Type C Chopper.	220
4.16	MATLAB/Simulink Model of a Type C Chopper.	221
4.17	Simulation Waveform of a Type C Chopper.	222
4.18	Type D Chopper.	222
4.19	Performance of a Type D Chopper.	223
4.20	MATLAB/Simulink Model of a Type D Chopper.	223
4.21	Simulation Waveform of a Type D Chopper.	224
4.22	Type E Chopper.	224
4.23	MATLAB/Simulink Model of a Type E Chopper.	225
4.24	Simulation Waveform of a Type E Chopper.	226
4.25	MATLAB/Simulink Model of a Buck Converter.	227
4.26	Simulation Waveform of a Buck Converter.	227
4.27	MATLAB/Simulink Model of a Boost Converter.	228
4.28	Simulation Waveform of a Boost Converter.	229
4.29	MATLAB/Simulink Model of a Buck–Boost Converter.	230
4.30	Simulation Waveform of a Buck–Boost Converter.	231
4.31	MATLAB/Simulink Model of a Voltage-Commutated Chopper.	232
4.32	Simulation Waveform of a Voltage-Commutated Chopper.	233
4.33	MATLAB/Simulink Model of a Current-Commutated Chopper.	235
4.34	Simulation Waveform of a Current-Commutated Chopper.	236
4.35	Simulation Model of a Load-Commutated Chopper.	237
4.36	Simulation Waveform of a Load-Commutated Chopper.	237
4.37	Jones Chopper Circuit.	238
4.38	Morgan Chopper.	239
S1	DC-DC converter	246
S2	A Synchronous Buck Converter.	248
S3	DC-DC converter	249
5.1	Simulation Diagram of a Single-Phase Half-Bridge Inverter with R Load.	261
5.2	Simulation Waveform for a Single-Phase Half-Bridge Inverter with R Load.	262
5.3	Simulation Model of a Single-Phase Full-Bridge Inverter with R Load.	262
5.4	Simulation Waveform of a Single-Phase Full-Bridge Inverter with R Load.	263
5.5	Simulation Model of a Single-Phase Half-Bridge Inverter with RL Load.	264
5.6	Simulation Waveform of a Single-Phase Half-Bridge Inverter with RL Load.	265
5.7	Simulation Model of a McMurray Auxiliary Commutated Inverter.	266
5.8	Simulation Waveform of McMurray Auxiliary Commutated Inverter.	267
5.9	Simulation Model of a Modified McMurray Half-Bridge Inverter.	268
5.10	Simulation Waveform of Modified McMurray Full-Bridge Inverter.	268

5.11	Simulation Waveform of a Modified McMurray Half-Bridge Inverter.	269
5.12	Simulation Waveform of a Modified McMurray Full-Bridge Inverter.	270
5.13	Pulse Width Modulation.	272
5.14	Pulse Width Modulation with Reference Signal.	272
5.15	Simulation Model of a Three-Phase Bridge Inverter.	273
5.16	Simulation Model of Three-Phase Bridge Inverter 180° Conduction.	274
5.17	Simulation Model of Three-Phase Bridge Inverter 120° Conduction.	274
5.18	Simulation Model of a Single-Phase Capacitor-Commutated Current Source Inverter with R Load.	276
5.19	Simulation Waveform for a Single-Phase Capacitor-Commutated Current Source Inverter with R Load.	277
5.20	Simulation Model of a Series Resonant Inverter.	278
5.21	Simulation Waveform of a Series Resonant Inverter.	279
5.22	Parallel Resonant Inverter.	279
5.23	ZVS and ZCS-PWM Converters.	280
6.1	Circuit of a Single-Phase Half-Wave AC Voltage Controller with R Load.	292
6.2	Waveform of a Single-Phase Half-Wave AC Voltage Controller.	294
6.3	MATLAB Model of a Single-Phase Half-Wave AC Voltage Controller with R Load.	295
6.4	Simulation Waveform of a Single-Phase Half-Wave AC Voltage Controller with R Load.	296
6.5	Circuit of a Single-Phase Full-Wave AC Voltage Controller with R Load.	297
6.6	Waveform of a Single-Phase Full-Wave AC Voltage Controller with R Load.	297
6.7	MATLAB Model of Single-Phase Full-Wave AC Voltage Controller with R Load.	298
6.8	Simulation Waveform of a Single-Phase Full-Wave AC Voltage Controller with R Load.	299
6.9	Circuit of a Single-Phase Full-Wave AC Voltage Controller with RL Load.	299
6.10	Waveform of a Single-Phase Full-Wave AC Voltage Controller with RL Load.	300
6.11	MATLAB Model of a Single-Phase Full-Wave AC Voltage Controller with RL Load.	301
6.12	Simulation Waveform of a Single-Phase Full-Wave AC Voltage Controller with RL Load.	302
6.13	Single-Phase Cycloconverter.	303
6.14	Waveform of a Single-Phase to Single-Phase Cycloconverter.	304
6.15	MATLAB Model of a Single-Phase to Single-Phase Cycloconverter.	305
6.16	Simulation Waveform of a Single-Phase to Single-Phase Step-Up Cycloconverter.	306
6.17	Simulation Waveform of a Single-Phase to Single-Phase Step-Down Cycloconverter.	306
6.18	3Φ–1Φ Half-Wave Cycloconverter.	307
6.19	3Φ–1Φ Bridge Cycloconverter.	308
6.20	Waveforms of 3Φ–1Φ Cycloconverter.	308

6.21	MATLAB Model of a Three-Phase to Single-Phase Cycloconverter.	310
6.22	Simulation Waveform of a Three-Phase to Single-Phase Cycloconverter.	310
6.23	3Φ–3Φ Half-Wave Cycloconverter.	311
6.24	3Φ–3Φ Bridge Cycloconverter.	312
6.25	Noncirculating Mode.	314
6.26	Circulating Mode.	315
6.27	MATLAB model of a Three-Phase to Three-Phase Cycloconverter.	316
6.28	Simulation Waveform of a Three-Phase to Three-Phase Cycloconverter.	317
6.29	Schematic of Matrix Converter.	318
7.1	Fuzzy Logic Toolbox.	342
7.2	FIS Editor.	343
7.3	Membership Function.	343
7.4	Rule Editor.	344
7.5	Rule Viewer.	344
7.6	Surface Viewer.	345
7.7	Membership Function for Input and Output of FLC.	345
7.8	Block Diagram of the Buck Converter.	347
7.9	Block Diagram of the FLC.	348
7.10	Membership Functions for the Fuzzy Model (a) Input ΔP_{pv}, (b) Input ΔV_{pv}, and (c) Output ΔV_{ref}.	349
7.11	FLC Simulink Model.	351
7.12	PV Source Block.	351
7.13	Subsystem Block.	352
7.14	Power Characteristic Curve with FLC.	352
7.15	Overview of Neural Network.	353
7.16	A Simple Neuron.	354
7.17	A Basic Artificial Neuron.	354
7.18	Basic Structure of a Neural Network.	355
7.19	Overall Simulink Model.	358
7.20	Structure of the Neural Network.	358
7.21	Structure of the First Layer of Neural Network.	359
7.22	Structure of the Second Layer of Neural Network.	359
7.23	Training Curve.	360
7.24	Voltage Curve for Neural Network.	361
7.25	Current Curve for Neural Network.	361
7.26	Power Curve for Neural Network.	362
8.1	Block Diagram of Offline UPS.	374
8.2	Block Diagram of Online UPS.	375
8.3	Block Diagram of Line Interactive UPS.	376

8.4	Forward-Mode SMPS.	378
8.5	Flyback-Mode SMPS.	379
8.6	VAR Compensators.	381
9.1	Basic Block Diagram of an Electrical Drive System.	389
9.2	Separately Excited DC Motor.	390
9.3	Steady State Operation of a DC Motor.	393
9.4	Speed–Torque Curve.	394
9.5	Four Quadrant Operation of a DC Motor.	395
9.6	Basic Thyristor-Based Drive.	396
9.7	Single-Phase DC Drive.	397
9.8	Three-Phase DC Drive.	398
9.9	Simulation Model of a Single-Phase DC Drive.	398
9.10	Speed–Torque Curves of a Single-Phase DC Drive.	399
9.11	Output Waveform of a Single-Phase DC Drive.	399
9.12	Simulation Model of a Three-Phase DC Drive.	400
9.13	Output Waveform of a Three-Phase AC Drive.	400
9.14	Speed–Torque Curves of a Three-Phase DC Drive.	401
9.15	Concept of Reversal.	401
9.16	DC–DC Chopper Drive.	402
9.17	Chopper-Fed Drive.	403
9.18	Speed–Torque Characteristics.	403
9.19	Speed–Torque Characteristics of an Induction Motor.	405
9.20	Induction Motor Drive.	406
9.21	VSI-Fed Induction Motor.	407
9.22	Induction Motor Drive.	408
9.23	Simulink Model of a Synchronous Motor-Fed Drive.	409
9.24	Waveform of a Synchronous Motor-Fed Drive.	410
9.25	Speed–Torque Characteristics of a Synchronous Motor-Fed Drive.	410

Tables

1.1	Installing and activating using an Internet connection	3
1.2	Installing and activating without using an Internet connection	3
1.3	Array Functions	24
1.4	Simulink Features	33
2.1	Power Electronic Converters	64
2.2	Characteristics of Various Types of Diodes	76
2.3	Applications of Power Semiconductor Devices	121
2.4	Power Electronic Systems	124
4.1	Features of Step-down and Step-up Choppers	215
7.1	Fuzzy Rule Base Table	346
7.2	Rule Base for the Fuzzy Model	349
7.3	Training Data	359
7.4	Results of ANN	360

Acknowledgements

The authors are thankful to the Almighty for giving them the strength to persevere and for their achievements.

The authors owe gratitude to Shri L Gopalakrishnan, Managing Trustee, PSG Institutions, Dr K. V. Kuppusamy, Managing Trustee, RVS Institutions, Dr R. Rudramoorthy, Principal, PSG College of Technology, Coimbatore, India and Dr V. Gunaraj, Principal, RVS College of Engineering and Technology, Coimbatore for their whole-hearted cooperation and encouragement for this successful endeavor.

Dr L. Ashok Kumar takes this opportunity to acknowledge many people who helped him in completing this book. This book would not have come to its completion without the help of my students, my department staff and my institute, and especially my project staff. I am thankful to all my students doing their projects and research work with me. The writing of this book would have been impossible without the support of my family members; parents and sisters. Most importantly, I am very grateful to my wife, Y. Uma Maheswari, for her constant support during the writing: without her nothing would have been possible. I would like to express my special gratitude to my daughter A. K. Sangamithra for her smiling face and support, which helped a lot in completing this work.

Professor A. Kalaiarasi would like to acknowledge her deep sense of gratitude to her family, specially her daughter V. K. Meha for the patience and endless support in completing this book. Most importantly, I take this opportunity to thank my supervisor Dr L. Ashok Kumar for giving me a chance to work on this writing alongside him. This acknowledgment would not be complete without expressing my gratitude to my friends and students for their constant support.

We thank the personnel at Cambridge University Press, who produced the book. In addition, special thanks to Rachna Sehgal, the Associate Commissioning Editor for this book.

Preface

Unlike electronic systems that transmit and process signals and data, power electronics involves the processing of substantial amounts of electrical energy. The power range begins with tens of watts to a few hundred watts in the common AC/DC converter (rectifier) used in consumer electronic devices such as battery chargers, personal computers or television sets. In the industry, a common application such as the Variable Speed Drive (VSD), which is used to control an induction motor, has a power range that starts from a few hundred watts and can go upto tens of megawatts.

Beginnning with high-vacuum and gas-filled diode thermionic rectifiers, and triggered devices such as the thyratron and ignitron, the field has evolved to a stage where it can be said that today power electronics is the application of solid-state electronics to the control and conversion of electric power. Steady improvements in the voltage and current handling capacity of solid state devices have made this possible.

During the past few decades the technical aspect of power electronics and drives has gone through significant scientific progression. Many inventions in strategy, apparatus, circuit, control and systems have made power electronics a formidable technology during this period. MATLAB helps learners understand the field because it integrates computation, visualization, and programming in an easy-to-use environment where problems and solutions are expressed in familiar mathematical notation. This book introduces this promising technology with MATLAB/SIMULÏNK in Power Electronic Circuits. It addresses the progress of early semiconductor devices and then moves on to introduce recently developed high-performance power semiconductor switching devices and their applications using MATLAB/SIMULÏNK. It also gives a condensed review of the current power electronic circuits and their outcomes.

The ultimate purpose of this book is to help engineering students engage in energy conversion on the up-to-date electronic applications. It covers almost all features and facilities of MATLAB/SIMULINK and their influence in Power Electronic Circuits like Rectifiers, Inverters, Converters, Choppers, Cyclo converters etc.

The book is recommended as an introductory handbook for starters and as a text book for undergraduate and postgraduate engineering students. It can also serve as a reference book for researchers who need preliminary knowledge of the design and investigation of Power Electronic Circuits.

1

Introduction to MATLAB

Learning Objectives

- ✓ To introduce the fundamental concepts of MATLAB programming
- ✓ To learn the basic steps in installation
- ✓ To understand the various functions involved in arithmetic and scientific concepts
- ✓ To discover the concepts of matrix and vectors, arrays and plotting
- ✓ To introduce the steps involved in circuit modelling and simulation
- ✓ To examine the basic operation of an electric circuit
- ✓ To differentiate the merits and demerits of MATLAB

1.1 Introduction and Outlook

The MATrix LABoratoy (MATLAB) is a high performance interactive multiparadigm numerical computing software system developed by MathWorks. Cleve Moler started developing MATLAB in the late 1960s and it was rewritten in C in 1984. MATLAB was first adopted by researchers and practitioners in control engineering; it has now spread to all domains. The MATLAB function is built roughly around the MATLAB language and the main use of MATLAB is the usage of the command window for execution of text files that includes functions or scripts. MATLAB provides a development environment for managing various sets of files, codes and multiple datasets. It is used to solve problems numerically; MATLAB is the best interactive tool for exploring the various levels of iterations, design, analysis and problem solving. In addition, MATLAB, which is a contemporary programming language, has its own stylish data structures that also contain built-in editing and debugging tools, and supports object-oriented programming.

MATLAB can handle plain numerical expressions and mathematical formulas. It integrates various platforms of computation, visualization, linear algebra, filtering fourier analysis, statistics and numerical integration in the programming environment. MATLAB is exclusively intended for matrix computations that involve explaining linear systems, computing eigenvectors, factoring matrices and arrays. Adding up, it has a variety of custom graphics capabilities with manipulator boundaries and can be extended over lists inscribed in its particular program design.

The extended capability of MATLAB enables it to provide solutions for non-linear problems, such as the elucidation of ordinary differential equations. Thus, it is better that MATLAB produces fairly accurate solutions rather than exact solutions. As a result, MATLAB is a tool designed for different tasks and therefore cannot be directly compared with other mathematical tools. The various functionalities of MATLAB are used to integrate it with other peripheral applications like FORTAN, C, C++, JAVA.

Nowadays, MATLAB is trending in education and is quite popular among scientists as it is the best language for encoding. Some applications are packed with reference to the tool box. The tool boxes include optimization, simulation, signal processing, control systems and all applied fields of engineering. These are the unique factors that make MATLAB a tremendous tool for education and investigation.

1.2 How to Start with MATLAB?

Before starting with MATLAB, installation and activation of the tool is very important. Using the MathWorks installer, installation and activation can be done in some of the following operating systems:

- Microsoft Windows operating systems (32 or 64 bit)
- Linux operating systems (32 or 64 bit)
- Mac operating systems

The installation and activation process can be accomplished in any of the following two ways:

- Using an Internet connection
- Without using an Internet connection

1.2.1 Installing and activation

Before installing, proceed with the steps shown in Tables 1.1 and 1.2.

Table 1.1 Installing and activating using an Internet connection

Step	
Step 1	Commence the installer
Step 2	Select 'Installation using Internet'
Step 3	Evaluate the agreement of software license
Step 4	MathWorks account login
Step 5	Choose the license to be installed
Step 6	Select the type of installation
Step 7	Mention the installation folder
Step 8	Select the products to be installed
Step 9	Select the installation options
Step 10	Confirm the choice of selection
Step 11	Installation is complete
Step 12	Activate your installation
Step 13	Mention license file path
Step 14	Activation complete

Table 1.2 Installing and activating without using an Internet connection

Step	
Step 1	Commence the installer
Step 2	Select 'Installation without using Internet'
Step 3	Review the license agreement
Step 4	Specify the file installation key
Step 5	Select the type of installation
Step 6	Mention the installation folder
Step 7	Select the products to be installed
Step 8	Select the installation options
Step 9	Confirm the choice of selection
Step 10	Installation is complete
Step 11	Activate your installation
Step 12	Mention license file path
Step 13	Activation complete

Once the installation is completed we can start the program using the following steps:

- GOTO start menu
- Double click MATLAB icon

The desktop then appears with the default layout as shown in Figure 1.1, which is the startup page of MATLAB.

From Figure 1.1, we can see some of the default panels that include: Start button, command window, current folder, workspace and command history.

Start button   We can single click the button for better access of tools and functions.

Command window In the command window, we can enter commands at the command line, which is indicated by *fx>>* (This is the command prompt).

Current folder We can access all the MATLAB files from the directory.

Workspace We can explore all the data sets that are imported from the files.

Command history We can view or analyze the commands given in the command window.

As we start up with MATLAB, the user can type some of the commands, such as *helpwin*, *helpdesk* and *demo* for better understanding of MATLAB functions and operations.

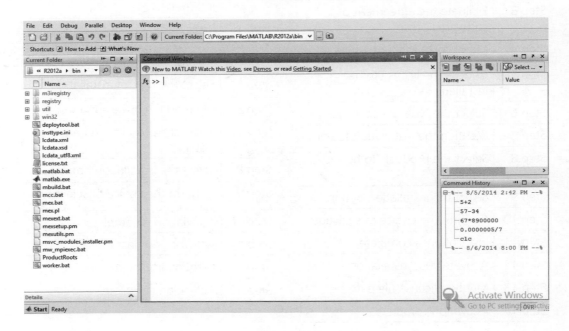

Figure 1.1 Startup Page of MATLAB.

1.3 MATLAB: A Calculator

When we start up with MATLAB, we come across a notation or a prompt sign that prompts us to start,

>> _

Since MATLAB is an inferred type language, the variables can be assigned using the operator '=' without the declaration of their type. Some examples of computation involving variables and constants are shown here:

Example 1

```
>> x = 100
```

MATLAB responds with

```
x =
100
```

Example 2

```
>> x = "hello"
```

MATLAB responds with

```
x = "hello"
Error: The input character is not valid in MATLAB
statements or expressions.
```

Example 3

```
>> x = pi
```

MATLAB responds with

```
x =
3.1416
```

1.3.1 Basic arithmetic operations

To perform basic arithmetic operations, MATLAB can be used as an evaluator of expressions. To execute the mathematical expression, we need to type the numeric constants into the command window after the prompt and press ENTER. MATLAB will print the result back in the command window. When an output variable is not assigned, MATLAB utilizes the default variable 'ans' to display the results of the operation.

Example 4 To perform addition

```
>> 5+2
```

MATLAB responds with

```
ans =
7
```

Example 5 To perform subtraction

```
>>57-34
```

MATLAB responds with

    ans =
    23

Example 6 To perform multiplication

    >>67 * 8900000

MATLAB responds with

    ans =
    596300000

Example 7 To perform division

    >>0.0000005/7

MATLAB responds with

    ans =
    7.1429e-08

Example 8 To perform trigonometric calculations

    >>cos(pi/2)

MATLAB responds with

    ans =
    6.1232e-17

In Figure 1.2, the mathematical computations of arithmetic operations are shown in the MATLAB window. Thus, it can be clearly seen that the MATLAB tool can be used purely as a calculator that performs mathematical and scientific operations.

1.3.2 Assigning values to variables

The alternate method of performing arithmetic operations in MATLAB is by assigning a suitable value to the variables. To create a variable using MATLAB, assign the variables say, for example, a, b and c with some values.

The basic syntax of assigning variables is as follows

    variable name = a value

Here the value can be a function call, operator, variable or a numerical value. Consider the following examples of assigning values to the variables.

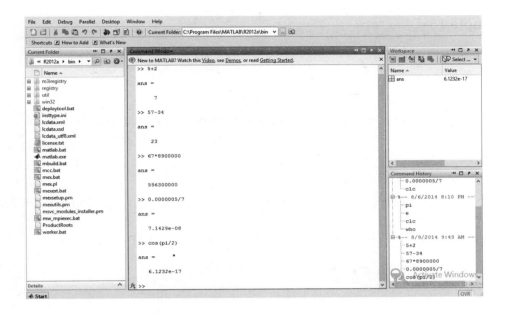

Figure 1.2 MATLAB as a Calculator.

Example 9

```
>> a = 2
   a = 2
```

Example 10

```
>> a = 6
   a = 6
>> b = a^2
   b = 36
```

Example 11

```
>> b = pi/6
   b = 0.523626
```

Example 12

```
>> c = b + a^0.25
   c = 1.7128
```

In Example 12, it is noted that the variable which is named c exists only as a numerical value.

Example 13

```
>>who
Your variables are
a    ans    b    c
```

The 'who' command represents the list of active variables that are currently available.

Example 14

```
>> clc
```

The 'clc' command clears the screen. It refreshes or empties the command window for a new set of operations.

Figure 1.3 shows how to assign values to variables in the MATLAB window.

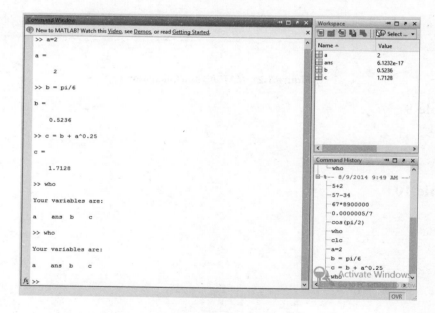

Figure 1.3 Working with Variables.

Some of the other tasks available in MATLAB computing are overwriting of variables, controlling the hierarchy, controlling the precision of floating points, multiple statements and display of error message.

Example 15

```
Overwriting of variables
    >> a = 7;
```

```
>> a = a + 4
a =
    11
```

The purpose of the semicolon (;) is to suppress the result after execution.

Example 16

```
Controlling the hierarchy
>>a = (1 + 2) * 3
a =
    9
>> a = 1 + 2 * 3
a =
    7
```

Here similar expressions produce different results. To avoid erroneous results, MATLAB considers hierarchy of mathematical operations to perform calculations.

Example 17

Controlling the precision of the floating number

The length of the floating number of the assigned variable can be increased or decreased using the syntax

$$\text{"format short"}$$
$$\text{"format long"}$$

```
>> 1/5987666555
ans =
   1.670099680425508e-10
>> format short
>> x = 1/56565656565
x =
   1.7679e-11
>> format long
>> x = 1/56565656565
x =
   1.767857142877663e-11
```

Example 18

```
Multiple statements
    >> a = 4;
    >> b = cos(a) * sin(a)
       b =
           0.494679123311691
```

The execution of multiple statements includes the execution of all linear expressions separated by commas (,) or semicolons (;).

Example 19

```
Display of error messages
    >> x = 12; y = 32;
       z = x + w
       Undefined function or variable 'w'.
```

When an undefined variable is being executed, the MATLAB responds with an error stating 'undefined function or variable'.

To terminate the session, we use the command called 'quit' in the command window to exit the application.

1.4 Basic Features of MATLAB

1.4.1 Investigation of a MATLAB function

In technical computing, having a large set of predefined mathematical functions is very important. Since MATLAB is an interactive tool for computation, it stores a large set of mathematical functions in the inbuilt mode. These stored functions are called built-in functions. Most mathematics standards are executed using the built-in functions. A set of mathematical functions collectively forms a MATLAB program. Hence, the MATLAB programs are stored in the form of plain text as files with an extension '.m'. Such files are called m-files. The unique quality of MATLAB is that the files can be used interactively, which also favours the process of debugging the files for huge programs.

1.4.2 Mathematical functions

Like BASIC language, the large set of various arithmetic operations can be assigned to the variables that are defined. The basic syntax of mathematical functions are listed here:

Exponential and Logarithmic Functions

exp (*x*)   Exponential; e^x
log (*x*)   Natural logarithm; ln(x)
log10 (*x*)   Common (base 10) logarithm; log(x) = log10(x)
sqrt (*x*)   Square root; √x

Trigonometric Functions

acos (*x*)   Inverse cosine; arcos x = cos − 1 (x)
asin (*x*)   Inverse sine; arcsin x = sin − 1 (x)
atan (*x*)   Inverse tangent; arctan x = tan − 1 (x)
cos (*x*)   Cosine; cos(x)
sin (*x*)   Sine; sin(x)
tan (*x*)   Tangent; tan(x)

Some of the other types of trigonometric functions include *acot* (*x*), *acsc* (*x*), *asec* (*x*), *atan2*(*y, x*), *cot* (*x*), *csc* (*x*), and *sec* (*x*).

Hyperbolic Functions

acosh (*x*)   Inverse hyperbolic cosine; cosh − 1 (x)
acoth (*x*)   Inverse hyperbolic cotangent; coth − 1 (x)
cosh (*x*)   Hyperbolic cosine; cosh(x)
coth (*x*)   Hyperbolic cotangent; cosh(x)/sinh(x)
sech (*x*)   Hyperbolic secant; 1/cosh(x)
sinh (*x*)   Hyperbolic sine; sinh(x)

Other types of hyperbolic functions include *acoth* (*x*), *acsch* (*x*), *asech* (*x*), *atanh* (*x*), *asinh* (*x*), *csch* (*x*) and *tanh* (*x*).

Complex Functions

abs (*x*)   Absolute value; |x|
angle (*x*)   Complex number angle x
conj (*x*)   Complex conjugate of x
imag (*x*)   Imaginary part of a complex number x
real (*x*)   Real part of a complex number x

Statistical Functions

erf (*x*)	Compute error function erf (x)
mean	Compute mean or average
median	Compute median
std	Compute standard deviation

Random Number Functions

rand	Generates uniformly distributed random numbers between 0 and 1
randn	Generates normally distributed random numbers

Numeric Functions

ceil	Rounds to the nearest integer towards •
fix	Rounds values to the nearest integer towards zero
floor	Rounds to the nearest integer towards •
round	Rounds to the nearest integer
sign	Signup function

String Functions

findstr	Occurrence of a string
strcmp	String comparison
char	String array character

Following is a typical example that relates to the basic mathematical functions.

Example 20

Solve for the expression $g = \sin(x) + \sin(y)(\log x) + e^1$ with values of $x = 3$, $y = 7$ and $z = 6$.

```
>> x = 3; y = 7; z = 6;
>> g = sin(x) + sin(y) * log(x) + exp(1)
   g =
   3.581175387361639
```

1.4.3 Vector and matrix operations

1.4.3.1 Vector Operations

MATLAB is an easy to understand tool that helps in controlling matrices and vector ideas. The essential functions of MATLAB rotate around the use of vectors and a vector is characterized by putting a succession of numbers inside the matrices, though in reality, a system could be characterized as the fundamental data set component in MATLAB.

A column vector can be characterized with an exhibit of $1 \times n$ measurements. The estimations of a vector can be encased inside a square section and may be divided by commas or by spaces. Consider the following example of how to assign a name to a row vector.

Example 21

```
A>> v = [1 2 3 4 5 6 7 8 9] click ENTER
v =
     1    2    3    4    5    6    7    8    9
```

A column vector can also be characterized with an exhibit of $m \times 1$ measurements. Like the line vector, the estimates can be encased inside a square section divided by spaces or with commas. Column vectors can even be made by utilizing semicolons (;) inside the square sections to divide the values. To make a section vector, simply take the transpose of the row vector utilizing the image ('). Following are illustrations for a segment vector.

Example 22

```
>> v = [-34 56 -445 54 -2 -5]'    click ENTER
v =
   -34
    56
  -445
    54
    -2
    -5
```

Example 23

```
>> v = [1;7;9;67;9]
v =
     1
     7
     9
    67
     9
```

A vector can be of any length. Following is an exchange method for characterizing a vector for a set of qualities.

Example 24

```
>> v = [1:9]
v =
     1    2    3    4    5    6    7    8    9
```

Another method of manipulating vectors is by the addition of the vector with a scaling. In this method, to get a component of a vector, we utilize the image colon (:). Assume, for example that we need to characterize a vector with a beginning quality and the closure esteem along with the addition esteem. The vector could be written as follows.

Example 25

```
>> v = [1:.20:7]
v =
  Columns 1 through 3
   1.000000000000000   1.200000000000000   1.400000000000000
  Columns 4 through 6
   1.600000000000000   1.800000000000000   2.000000000000000
  Columns 7 through 9
   2.200000000000000   2.400000000000000   2.600000000000000
  Columns 10 through 12
   2.800000000000000   3.000000000000000   3.200000000000000
  Columns 13 through 15
   3.400000000000000   3.600000000000000   3.800000000000000
  Columns 16 through 18
   4.000000000000000   4.199999999999999   4.400000000000000
  Columns 19 through 21
   4.600000000000000   4.800000000000000   5.000000000000000
  Columns 22 through 24
   5.200000000000000   5.400000000000000   5.600000000000000
  Columns 25 through 27
   5.800000000000000   6.000000000000000   6.200000000000000
```

```
Columns 28 through 30
   6.400000000000000    6.600000000000000    6.800000000000000
Column 31
   7.000000000000000
```

From this illustration, it can be seen that a line of vectors beginning from 1 to 7 with increment in the estimations of .20 is continuously registered. MATLAB executes the entries that are needed faultlessly. Subsequently, a set of 31 sections are continuously declared to characterize the given vector.

The following program accesses elements within the vector, just type

Example 26

```
>> v(4)
ans =
    1.600000000000000
```

The result brings out the estimation of the vector created in Example 25, that is, the fourth entrance of the vector has been printed. The other values are excluded. Therefore, the result is continuously printed with the ans.

The following is a program that signifies the end value of the vector

Example 27

```
>> v(1, end)
ans =
    7
```

The following program tracks the recent or the past result of a vector.

Example 28

```
>> ans'
ans =
    1.600000000000000
```

A simple example for operating with vector is shown here.

Column vector:

Example 29

```
>> v = [1:5:8]; u = [3:6:4];
>> y = 2 * (u + v)
```

```
y =
    8    18
```

Row vector:

Example 30

```
>> v = [1:5:8]'; u = [3:6:4]';
>> y = 5 * (u + v)
y =
    20
    45
```

The vector can also be represented in a simpler form.

Example 31

```
>> v(:) for column vector
>> v(1:end) for row vector
```

MATLAB, as a matter of course, stores all the outputs that have been executed, which is convenient when substantial set of operations or reckonings are continuously done.

1.4.3.2 Matrix operation

To perform matrix operation, a system in MATLAB could be characterized as a show of numbers. A network is like a vector with a gathering of column and segment vectors.

The following examples represent a 3 × 3 matrix as rows of column vectors or columns

Example 32

```
>> x = [1 4 6; 2 4 5; 5 6 7]
x =
    1    4    6
    2    4    5
    5    6    7
```

The transpose of this vector can be obtained by using the symbol (');

Example 33

```
>> x = [1 4 6; 2 4 5; 5 6 7]'
x =
     1     2     5
     4     4     6
     6     5     7
```

Henceforth, to manage matrix operations in MATLAB, the notations to be considered are square section, commas and semicolon as said previously for vector operations.

To extract a particular element in a matrix, we can perform the following operation.

Example 34

```
>> x(1,3)
ans =
     5
```

The element, being located in row 1 and column 3 of x in Example 33, is executed.

1.4.3.2.1 Indexing of a matrix

Indexing of a network can be done in different ways as for the vector. The components could be named with i and j in it for a column and a section for matrix x. A component can be denoted as $x(i, j)$ or x_{ij}. The record of a network is parted into two. The main file could be the line number and the second one could be the segment number.

Consider a matrix

```
x =
     1     2     5
     4     4     6
     6     5     7
```

The following example shows us how to replace, say, the second column and second row value 4, by any other value, say 6.

Example 35

```
>>x(2,2) = 6
x =
```

```
    1   2   5
    4   6   6
    6   5   7
```

Zero and negative qualities are not acknowledged though the components can be obtained as $x(i, j)$ where $i \geq 1$ and j

1.4.3.2.2 Inverse of a matrix

Example 36

Consider the matrix

```
x =
    1   2   5
    4   6   6
    6   5   7
```

The following example shows how we can operate with a matrix in terms of its inverse.

Example 37

```
>> inv(x)
ans =
   -0.230769230769231  -0.211538461538462   0.346153846153846
   -0.153846153846154   0.442307692307692  -0.269230769230769
    0.307692307692308  -0.134615384615385   0.038461538461538
```

To calculate the eigenvalues of the matrix x, we can perform the following operation.

Example 38

```
x =
    1   2   5
    4   6   6
    6   5   7
>> eig(x)
ans =
   14.645217647029696
   -2.234341959325249
    1.589124312295557
```

Let us see how we can manipulate the eigenvector and the eigenvalues.

Example 39

```
>> [v,e] = eig(x)
v =
  -0.345702432092625  -0.842635176101989   0.362970606773015
  -0.637690615601249   0.017156405528112  -0.849621987304843
  -0.688360739161774   0.538211499082772   0.382615756741668
e =
  14.645217647029696                   0                  0
                   0  -2.234341959325249                  0
                   0                   0  1.589124312295557
```

To obtain the diagonal values, we can perform the following operation.

Example 40

```
>> diag(x)
ans =
     1
     6
     7
```

The other subroutines under matrix operations are the approximation values of linear expressions.

For example, to obtain the solution for $Ax = B$, we can perform the following operation.

Example 41

```
>> v=[1 2 3]; x=[1.428  7.926  4.856];
>> A = x/v
A =
   1.428571428571429
   2.428571428571428
   2.642857142857143
>> B = [ 1 2 4];
>> A = B/v
>> A = B/v
```

```
A =
   1.214285714285714
>> A * B
ans =
        1.214285714285714   2.428571428571428   4.857142857142857
>> A1 = v/B
A1 =
        0.809523809523809
>>   A1 * B
ans =
        0.809523809523809   1.619047619047619   3.238095238095238
```

1.4.3.2.3 Concatenating of matrices

Concatenation includes the creation of submatrices.

Example 42

```
x =
     1     2     5
     4     6     6
     6     5     7
```

The submatrix would be

```
>> y = [x 10 * x; -x [1 0 0; 0 1 0; 0 0 1]]
y =
     1     2     5    10    20    50
     4     6     6    40    60    60
     6     5     7    60    50    70
    -1    -2    -5     1     0     0
    -4    -6    -6     0     1     0
    -6    -5    -7     0     0     1
```

1.4.3.2.4 To delete a row or a column

Use the empty vector to delete any row or a column.

Example 43

```
>> x(2,:) = []
x =
     1     2     5
     6     5     7
```

Other matrix generators include functions, such as eye, zeros and ones. The intellectual properties of matrices include hilb, invhilb, magic, pascal, toeplitz, vander and Wilkinson.

There is also the whos command which will identify the set of variables in the workspace.

Example 44

```
>> whos
  Name      Size       Bytes    Class       Attributes

  A         1×1            8    double
  A1        1×1            8    double
  B         1×3           24    double
  a         1×1            8    double
  ans       1×3           24    double
  b         1×1            8    double
  c         1×1            8    double
  e         3×3           72    double
  g         1×1            8    double
  u         1×1            8    double
  v         1×3           24    double
  x         2×3           48    double
  y         6×6          288    double
  z         1×1            8    double
```

1.4.4 Arrays

In MATLAB, the vector can be considered to be an array of single dimension and the matrix an array of two dimensions. The variables are multidimensional.

The basic type of an array includes two types:

A double argument creates a rectangular array.

Example 45

```
>> ones(8)
ans =
    1   1   1   1   1   1   1   1
    1   1   1   1   1   1   1   1
    1   1   1   1   1   1   1   1
    1   1   1   1   1   1   1   1
    1   1   1   1   1   1   1   1
    1   1   1   1   1   1   1   1
    1   1   1   1   1   1   1   1
    1   1   1   1   1   1   1   1
```

A single argument creates a square array.

Example 46

```
>> zeros(1,8)
ans =
    0   0   0   0   0   0   0   0
```

The function 'eye' represents the identity matrix.

Example 47

```
>> eye(4)
ans =
    1   0   0   0
    0   1   0   0
    0   0   1   0
    0   0   0   1
```

The function 'random' creates a Example 48 distributed random numbers.

Example 48

```
>> rand (3,8)
ans =
```

```
Columns 1 through 3
0.814723686393179   0.913375856139019   0.278498218867048
0.905791937075619   0.632359246225410   0.546881519204984
0.126986816293506   0.097540404999410   0.957506835434298
Columns 4 through 6
0.964888535199277   0.957166948242946   0.141886338627215
0.157613081677548   0.485375648722841   0.421761282626275
0.970592781760616   0.800280468888800   0.915735525189067
Columns 7 through 8
0.792207329559554   0.035711678574190
0.959492426392903   0.849129305868777
0.655740699156587   0.933993247757551
```

These are the set of special arrays that are especially used by MATLAB functionalities.

1.4.4.1 Multidimensional arrays

Multidimensional clusters in MATLAB have more than two measurements. They are expansions of the typical two-dimensional matrix. A two-dimensional cluster can be converted into a multidimensional matrix.

Example 49

```
>> x = [2 4 5; 3 5 6; 5 6 7]
x =
     2     4     5
     3     5     6
     5     6     7
```

To improve the dimension of the array, we do the following operation.

```
>> x(:, :, 2) = [ 1 2 3; 4 5 6; 7 8 9]
x(:, :, 1) =
     2     4     5
     3     5     6
     5     6     7
```

```
x(:, :, 2) =
     1     2     3
     4     5     6
     7     8     9
```

Table 1.3 Array Functions

length	Vector length
ndims	Array dimensions count
numel	Array elements count
size	Dimension of array
iscolumn	Identify column vector
isempty	Identify empty vector
circshift	Shift array circularly
flipdim	Flip array along specified dimension
flipud	Flip matrix up to down
permute	Rearrange dimensions of N-D array
repmat	Replicate and tile array

Multidimensional arrays can also be implemented for special arrays.

Table 1.3 represents the list of array functions available in MATLAB.

Other types of array functions include *blkdiag, ismatrix, isrow, isscalar, isvector, ctranspose, diag, fliplr, ipermute, reshape, rot90, issorted, sort, sortrows, shiftdim, squeeze, transpose, vectorize* and *ctranspose*.

1.4.5 Basic plotting

Plotting a chart utilizing the data sets within MATLAB is simple. With a few changes, the required vectors can be plotted utilizing the implicit capacity. Straightforward charts can be plotted using clusters of matrices.

Example 50

```
>> x = [2 4 6 8 10 12]; y = [-1 3 5 -2 12 6];
Plot (x,y)
```

The MATLAB responds with the result as shown in Figure 1.4.

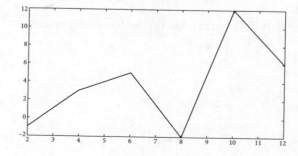

Figure 1.4 Basic Plotting 1.

As a second example, the following function can be plotted.

Example 51

$$y = x/(1 + x^2) \text{ for } -5 < x < 5.$$

To plot, the following operation can be used.

```
>> x = linespace(-5,5,100);
>> y=x/(1+x*x);
>> plot(x,y)
```

MATLAB opens up a graphics window and displays the result shown in Figure 1.5.

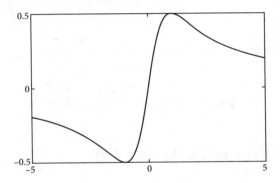

Figure 1.5 Basic Plotting 2.

1.4.5.1 Labelling and annotations in plotting

While plotting a set of variables, we need to mention what the x-axis and the y-axis denote; moreover, the values on the x- and y-axis need to be marked. These along with the title of the plot might be executed utilizing the accompanying operation (see Figure 1.6).

Example 52

```
>> x = [1 2 3 4 5 6 7 8 9]; y = [3 -1 2 4 5 1 5 -9 1];
>> title('Mean versus time')
>>xlabel('time')
>>ylabel('average')
>>plot(x,y)
```

In MATLAB, the plot can be arranged in many ways according to the client's preference and to enhance the nature of the data analysis.

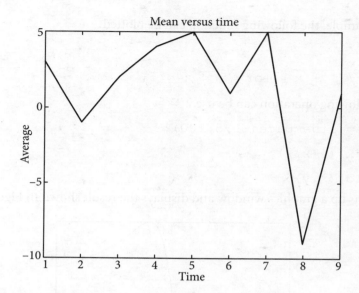

Figure 1.6 Plot with Label and Annotation.

1.5 Programming with MATLAB

All MATLAB commands are executed using the prompt where it provides practical solutions even for tedious problems. Programming with MATLAB can be executed in the editor window.

The MATLAB editor is the editor window, which is used to create .m files. To start up with programming:

Goto file menu> New> Script

Scripts Script documents are system records with .m expansion. In these documents, we compose arrangement of orders, which are to be executed at the same time. Without any acknowledgements, the scripts are placed on the work space.

Or

Goto file menu> New> Function

Function Functions are program records with .m extension.

1.5.1 Creating M-files

There are two methods of writing files in MATLAB

Figure 1.7 shows the MATLAB editor window.

Figure 1.7 Editor Window.

A script document contains different succeeding outlines of MATLAB instructions and calling of functions. The script can be run by writing its name on the command line. One can sort in all the orders in a content record having an expansion of 'm'. This 'script' record must be placed in the current work index. When the name of this 'm' document is entered at brief level, MATLAB peruses the substance of this record and executes each order in this document one by one consecutively. Indeed, what is made is a MATLAB project called 'm-code'.

1.5.2 M-file functions

A function can incorporate legitimate MATLAB outflows, control streams, remarks, clear lines and settled functions. Any variable that we create inside a capacity is put away inside a workspace particular to that function, which is partitioned from the base workspace. Functions end with either an end declaration, the end of the document, or the definition line for an alternate capacity, whichever starts things out. The end explanation is used when a function in the document contains a settled function.

Example 53

```
>> f = factorial(8)
f =
    40320
```

System documents can contain various functions. The principle function is the fundamental function and is the function that MATLAB partners with the document name. Resulting functions that are not settled are called local functions. They are only accessible to different functions inside the same record.

Listed here are different types of file functions

Functions and Scripts

depdir	Lists dependent folders for a function or for a p-file
echo	Displays the statements of a function during the execution of a function
function	Function declaration

input	Requests the user for an input
mfilename	The file name of the currently running function
onCleanup	Cleans up the task after completion of the execution
script	Sequence of the MATLAB statements in the file
syntax	To call a MATLAB function

Other types of functions include *depfun, end, input name, input parser, namelengthmax, nargchk, nargin, narginchk, nargout, nargoutchk, pcode, varargin* and *varargout*.

Evaluation

ans	The most latest answer
builtin	The built-in function executed from overloaded method
pause	Stop execution in the interim
run	Runs the script that is not on the current path

Other evaluation functions include *arrayfun, assert, cellfun, echo, evalc, evalin, feval, isvarname, script, structfun, symvar, tic* and *toc*.

Variables and Functions in Memory

global	Declaration of global variables
memory	Displays the memory information
mlock	Prevents clearing function from memory
pack	Consolidates workspace memory
persistent	Defines persistent variable

Other types of evaluation functions include *ans, assignin, clearvars, datatipinfo, genvarname, inmem, isglobal* and *mislocked*.

1.5.3 Control structures and operators

MATLAB is also a programming language. Like other machine programming language, MATLAB has some decision making structures for controlling command execution. These decision making or control flow structures incorporate for loops and if-else-end developments. Control flow structures are frequently utilized within script m-files and m-files. By making a file with the augmentation .m, we can without much effort compose and run programs. We do not have to assemble the system since MATLAB is an interpretative language. MATLAB has thousand of functions, and we can include our own particular function utilizing m-files. MATLAB gives a few instruments that can be utilized to control the flow of a project.

The four control flow structures of MATLAB are the if statement, the for loop, the while loop, and the switch statement.

The if-else-end statement:

if ... elseif ... else ... end

Example 54

```
discr = b^2-4ac
if discr < 10
disp('Warning: discriminant is negative, roots are real');
else
disp('Warning: undefined');
end
```

Example 54(a)

```
>>  m = 1:5
for n = 1:5
    if m == n
        a(m,n) = 2;
    elseif abs(m - n) == 2
        a(m,n) = 1;
    else
        a(m,n) = 0;
    end
end
end
```

Result:

```
m =
     1     2     3     4     5
```

Example 54(b)

```
>> t = rand(1);
>> if t > 0.75
    s = 0;
elseif t < 0.25
    s = 1;
else
    s = 1 - 2 * (t - 0.25);
end
```

Result:

```
>> s
s =
    0
>> t
t =
    0.9058
```

The for...end statement:

for variable = expression
....
End

Example 55

```
for ii = 1:5
x = ii * ii
end
```

The while...end statement:

while expression
....
end

Example 56

```
x = 1
while x <= 10
```

```
x = 39 * x
end
```

Example 56(a)
```
Discr = b^2-4*a*c ;
while discr >= 20
Disp()
End
```

Example 56(b)
```
x = 7, y = 10;
while x <= 7, y <= 10
x = y * 20
End
```

To display the results on the window use the following set of commands:
- Open a file using fopen
- Write the output using fprintf
- Close the file using fclose

1.5.4 Debugging M-files

MATLAB can also find sentence errors during compilation. These errors are typically simple to settle. MATLAB can likewise experience errors during a run time.

Listed are the set of debugging commands:

dbstop	Set breakpoint
dbclear	Remove breakpoint
dbcont	Resume execution
dbdown	Change local workspace context
dbstack	List who called whom
dbstatus	List all breakpoints
dbstep	Execute one or more lines
dbtype	List M-files with line numbers
dbup	Change local workspace context
dbquit	Quit debug mode

During an error in an m-function, we can utilize the debugging commands to set breakpoints to help debug the error. Some essential points that we need to remember about MATLAB's debugging orders are:

- the debugging orders deal with m-functions, not scripts.
- the m-file breakpoint data are nearly connected with the arranged m-file. If by chance the m-file is cleared either by altering or by the acceptable command, all m-file breakpoints are erased. We can utilize this strategy when it is important to clear various breakpoints from one or more m-documents.

1.5.5 Creating plots

MATLAB provides numerous strategies to plot numerical information. Graphical abilities of MATLAB incorporate plotting apparatuses, standard plotting functions, realistic control and data analysis instruments and devices for printing and trading representation to standard configurations. Typical Math Toolbox extends these graphical abilities and makes it possible to plot typical capacities (see Figure 1.8).

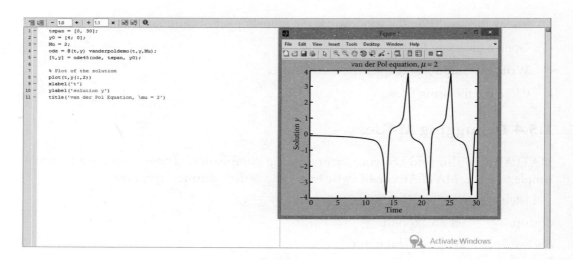

Figure 1.8 Plotting.

ezplot: To create 2D plots of symbolic expressions, equations or functions
ezplot3:   To create 3D parametric plots
ezpolar: To create plots in polar coordinates
ezsurf: To create surface plots
ezcontour: To create contour plots
ezmesh: To create mesh plots

1.6 Circuit Descriptions

MATLAB Simulink provides segment libraries and investigation instruments for displaying and modelling electrical systems. The libraries offer models of electrical segments, including three-phase machines, electric drives and parts for applications, for example, flexible AC transmission systems (FACTS) and renewable systems. Total Harmonic Distortion (THD), load flow and other key electrical power system investigations are mechanized.

MATLAB Simulink models can be utilized to create control systems and test system level execution. We can parameterize models utilizing MATLAB variables and representations, and outline control systems for a given electrical power system in Simulink. To convey models to other recreation situations, including Harware In the Loop (HIL) systems, Simpowersystems helps C-code era.

1.6.1 Format and layout

Simulink is a piece outline environment for multidomain recreation and model-based design. It underpins simulation, programmed code era and consistent test and check of installed systems. Simulink provides a graphical proofreader, adaptable block libraries and solvers for demonstrating and resenacting element systems. In Simulink, it is exceptionally easy to speak to and afterwards recreate a scientific model language to a physical system. Models are enunciated graphically in Simulink as block outlines. Wide exhibits of blocks are accessible to the client in given libraries to access different phenomena and models in varied configurations. One of the essential advantages of utilizing Simulink for the examination of element systems is that it permits us to rapidly examine the reaction of confounded systems that may be restrictively hard to break down systematically. Simulink can numerically estimate the answers for scientific models that are difficult to solve. Simulink exhibits how to determine a numerical model and then actualize that model. Some of the features of Simulink are listed in Table 1.4.

Table 1.4 Simulink Features

Building the model	Models various level subsystems with predefined library pieces
Recreating the model	Simulates the element conduct of a given system; perspectives come about as the model runs
Examining simulation results	Views recreation and debugs the model
Overseeing projects	Easily oversees records, parts and a lot of other information
Associating with hardware	Connects the model to equipment for ongoing testing and inserts into system org

1.6.2 Electrical circuit description

MATLAB Simulink favours segment libraries and investigation apparatuses for demonstrating and recreating systems. The libraries offer models of electrical power parts, including three-stage machines, electric drives and fragments for applications, say for instance: FACTS and renewable energy schemas; THD analysis, load flow investigation and other key electrical power system investigations can be customized. MATLAB Simulink models could be used to make power electronics schemas and test system level execution. We can parameterize the models using MATLAB variables and affirmations, and set up equipment structures for our electrical power structure in Simulink. MATLAB Simulink can incorporate mechanical, pneumatic and diverse parts to the model using Simscape and test them together in a single window.

1.6.3 Simulink library browser

To get exposed to Simulink,

   Open library browser> click new Simulink model

On the left window panel, shown in Figure 1.9, the library is categorized into subsystems; clicking each one will display the blocks on the right window.

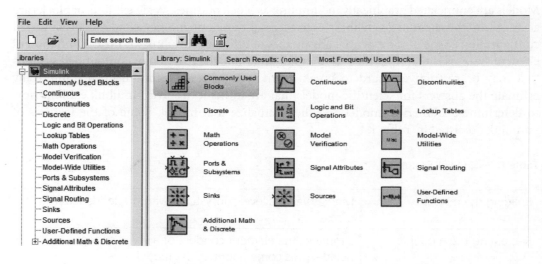

Figure 1.9   Simulink Library Browser.

1.6.3.1 Building the model

Simulink contributes an established set of predefined hinders, which can be joined together to make a complete system. Various tools are available for progressive displaying, information administration and subsystem customization, which empowers one to access to even the most mind-boggling system succinctly and exactly. The Simulink Library Browser includes specialized components for aerospace, communications, control system tool box, optimization, and many other applications. MATLAB can also be used for electrical, mechanical and hydraulic components; that is, we can display physical systems with mechanical, electrical and water-powered segments. We can build reconstructed works by utilizing these pieces or by consolidating using MATLAB C, Ada and Fortran in the model.

1.6.3.2 Circuit elements

Electrical components can be considered as reflections related to their corresponding electrical parts, for example, resistors, capacitors and inductors, utilized within the gambit of electrical systems. Any electrical system consists of various, interconnected electrical components in a schematic chart or circuit graph, each of which influences the voltage in the system or current through the system. These perfect electrical components relate to genuine, physical, electrical or electronic segments; however, they do not exist physically and are expected to have perfect properties. A model can be assembled by dragging the block from the Simulink Library browser into the Simulink editor window. Using connection lines, the block can be interlinked with each other.

For example, to build a basic electrical circuit as shown in Figure 1.14, the given steps can be followed:

First, open Simulink and open a new model window.

Then, drag the selected blocks one by one into the model window and place them approximately as shown in Figures 1.10–1.13.

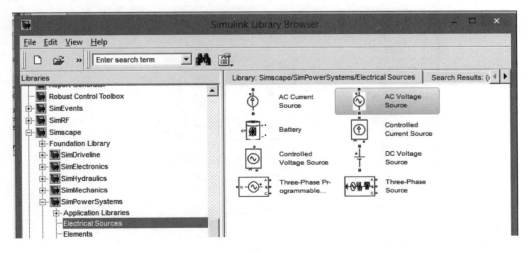

Figure 1.10 Electrical Sources.

Open Simulink Library Browser>Libraries>Simscape>SimPowerSystems

To select AC voltage source:

Goto SimPower Systems>Application libraries>Electrical Sources> AC Voltage Source

To select AC current source:

Goto SimPower Systems>Application libraries>Electrical Sources>AC Current Source

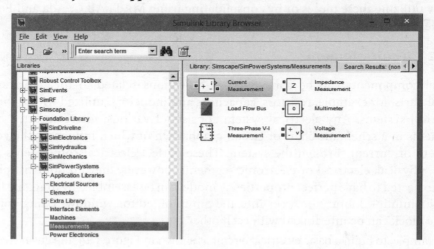

Figure 1.11 Measurement.

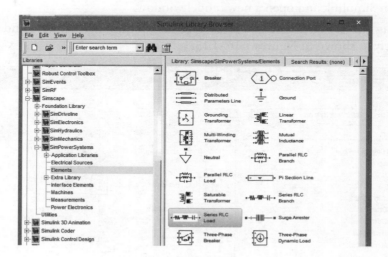

Figure 1.12 Elements.

To measure voltage and current values:

Goto SimPower Systems>Extra Library>Measurements>Current Measurement

Goto SimPower Systems>Extra Library>Measurements>Voltage Measurement

Goto SimPower Systems>Application libraries>Elements>Series RLC Load

Goto Simulink>Commonly used Blocks>Scope

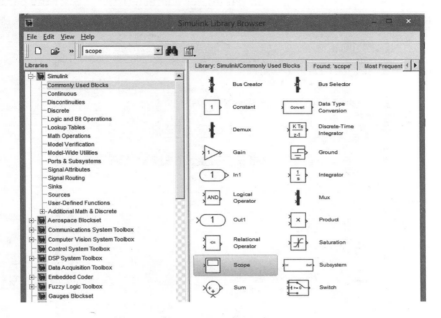

Figure 1.13 Commonly Used Blocks.

Drag the selected blocks into the Simulink Editor and connect all the blocks using the pointer to make it a closed circuit. The Simulink Editor provides a finished control over what we see and use inside the model. Similarly, we can add orders and submenus to the manager and setting menus by double clicking each piece. We can also add a custom interface to a subsystem or model by utilizing a cover that shrouds the subsystem's substance and furnishes the subsystem with its symbol and parameter dialog box.

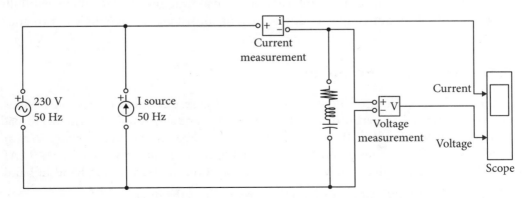

Figure 1.14 Simple Electrical Circuit.

The Simulink model contains both indicators and parameters. Parameters are coefficients that characterize system elements and conduct as shown in Figure 1.15. Simulink helps focusing on the accompanying sign and parameter characteristics. To set the parameters, double click the block and determine the traits; click OK afterwards.

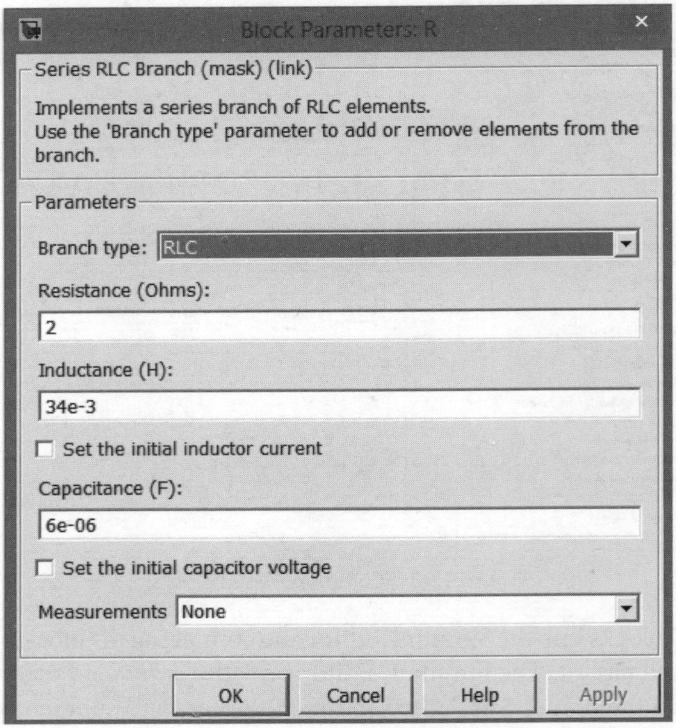

Figure 1.15 Parameters.

If we choose not to determine information traits, Simulink decides them naturally through spread calculations and checking behaviour consistency to guarantee information accuracy.

1.6.3.3 Simulating the model

To simulate the behaviour of the structure and see the results, focus run time decisions, including the sort and properties of the solver, model start and stop times, and whether to load or extra propagation input. Different combinations could be saved with the model. We can run the changes spontaneously from the Simulink Editor or proficiently from the MATLAB summon line. It is possible to have different propagation modes working hand in hand.

- Normal (the default), which interpretively reproduces the model

- Accelerator, which increases generation execution by making and executing requested target code; yet gives the flexibility to change model parameters in the midst of run time.
- Rapid Accelerator, which can reproduce shows faster than in the Accelerator mode by making an executable that can run outside Simulink on a second changing focus.

To reduce the time required to run different models, we can run the models in parallel on a multifocus machine or machine bundle. While running a model, we can break down the model using MATLAB and Simulink. Simulink fuses debugging instruments to help us understand the proliferation conduct. We can picture the run time lead by study markers with the showcases and degrees given in Simulink. With the Simulink debugger, we can wander through a proliferation system without a moment's delay and break down the eventual outcomes of executing the technique.

1.6.4 Circuit elements

Electrical components, for example, resistors, capacitors and inductors, are utilized in electrical systems. An electrical system might be broken down into different, interconnected electrical components in a schematic graph or circuit outline, each of which influences the voltage in the system or current through the system. These perfect electrical components shown in Figure 1.16 represents physical, electrical or electronic segments; yet they do not exist physically and are accepted to have perfect properties.

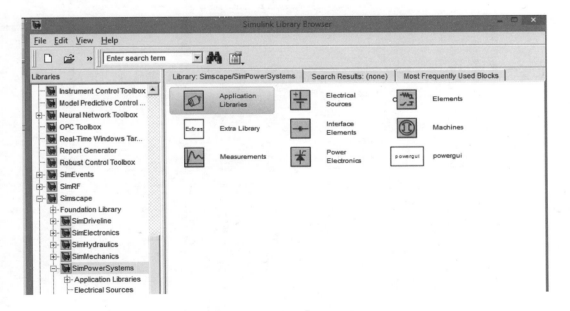

Figure 1.16   Circuit Elements.

1.6.4.1 Active elements

For a circuit to be legitimately called electronic, it must contain no less than one active device. Transistors and Integrated Circuits are some of the examples of active elements.

1.6.4.1.1 Dependent sources

In the hypothesis of electrical systems, a ward source is a voltage source or a current source whose worth lies on a voltage or current in the system.

1.6.4.1.2 Independent sources

An autonomous voltage source maintains a voltage (altered or shifting with time), which is not influenced by any other quantity. Similarly, an autonomous current source maintains a current (settled or time-fluctuating), which is unaffected by any other quantity.

1.6.4.2 Passive elements

A passive element is an element of the electrical circuit that does not create power, such as a capacitor, an inductor, a resistor or a memristor. Resistors, capacitors, inductors, transformers and even diodes are considered passive devices (Figure 1.17).

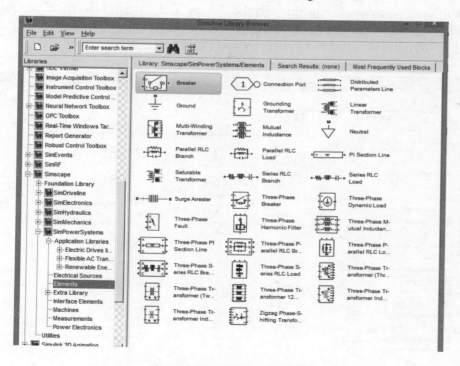

Figure 1.17 Passive Elements.

1.6.5 DC analysis

DC circuit measurements and calculations can be made easier with the help of MATLAB/Simulink tools. Even complex circuits can be derived and solved with minimum time consumption, which will be a credential for MATLAB users.

A simple DC circuit is derived here as an example:

Example 57

For Figure 1.18, which is a DC circuit, derive the loop currents I_1 and I_2 using mesh analysis.

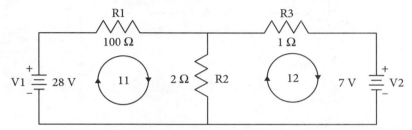

Figure 1.18 DC Analysis.

Compare the results using conventional and Simulink methods.

Solution

Conventional Method:

Using Kirchoff's Voltage Law (KVL),

Loop 1:

$$28 = 4I_1 + 2(I_1 - I_2)$$

$$28 = 4I_1 + 2I_1 - 2I_2$$

$$28 = 6I_1 - 2I_2 \tag{1}$$

Loop 2:

$$7 = 1I_2 + 2(I_2 - I_1)$$

$$7 = 1I_2 + 2I_2 - 2I_1$$

$$7 = -2I_1 + 3I_2 \tag{2}$$

Solving (1) and (2),
The loop currents $I_1 = 5$ A and $I_2 = 1$ A

Simulink Method:

Refer to Figure 1.19. Thus, from the conventional and Simulink methods, the loop currents derived from the DC circuit are the same.

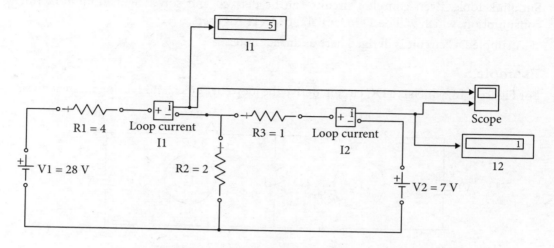

Figure 1.19 Simulink Model.

1.6.6 AC analysis

Similar to DC circuits, analysis of ac circuits can also be done in a simple manner. Consider the example as shown in Figure 1.20.

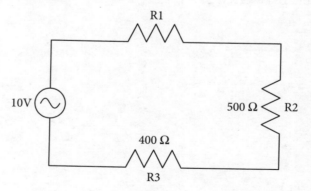

Figure 1.20 AC Analysis.

Example 58

For the circuit shown in Figure 1.20, calculate the total current flow and the voltage across each resistor.

Compare the results using Simulink method.

Conventional Method:

The total resistance value in the closed circuit is

$$R_{total} = R_1 + R_2 + R_3$$
$$= 100 + 500 + 400$$
$$= 1000 \, \Omega$$

$$I_{total} = V/R_{total} = 10/1000$$
$$= 10 \, mA$$

The total current flow in the circuit $I = 10$ mA

Voltage across resistor R_1:

$$V_{R1} = I_{total} \cdot R_1$$
$$= 10 \, mA \times 100 \, \Omega$$
$$= 1 \, V$$

Voltage across resistor R_2:

$$V_{R1} = I_{total} \cdot R_2$$
$$= 10 \, mA \times 500 \, \Omega$$
$$= 5 \, V$$

Voltage across resistor R_3:

$$V_{R1} = I_{total} \cdot R_3$$
$$= 10 \, mA \times 400 \, \Omega$$
$$= 4 \, V$$

The voltages across the resistors R_1, R_2, and R_3 are 1 V, 5 V, and 4 V.

Simulink Method

From the simulation results (Figures 1.21 and 1.22), it is proved that the values obtained using the conventional method and the Simulink method are the same. Hence, the ac circuit is verified.

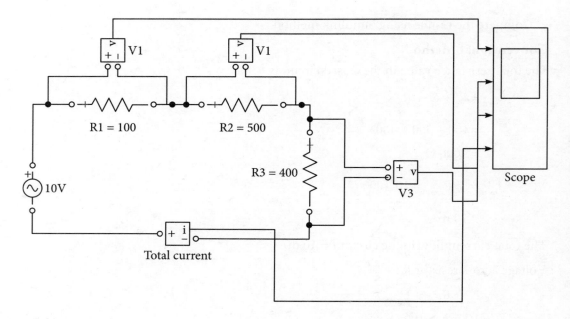

Figure 1.21 Simulink Model.

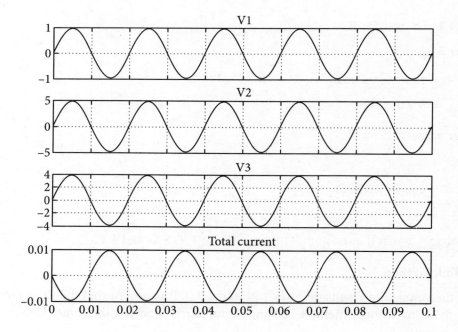

Figure 1.22 Scope Output.

1.7 Examples of MATLAB Simulations

Listed are some samples of MATLAB simulation circuits and their output responses.

1.7.1 Steady state analysis of a linear circuit

Example 59

Simulation Model

Circuit Description

This linear system shown in Figure 1.23 consists of different inductor currents and capacitor voltages, with two input sources (V_s, I_s) and two sets of outputs: current and voltage measurements. When an Z source filter is allied with a feed of inductive source, harmonic current is injected at the bus bar. An impedance measurement block is used to work out the impedance in opposition to frequency of the circuit.

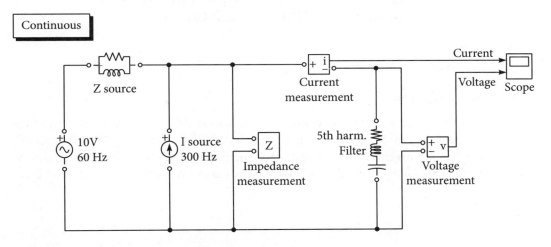

Figure 1.23 Steady State Analysis of a Linear Circuit with RLC.

Scope Output

Save the Simulink block, compile for any errors and then click run. The response of the voltage and current across the fifth harmonic filter can be viewed in the scope, which is shown in Figure 1.24.

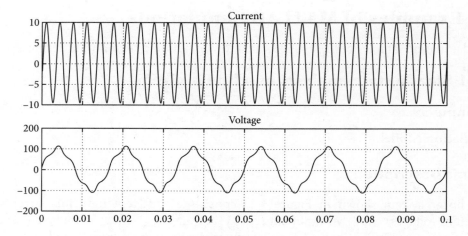

Figure 1.24 Scope Output for Steady State Analysis of a Linear Circuit.

1.7.2 Resonant switch converter using metal oxide semiconductor field effect transistor (MOSFET)

Example 60

Simulation Model

Circuit Description

The MOSFET in Figure 1.25 is utilized within a resonant switch converter topology and this topology portrays the device current moving through the resonant tank (L_r, C_r), subsequently bringing about zero current turn off. The design is demonstrated as a perfect current source, where the MOSFET is determined by a 2 MHz pulse generator with 20 per cent duty cycle.

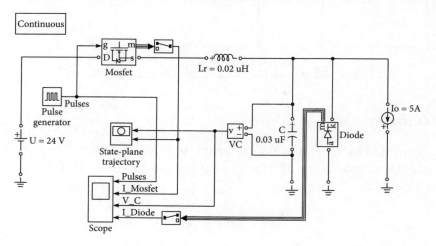

Figure 1.25 Resonant Switch Converter using MOSFET.

To stop from falling a current source and an inductance (L_r), a 100 kohm resistive snubber is joined over the MOSFET. The circuit can likewise be made discrete by selecting a ceaseless electrical model.

Scope Output

Save the Simulink block, compile for any errors and then click run. The response of the voltage and current across the semiconductor switches can be viewed in the scope, which is shown in Figure 1.26.

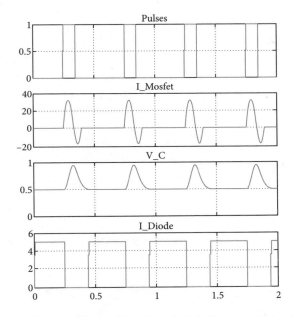

Figure 1.26 Scope Output of Resonant Switch Converter Using MOSFET.

1.7.3 Gate turn off (GTO) thyristor-based converter

Example 61

Simulink Model

Circuit Description

A captivated snubber circuit shown in Figure 1.27 comprising a D_s, R_s, and C_s joined over the GTO. The snubber circuit inductance L_s is additionally taken care of. The working frequency is 1 kHz and the duty cycle is 60 per cent; with perfect switches, the hypothetical estimation of the mean load current can be endured.

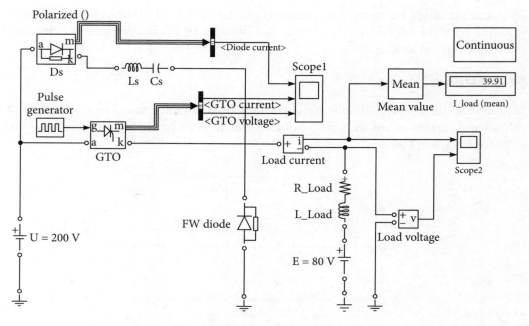

Figure 1.27 GTO-based Converter.

Scope Output

Save the Simulink block, compile for any errors and then click run. The simulation results at the load is shown in the scope (Figure 1.28).

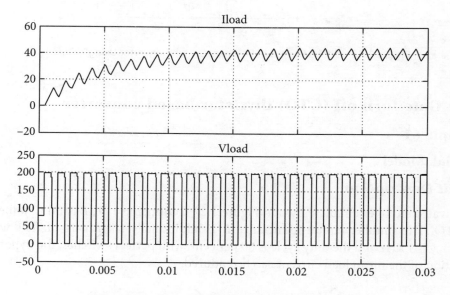

Figure 1.28 Scope Output of GTO-based Converter.

1.7.4 Regulation of zener diode

Example 62

Simulation Model

Circuit Description

Zener diodes shown in Figure 1.29 are usually utilized in applications as voltage controllers. The circuit demonstrates an AC source sustained to a step down transformer. The output of the transformer is then rectified by utilizing a diode extension and smoothened utilizing a capacitive channel. The Zener diode then acts to direct the output voltage to the Zener voltage 10 V. The input current in the Zener is constrained by the resistor (R) point of confinement to admissible qualities. The programmable voltage source is set up to build its output voltage at 0.1 s. As the source voltage builds, so does the voltage connected at the input of the Zener. Nonetheless, the Zener can direct the output only as long as the length of its input current is beneath the greatest defined quantity. The current increments as we expand the source voltage, and the Zener at last falls flat. With the Zener behaving like an open circuit at fault, voltage regulation is lost and the output of the capacitive channel gets connected to the load.

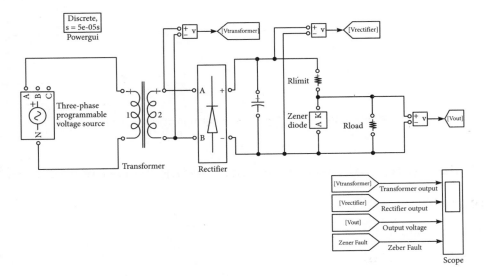

Figure 1.29 Regulation of Zener Diode.

Scope Output

Save the Simulink block, compile for any errors and then click run. The response of the Zener fault, transformer response, rectifier ouptut, and voltage levels can be viewed in the scope, which is shown in Figure 1.30.

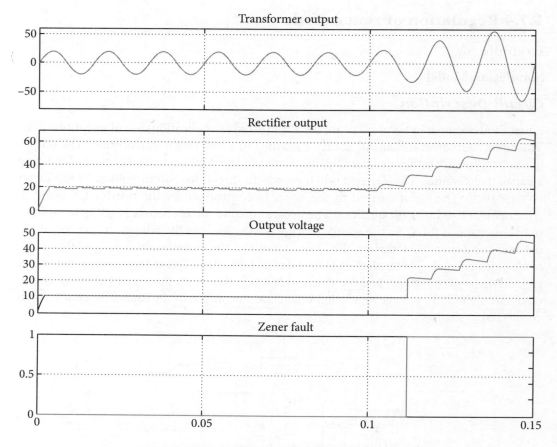

Figure 1.30 Scope Output of Regulation of Zener Diode.

1.7.5 Regulation of pulse generator using thyristor converter

Example 63

Simulation Model

Circuit Description

A pulse is synchronized on a voltage source where it is generated. The DC motor shown in Figure 1.31 representing a basic RL-E model has an inductive three-phase source through a six-pulse thyristor span. The converter output current is controlled by a Proportional Integral (PI) current controller assembled with Simulink. A step sign is connected to the reference input to test the element reaction of the current controller.

Introduction to MATLAB 51

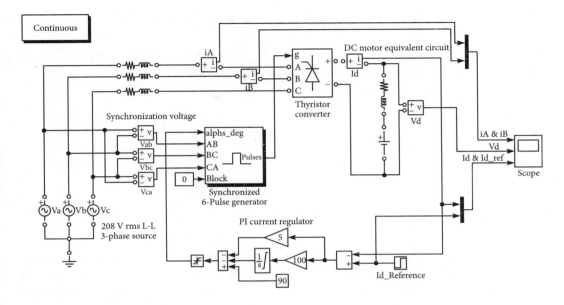

Figure 1.31 Regulation of Pulse Generator Using Thyristor Converter.

Scope Output

Save the Simulink block, compile for any errors and then click run. The response of the voltage and current measurements can be viewed in the scope which is shown in Figure 1.32.

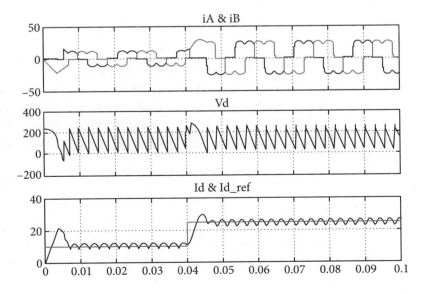

Figure 1.32 Scope Output of Regulation of Pulse Generator Using Thyristor Converter.

1.8 Other Types Circuit Simulators

MATLAB is very expedient for mathematical analysis and linear algebra. It is often used in the engineering field. Other types of circuit simulators include:

- PSpice
- LabVIEW
- PSIM
- Scilab
- VisSim

1.8.1 PSpice

PSpice is a general purpose software package that simulates completely different circuits and performs varied analysis of electrical and electronic circuits together with time-domain response, small signal rate response, total power indulgence, nodal voltages and division current in a circuit, transitory analysis, determination of use of transistors, determinations of shift response, etc. This software package contains models for various circuit parts. PSpice has the following limitations:

- Certain versions of PSpice is constrained to investigate circuits only up to 10 transistors exclusively.
- PSpice does not support repetitious methodology of answers.
- Distortion analysis is not attainable.

1.8.2 LabVIEW

LabVIEW is a completely featured programming language created by National Instruments. It is a graphical language, which is quite distinctive within the technique by which the code is prepared and saved. There is no text-based code in and of it; however, the data flows through the program. Therefore, LabVIEW is preferred by those who like to usually visualize the information flow instead of using a text-based typical programming language to understand a task.

1.8.3 PSIM

Physical Security Information Management (PSIM) is a class of computer code that has a stage and application, created by middleware designers and developed to integrate numerous independent protection applications and strategy to manage them through one inclusive computer program. PSIM is an electronic circuit simulation computer code package, designed specifically to be used in power electronics and motor drive simulations; however, it can simulate any electronic circuit. PSIM uses nodal analysis and the tetragon rule integration because of the basis of its simulation rules. It provides a schematic capture interface and a wave reader Sim view. PSIM has many modules that stretch its practicality into specific areas of circuit simulation and style including: management theory, electrical motors and wind turbines. It is employed by businesses for analysis and merchandise development and by instructional establishments for analysis and teaching.

1.8.4 Scilab

Scilab is an open-source, cross-stage numerical methodology package. It is utilized for sign methodology, factual investigation, picture transforming, liquid progress models, numerical advancement and demonstration and typical controls. The language gives a programming environment, with grids in view of the principle learning sort. By system based processing, element composing and programmed memory administration, a few numerical issues are communicated among the reduced scope of code lines, as contrasted with FORTRAN, C or C++. This empowers clients to rapidly build models for an assortment of scientific issues. Although the language gives straightforward lattice operations, Scilab gives gigantic records of abnormal state operation, for example, in three-dimensional modelling.

1.8.5 VisSim

VisSim is a pictorial slab diagram language that can be used for dynamic system simulation and for model-based design of embedded systems. The software was developed by the Visual Solutions of Westford, Massachusetts. VisSim is engaged with the paradigm of graphical data flow to implement differential equations on dynamic systems. The modeling of startup sequencing and serial protocol decoding is much easier in VisSim. Some of the demerits include:

- Coding requirements are significant
- Difficulty in usage
- Less intuitive than other models
- Learning curve may be steeper
- Runtimes can be slow, especially in 3D mode or for large network
- Occasionally get cryptic german error messages

1.9 Merits and Demerits of MATLAB

1.9.1 Merits

MATLAB is an interpret language for arithmetical working. It allows one to execute numerical calculations and envisages the outcome without the need for intricate and time-consuming encoding. MATLAB allows the users to precisely resolve difficulties, create graphics directly and construct rules competently.

- MATLAB lets you do convoluted calculations promptly, without writing a complete program or worksheet.
- Plenty of workbench and add-ons are accessible to do a wide ranges of jobs.
- MATLAB's functionality can be significantly stretched out by adding a number of toolboxes.
- MATLAB lets us work in matrices without difficulty.
- MATLAB is relatively vector operated.
- Plotting can be done easily using graphical interactive tools.

1.9.2 Demerits

MATLAB is a highly interpreted language that utilizes a large amount of memory, which slows down the operating speed, which in turn will affect real-time applications and makes it complicated.

Summary

After reading this chapter, the beginner will be able to compute numerical concepts from basics to the complexities. One can easily understand the design concepts of various fields using MATLAB and will be able to excel in various levels of programming in MATLAB. The programming sequences are given step by step which can facilitate better programming.

Review Questions

1. Explain the arithmetic models used in MATLAB.
2. Compare the different simulation module softwares and comment.
3. Explain the concepts of vectors and matrices used in MATLAB.
4. Explain the various control structures and operators used in MATLAB with examples.
5. List out the merits and demerits of MATLAB.

6. Explain the concept of DC and AC analysis of electric circuits using MATLAB.
7. Explain the plotting methods used in MATLAB with relevant examples.
8. Design a simple electric circuit using MATLAB and compare the results using theoretical calculations.
9. For a sphere having a radius of 15 cm, determine the length of the sides of the cube with the same surface area as the sphere; also determine the length of the sides of the cube, which has the same volume as that of the sphere.
10. A ball is being dropped at a height of $h = 2$ m, where the velocity is given by $v^2 = 2gh$ and rebounds with a velocity which is 85 per cent of the velocity impact. What is the height after the eighth bounce when the ball rebounds to a height of $h = v^2/2g$?

Practice Questions

1. Compute: $1 - 345$.
2. Compute to 15 digits:
   (a) $\cosh(0.1)$
   (b) $\ln(5)$
   (c) $\arcsin(1/2)$
3. Simplify the following expressions:
   (a) $5/(1 + 4/(1 + 1a))$
   (b) $\sin 2x - \tan 2x$
4. Compute 337777789701 both as an approximate floating point number and as an exact integer.
5. Try to solve the system of linear equations
   $3x - 1y + 8z = 7$
   $2x - 3y + 7z = -1$
   $x - 6y + z = 3.$
   Cross check the answer using matrix functions.
6. Use plot or ezplot, to graph the following functions:
   (a) $x = a^2 - a$ for $-4 \le a \le 4$
   (b) $y = \cos(x/4)$ for $-\pi \le x \le \pi$, $-10 \le y \le 10$
   (c) $y = e - x^2$ and $y = x^4 - x^2$ for $-2 \le x \le 2$
7. Plot the functions x^3 and $7x$ on the same graph and determine how many times their graphs intersect.
8. Plot the following surfaces:
   (a) $z = \tan x \tan y$ for $-3\pi \le x \le 3\pi$ and $-3\pi \le y \le 3\pi$
   (b) $z = (x^2 + y^2) \sinh(x^2 + y^2)$ for $-1 \le x \le 1$ and $-1 \le y \le 1$
9. Enter the matrix M by
   $> X = [1, 3, 4, 6; 2, 4, 0, -1; 0, -2, 3, -1; -1, 2, -5, 1]$
   and also the matrix
   $Y =$

$$\begin{matrix} 1 & 2 & 5 \\ 4 & 4 & 6 \\ 1 & 1 & 1 \\ 6 & 5 & 7 \end{matrix}$$

Multiply X and Y using X * Y. Can the order of multiplication be switched?

10. Set up a vector x, which contains the values from zero to one in steps of one-fifth.
11. Debug the code that is supposed to set up the function f(x) = x² cos(x + 1) on the grid x = 0 to 5 in steps of 0.155 and give the value of the function at x = 1 and x = 3.
12. For Figure 1.18, verify the superposition theorem using both conventional and Simulink method.
13. Calculate the integration of matrices using MATLAB for (cos^3(x)*sin^10(x) ranging from 0 to π/2.
14. Consider two one-dimensional arrays namely a and b using the function interleave for c, at the instance where a = [2, 4, 5, 3, 4] and b = [6, 7, 2, 5] and c = [−1, −5, 3, 5, −7].
15. Write a MATLAB function to plot the graph of an equilateral triangle with the vertices (a, c) and (a, b).
16. Solve the following equations using the functions solve or f zero.
    (a) 9x + 4 = 0
    (b) 8x + 3 = 0 (numerical solution to 15 places)
    (c) x³ + px+ q = 0 (solve for x in terms of p and q)
    (d) eˣ = 8x − 4 (all real solutions). It helps to draw a picture first.
17. Solve the following question using contour function:

    Plot the level curves of the function f (x, y) = 3y + y³ − x³ in the region where x and y are between −1 and 1
18. Determine the derivatives of the given functions.
    (a) f(x) = 6x³ − 5x² + 2x − 3.
    (b) $f(x) = \dfrac{2X - 1}{X^2 + 1}$
    (c) f(x) = sin(3x²+ 2)
    (d) f(x) = arcsin(2x + 3)
    (e) f(x) = √1 + x⁴
    (f) f(x) = xr
    (g) f(x) = arctan(x² + 1)
19. For the following integrals, check the results by x² differentiating.

    (a) $\int_0^{\pi/2} \cos x \, dx$

    (b) xsin(x²) dx

    (c) sin(3x)√1 − cos(3x) dx

    (d) $\int x^2 \sqrt{x} + 4 \, dx$

    (e) $\int_{-\infty}^{\infty} e - x^2 dx$.

20. Determine the Taylor polynomial for the following functions.
    (a) $f(x) = e^x$, $n = 7$, $c = 0$
    (b) $f(x) = \sin x$, $n = 5$ and 6, $c = 0$
    (c) $f(x) = \sin x$, $n = 6$, $c = 2$
    (d) $f(x) = \tan x$, $n = 7$, $c = 0$
    (e) $f(x) = \ln x$, $n = 5$, $c = 1$
    (f) $f(x) = \text{erf}(x)$, $n = 9$, $c = 0$

21. Plot the following equations:
    (a) $z = \sin x \sin y$ for $-3\pi \leq x \leq 3\pi$ and $-3\pi \leq y \leq 3\pi$
    (b) $z = (x^2 + y^2) \cos(x^2 + y^2)$ for $-1 \leq x \leq 1$ and $-1 \leq y \leq 1$

22. Plot the following surfaces filled with red circles.
    (a) $z = \sin x \sin y$ for $-3\pi \leq x \leq 3\pi$ and $-3\pi \leq y \leq 3\pi$
    (b) $z = (x^2 + y^2) \cos(x^2 + y^2)$ for $-1 \leq x \leq 1$ and $-1 \leq y \leq 1$

23. Write a function by taking an input string containing a text file to create a histogram of the number of occurrences of each letter from A to Z.

24. For the given vectors x and y of equal length, perform the following:
    - Calculate the cumulative integral of y and store the result in the variable integer
    - Store in fit after fitting a second-order line to the integral
    - Store in new integ after evaluating the original value of x
    - Draw the plot integ and new integ against x

25. You are given a sound called 'sound.wav'. Determine the duration of the sound (how many seconds it plays for) and store this duration. Then, plot the amplitude of the sound against time.

26. For the given function $f(x) = x^2$, write the code to plot the rotation around the x-axis.

27. Given a vector v, write a code that creates the following items:
    - vecA, which contains all the elements of v which are less than 9
    - vecB which is v in reverse order
    - vecC, which contains only the elements of v which are at odd indices
    - D, which is the average of the values in v
    - vecE, which is a vector of length 5 with random values between 5 and 7

28. Compute e^{14} and 3801π to 15 digits each. Which is greater?

29. The fractions 09/124, 183/400 and 24/75. Which of these is the best approximation to $\sqrt{7}$?

30. Plot the curve $3y + y^3 - x^3 = 5$.

31. Plot the level curve of the function $f(x, y) = y \ln x + x \ln y$ that contains the point (1, 1).

Multiple Choice Questions

1. State the full form for MATLAB?
    (a) Math Laboratory
    (b) Matrix Laboratory
    (c) Mathworks
    (d) Nothing (e) None of the above

2. The symbol that precedes all comments in MATLAB is
   (a) "   (b) %   (c) //   (d) <   (e) None of the above
3. The predefined variable that is not used in MATLAB is
   (a) pi   (b) inf   (c) i   (d) gravity   (e) j
4. The command that clears all the data stored in memeory is
   (a) clc   (b) clear   (c) delete   (d) deallocate   (e) None of the above
5. The character that is used in MATLAB, which represents their value in memory is
   (a) decima   (b) ASCII   (c) hex   (d) string   (e) None of the above
6. In MATLAB, which is not an aspect of for/while loop?
   (a) updat   (b) initializatio   (c) runner   (d) condition   (e) All are aspects of loops
7. To prevent unnecessary memory allocations and memory management, MATLAB uses
   (a) Vectors   (b) Scalars   (c) Matrix math   (d) Delayed copy   (e) Licenses
8. Which is the escape character that is used to print a newline using fprintf statement?
   (a) \t   (b) \nl   (c) \nxt   (d) \n   (e) None of the above
9. Which is the keyword that is used to move to the next iteration of a loop immediately?
   (a) update   (b) goto   (c) continue   (d) break   (e) None of the above
10. The syntax used to represent symbols x, y and z are
    (a) sym (x, y, z)   (b) syms x y z   (c) syms x, y, z   (d) sym x, y, z   (e) None of the above
11. Assuming that there is one element in the vector, the best way to access the first element in a vector named v is
    (a) v(0)   (b) v(1)   (c) v   (d) v(:, 0)   (e) None of the above
12. In MATLAB, which is the command used to see whether two elements are same or not?
    (a) !=   (b) ==   (c) is equal   (d) =   (e) None of the above
13. For a given vector [1 2 3 4; 11 24 92 100; 345 65 90 1], the value of 'a' when the command MATLAB>>[a,b] = size(vector) is
    (a) 1 2 3 4   (b) 12   (c) 1   (d) 4   (e) 3
14. Determine the answer for the code run:isnumeric(32)
    (a) 1   (b) 0   (c) 32   (d) yes   (e) true
15. Which is the command that is used to save a formatted string to memory, but that need not be printed out?
    (a) fprintf   (b) sprintf   (c) disp   (d) echo   (e) None of these
16. Executing in the command window, the following code a = (1:3)'; (a(1,2))' returns
    (a) error message   (b) 1   (c) 2   (d) 12   (e) None of the above
17. Executing in the command window, the following code a = ones(3,4); b = a(4,3), size(b); returns:
    (a) 1   (b) −1   (c) 11   (d) error message   (e) None of the above
18. Executing the command is keyword size returns 0, that is, size is not a MATLAB keyword. Given this information, which of the following statements shows the result of executing the following line in the command window size = (1:3)'; size(size)
    (a) wrong result   (b) 12   (c) 21   (d) 22   (e) None of the above

19. Executing in the command window, the code a = [1:2]'; size(a) returns
    (a) error message (b) 12 (c) 21 (d) 2 (e) None of the above
20. Executing in the command window, the code a = 1:2; size[a] returns
    (a) error message (b) 12 (c) 21 (d) 2 (e) None of the above
21. Which of the following statements is correct?
    (a) Function 'clear' clears only the scalar variables, but does not clear arrays
    (b) Function 'clc' clears all variables
    (c) Function 'clc' clears only constants
    (d) Function 'clear a' clears all variables of the type 'array'
    (e) None of the above
22. Executing in the command window, the code A = eye(1,10); size(A(2, 2)) returns
    (a) 1 (b) 0 (c) error message (d) 11 (e) None of the above

2

MATLAB Simulation of Power Semiconductor Devices

> **Learning Objectives**
> - To introduce the concept of simulation using MATLAB for power semiconductor devices
> - To enumerate the various features of power electronics and its applications
> - To represent the characteristics of the power semiconductor devices using simulation waveforms

2.1 Introduction and Outlook

A semiconductor is a robust device whose electrical conductivity is controlled over a wide range for superior or dynamic control. Semiconductor devices are electronic components that, like insulators, exploit the electronic properties of semiconductor resources, principally silicon, germanium, and gallium arsenide, as well as natural semiconductors. Semiconductors are financially and commercially attractive; these solid state devices have replaced vacuum tubes in most applications.

Semiconductors are remarkable in their innovation and financial viability and financially. Semiconductors are the crucial materials for many electrical devices, from machines to cell phones to advanced sound players. When the circuit is not connected, pure semiconductors and insulators have generally comparable electrical properties.

Semiconductors' inherent electrical properties are altered by adding impurities, a technique referred to as doping. In general, it is a best practice to imagine that the impure particles hold an electron that could flow freely. When the number of dopants increase, the semiconductor materials carry the power forward. The intersections of the semiconductors are doped with a definite kind of impurity, which may generate electrical fields imperative to the operation of the semiconductor. In addition, there would be a progressive change in the property of the semiconductor due to the application of the electric field. The ability to control the physical phenomenon in semiconductor materials statically through doping and the application of electrical fields has resulted in a wide range of semiconductor devices – in the same way as transistors, metal–oxide–semiconductor field-effect transistors (MOSFETs), insulated-gate bipolar transistors (IGBTs), integrated gate-commutated thyristors (IGCTs) and so on.

Semiconductor devices with progressively controlled physical phenomenon are the building blocks of integrated circuits. These 'dynamic' semiconductor devices are consolidated with adaptable uninvolved parts, in a similar way as capacitors and resistors to create power electronic devices. In some semiconductors, when an electron moves from the conduction region to the valence region, the energy levels below and above the band hole commonly discharge light. This is the principle behind the light-emitting diode (LED) and other semiconductor optical devices that have become economically so vital to our lives. In fact, semiconductors have led to photo detectors, fibre optic communications and even solar-based cells.

2.2 Why is Power Electronics Important?

Power electronics is the principle and technology behind solid state devices that controls and renovates electric power. It further refers to that area in electronic and electrical exchange that deals with planning and control, processing and integration of nonlinear, time variable power dispensation of electronic systems. Power electronics brings out essential technological opportunities that are presently enhancing the role and cost of power in all aspects from innovation to the end of use. It helps in improving the part and estimation of power.

Power electronics plays an important role in the following areas:

- Conversion of power from AC to DC, AC to AC, DC to AC and so on.
- Management of distortions, harmonics, voltage dips and over voltages, along lines enhancing the nature of power.
- Frequent control of electrical parameters such as voltage, current, impedance and phase.

2.3 Features of Power Electronics

Before power electronics became a specialized field, electrical transformation was normally dodged in applications. A definitive reason was the trouble in execution. The exchange occurred in the generator–engine set. The primary downside was less uniform quality, less transformation effectiveness, gigantic size, and noise contortion.

Nowadays, transformations are performed with semiconductor power devices, for instance, with thyristors and the transistor family in the early 1950s. Compared to electronic systems concerned with transmission and planning of signs and data, in power electronics noteworthy measures of electrical imperativeness are taken care of.

The concept of power electronics involves the examination of electronic circuits that are used to change and control the entire stream of electrical power. Power electronics methodology, as mentioned earlier, generally deals with solid state electric devices and circuits used in the field of control systems and electric power changing applications.

The field of power electronics covers a wide range of power from a few microwatts to thousands of watts. It investigates power, electronics and control.

- Power, in the sense of transmission, dissemination, and the continuous use of electrical power.
- Electronics, in the sense of analysis of power semiconductor devices and transformations in electric circuits at low power levels.
- Control systems, in the sense of investigation of the response or reaction and the steadiness of the circuit in a closed loop system.

Power electronics involves the utilization of hardware for control and transformation of large amounts of electrical power. The block diagram of a power electronics system is shown in Figure 2.1.

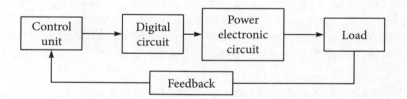

Figure 2.1 Basic Block Diagram of a Power Electronic System.

Initially, the primary source for power supply may be either AC or DC, which is focused throughout the application. The output response of the power electronic circuit may be variable AC or DC, or even a variable voltage and frequency, which is focused around the requirement. Hence, such circuits are referred to as power electronics transformation circuits. The power electronics circuits are depicted as electric power starting structural circuits. The investigation part includes the measurement of load signals in comparison with the input signals.

Electro magnetic interference (EMI) can develop from numerous sources, being either man-made or inherent. It can likewise have an assortment of qualities depending upon its source and the way the instrument is placed in the circuit. Man-made EMI, by and large, develops from different hardware circuits, although some EMI can emerge from the exchange between substantial streams, and so forth. Thus, EMI can develop from numerous sources—different conditions, lightning and other sorts of noise, all contribute.

Load circuits are controlled by advanced techniques conforming to the order of the signal. The role of filters in the circuit operation is very essential where control is necessary to reduce noise produced by the converter from critical mains or when there is breakdown. The significant order of power electronics systems is as follows:

- DC–DC transformation: Change and control voltage size
- AC–DC correction: Possibly control DC voltage and AC current
- DC–AC reversal: Produce sinusoid of controllable magnitude and frequency
- AC–AC transformation: Change and control voltage magnitude and frequency

Merits

- High productivity because of low adversity in semiconductor devices.
- High uniform quality of power electronic converter systems.
- Longer life and less support needed because there are no moving parts.
- Flexibility in operation.
- Fast dynamic reaction compared to electromechanical converter systems.
- Small size and less weight, therefore low establishment cost.

Demerits

- Circuits in power electronics systems tend to create sounds in the supply system and the load circuit.
- AC to DC and AC to AC converters work at low input power under certain working conditions.
- Regeneration of power is troublesome in power electronic converter systems.
- Power electronic controllers encompass low slide capacity.

2.4 Applications of Power Electronics

The applications of power electronics include battery chargers, AC connectors, light stabilizers, variable frequency drives and DC motor drives used to work pumps, assembling apparatus, fans and other power transmission structures that are used to interconnect electrical lattices. The aim of power electronic converters in all aspects and their purpose in various electronic devices are discussed in Table 2.1.

Table 2.1 Power Electronic Converters

S. No.	Converter	Description	Application	Commercial Use
1	DC–DC	To maintain voltage at a fixed value whatever the voltage level of the battery is	Used for electronic isolation and power factor correction	Mobile phones, personal digital assistants (PDAs)
2	AC–DC	Whenever the power electronic device is connected to the main supply, AC–DC conversion is required	Used for simple conversion from AC–DC; the voltage level will also be changed as part of the operation	Televisions and computers
3	AC–AC	Whenever the voltage level or the frequency level is to be changed, AC–AC conversion is required	Used for AC–AC conversion when frequency levels range from 50 Hz to 60 Hz in utility grids	Light dimmers and power adapters
4	DC–AC	Whenever the main supply fails, the inverter supplies on behalf of the mains from its storage battery	UPS or renewable energy systems	Emergency lighting systems

Power electronics circuits are used in pumps, blowers, power plant drives and also in business locales. Drive circuits are used to exchange power and for development control. For AC motors, the applications involve variable frequency drives, motor starters; they are also used in excitation systems. DC–DC converters are employed in electric vehicles; the role of DC–AC converters is to power the power motor. Electric trains use these power electronic circuits to get power and also for vector control using regulation rectifiers. Lift systems also use power electronic systems, such as thyristors, inverters, magnetomotors, or distinctive mixture schemas with a circuit regulation system and standard motors.

Power electronic circuits are reproduced using machine reenactment ventures, for instance, circuits are enacted using MATLAB before they are made, to test how the circuits behave under particular conditions. This makes the creation of circuits cost-effective and faster, which acts as an effective model for testing. Such accommodations can be furnished just with the offices of utilizing power semiconductor devices.

2.5 Power Semiconductor Devices in MATLAB/Simulink

In general, semiconductor devices are classified into four categories: two terminal devices, three terminal devices, four terminal devices, and multiterminal devices. Two terminal devices rely on the external power current to which the circuits are connected. They also direct the flow of current, allowing it to flow in one direction but not on the other. Three terminal devices are semiconductor devices whose state is not only dependent on the external power circuit but also on the signal of the driving terminal or gate (e.g., thyristor). Some frequently used two terminal and three terminal devices are explained in this chapter. The power semiconductor devices available for power electronics applications in MATLAB/Simulink are shown in Figure 2.2.

To select power semiconductor devices from the Simulink Library Browser, follow the given steps:

Goto

SimulinkLibraryBrowser>>SimScape>>SimPowerSystems>>PowerElectronics

The commonly used semiconductor switches that provide better outcome for power electronics are diodes, thyristors, IGBTs, MOSFETs, and gate turn offs (GTOs). An ideal switch would be one that serves as an alternate for all the switching devices in common. The purpose of the three-level bridge and universal bridge shown in Figure 2.2 are explained in the forthcoming chapters.

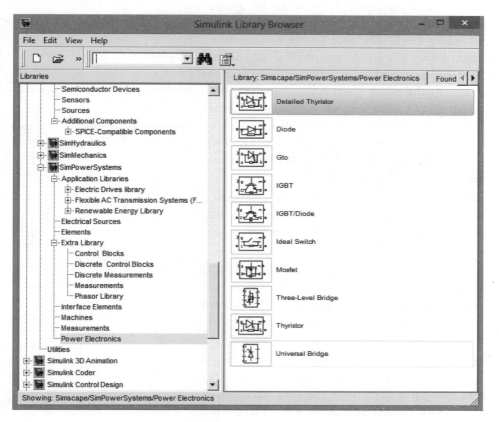

Figure 2.2 Power Semiconductor Devices.

2.5.1 Power diode and its characteristics

A power semiconductor device with two terminals, permitting the flow of current in only one direction is called a diode. The diode is said to be a restricted conductor with two terminals, namely anode, the positive terminal and cathode, the negative terminal as shown in Figure 2.3.

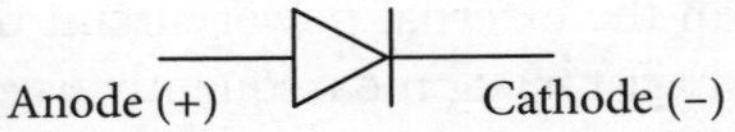

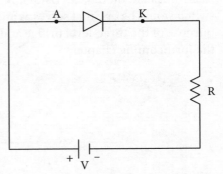

Figure 2.3 Symbol of Diode.

Power diodes are made of silicon p–n junction with two terminals, the anode and the cathode. The p–n junction is shaped by alloying, dispersion and epitaxial development. The diodes have the accompanying advantages such as high mechanical and warm uniform quality, high top converse voltage, low switch current, low forward-voltage drop, high proficiency, and conservativeness.

Figure 2.4 shows the schematic representation of a diode circuit. A diode is forward one-sided when the anode (A) is made positive with respect to the cathode (K). It directs completely when the diode voltage is more than the cut-in voltage (0.7 V for Si). A leading diode will have a little voltage drop crosswise over it. A diode is converse one-sided when the cathode is made positive with respect to the anode. At the point when it becomes converse one-sided, a little turn around current known as leakage current flow is present. This leakage current, shown in Figure 2.5, increments with expansion to the extent of the opposite voltage until heavy slide voltage reaches breakdown voltage.

Figure 2.4 Schematic Representation of a Diode Circuit.

In fact, a power diode is more intricate in structure and in operation than its low-power partners that must be altered to make it suitable for high-power applications.

Current passes through the diode when the anode terminal acts more positive than the cathode. Power diodes are used in rectifier circuits; they are also used as freewheeling diodes, for reverse voltage protection, and in voltage regulation circuits.

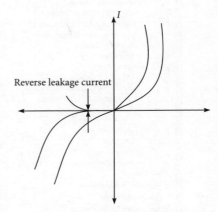

Figure 2.5 Reverse Leakage Current.

At low frequency and low current, the diode might be expected to behave as a perfect switch and the dynamic attributes of turn-on and turn-off qualities are not critical. Be that as it may, at high frequency and high current, the dynamic qualities assume an important part on the grounds that it increments power adversity and allows vast voltage spikes that might harm the device if appropriate assurance is not given to the device. The ON and OFF characteristics of a diode is shown in Figure 2.6

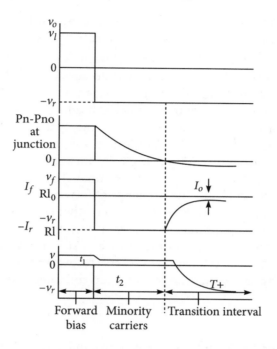

Figure 2.6 Diode On and Off Characteristics.

2.5.1.1 *Reverse recovery characteristic*

Reverse recovery characteristic is substantially more essential than forward recovery attributes on the grounds that it adds recovery troubles to the forward issues. At the point when diode is in forward conduction mode and after, its forward current is decreased to zero by applying reverse voltage. The diode keeps on leading because of minority bearers, which remains kept away in the *p–n* intersection and in the main part of the semiconductor material. The minority transporters take an opportunity to recombine with inverse charges and be zero. This time is known as the reverse recovery time. The reverse recovery time (t_{rr}) is measured from the beginning zero intersection of the diode current to 25 per cent of the greatest opposite current I_{rr}. T_{rr} has two segments, t1 and t2. t1 is as a consequence of charge storing in the exhaustion area of the intersection, that is, it is the time between the zero intersection and the top converse current I_{rr}. t2 is as a consequence of charge storing in the mass semiconductor material. The reverse recovery characteristic is shown in Figure 2.7.

$$T_{rr} = t_1 + t_2 \tag{2.1}$$

and

$$I_{rr} = t_1(di/dt). \tag{2.2}$$

The reverse recovery time relies on the intersection temperature, rate of fall of forward current and the greatness of forward current before recompense.

At the point when the diode is in the backward one-sided condition, the stream of leakage current is because of minority bearers. At that point, use of forward voltage would constrain the diode to convey current in the forward heading. But, a specific time known as forward conduction time, that is, turn-on time is required before all the transporters over the entire intersection can add to current flow. Typically, forward recovery time is not exactly the reverse recovery time. The forward recovery time restrains the rate of change of forward current and the exchanging speed. Reverse recovery charge Q_{RR} is the measure of charge bearers that stream over the diode in the opposite course because of the change of state from forward conduction to switch blocking condition. The estimation of reverse recovery charge Q_{RR} is resolved in the zone encased by the way of the opposite recovery current. Therefore,

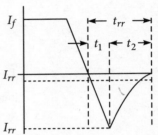

Figure 2.7 Reverse Recovery Characteristics.

$$Q_{RR} = 0.5 I_{RR} t_{RR} \tag{2.3}$$

The MATLAB/Simulink model of the diode is shown in Figure 2.8. To build a closed loop model of a power diode, the various blocks that are chosen from the Simulink Library Browser are as follows:

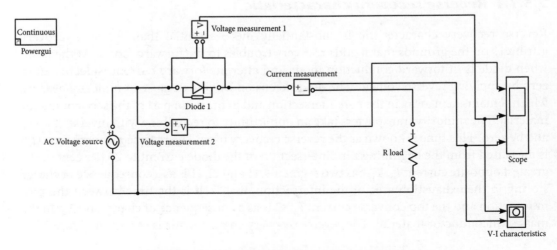

Figure 2.8 Simulink Model of a Diode (Forward Bias).

AC Voltage Source

SimulinkLibraryBrowser>>Simscape>>SimPowerSystems>>ElectricalSources>>ACVoltageSource

Diode

SimulinkLibraryBrowser>>Simscape>>SimPowerSystems>PowerElectronics>>Diode

Voltage and Current Measurement

SimulinkLibraryBrowser>>Simscape>>SimPowerSystems>Measurements>>

Voltage Measurements, Current Measurements, R Load

SimulinkLibraryBrowser>>Simscape>>SimPowerSystems>Elements>RLCBranch

Powergui

SimulinkLibraryBrowser>>Simscape>>SimPowerSystems>>Powergui

Scope

Simulink>>Sinks>>Scope

After dragging the aforementioned blocks, connect the blocks and make a closed loop circuit as shown in Figure 2.8. On researching the operation of the power diode, it leads only in one heading. The *V–I* characteristics of the diode is the curve between voltage over the diode and current through the diode. At the point when outside voltage is zero, the circuit is open and the potential obstruction does not permit current to flow. Along these lines, the circuit current is zero.

Forward Biasing

When the anode terminal of the diode is connected to the +ve terminal and cathode is connected to −ve terminal of the supply voltage as shown in Figure 2.8, forward biasing takes place. In this event, the anode terminal encounters a higher potential compared to the cathode terminal; the device is said to be forward inclined and a forward current will spill out of the anode to the cathode. This causes a little voltage drop over the device called the forward-voltage drop, which under ideal conditions is typically disregarded.

Example 1

The potential boundary is diminished when a diode is in the forward one-sided condition. At some forward voltage, the potential interference is complete and disposed of and current begins coursing through the diode further more in the circuit. At this point, the diode is said to be in the on state. The current increments with expanding forward voltage. The response of a diode can be demonstrated by setting the diode parameters as indicated in Figure 2.9.

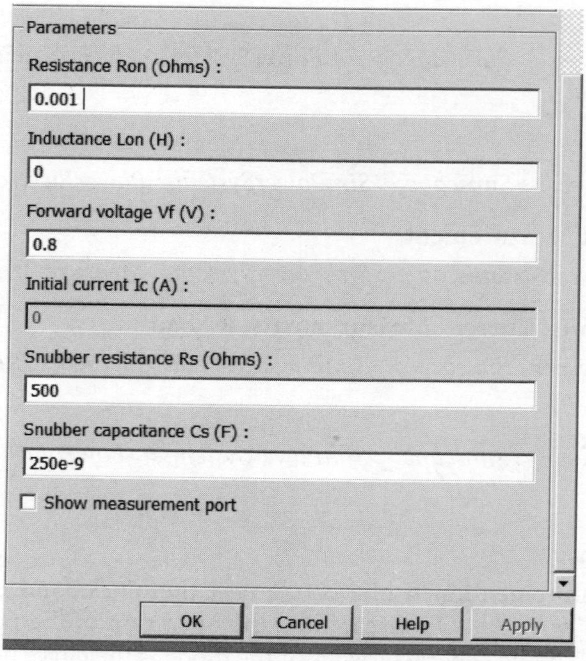

Figure 2.9 Parameters of a Diode.

Figure 2.10 displays the forward-biasing switching characteristics or the responses of the diode voltage and diode current.

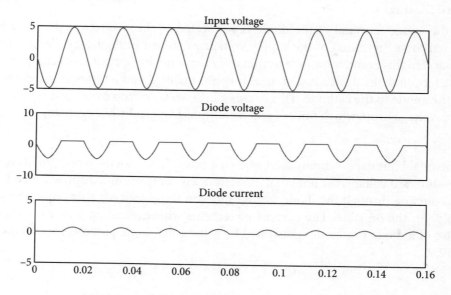

Figure 2.10 Switching Waveform of a Diode (Forward Bias).

The voltage and current relationships of any switch can be viewed using the *XY* graph available in Simulink>>Sinks. The *V–I* characteristics of an ideal diode under forward-biased condition is shown in Figure 2.11.

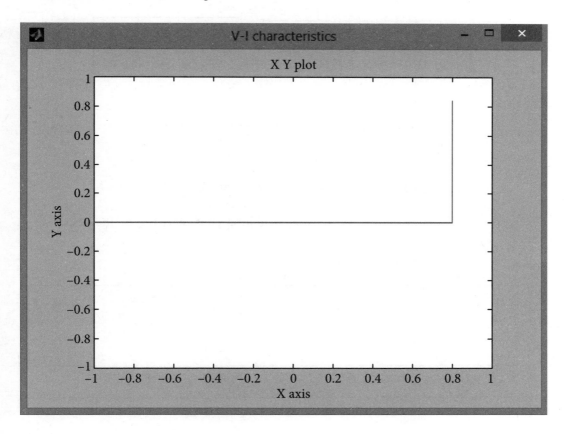

Figure 2.11 *V–I* Characteristics of a Diode (Forward Bias).

Reverse Biasing

The point when the cathode terminal of the diode is associated with the +ve terminal and the anode is associated with the −ve terminal of the supply voltage is known as reverse biasing; here the potential barrier over the intersection increases. By intricacy, when a diode is reverse biased, the diode experiences a slight current streaming in the inverse manner called the leakage current. Along these lines, the intersection safety gets to be high and a small current called reverse saturation current streams in the circuit. At this point, the diode is said to be in the off state. The reverse-biasing current is because of the minority charge carriers. The Simulink model of the reverse-biasing circuit of the power diode is shown in Figure 2.12.

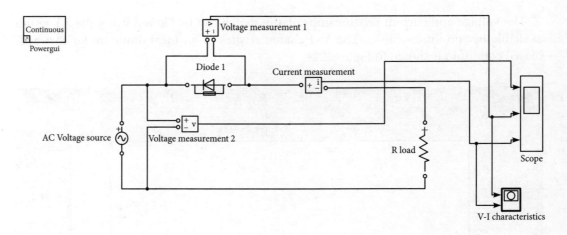

Figure 2.12 Simulink Model of a Diode (Reverse Bias).

Figure 2.13 displays the reverse-biasing switching characteristics or the responses of diode voltage and diode current.

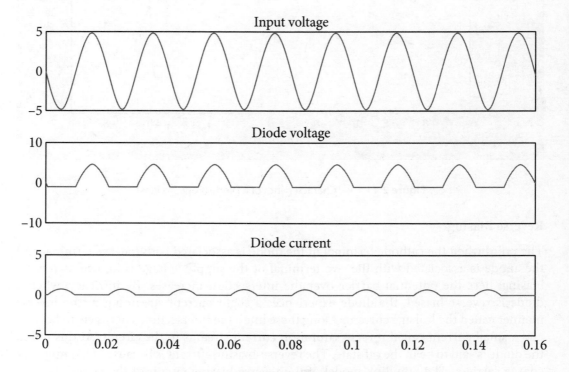

Figure 2.13 Switching Waveform of a Diode (Reverse Bias).

The *V–I* characteristics of an ideal diode under reverse-biased condition is shown in Figure 2.14.

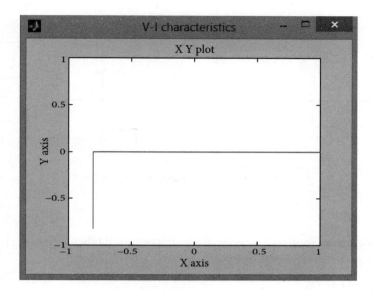

Figure 2.14 *V–I* Characteristics of a Diode (Reverse Bias).

Like the diode, all the other power semiconductor switches can be simulated and verified experimentally.

Note 1 All MATLAB/Simulink models simulated are set with default values. Users can modify their parameter set ups as per their application.

Note 2 Power semiconductor devices that are not supported by MATLAB are diode AC switch (DIAC), triode alternating current (TRIAC) and IGCT.

2.5.2 Zener diode

Zener diodes are fundamentally the same as the standard power diode; however they are intended to have a low reverse breakdown voltage that exploits this high turn around voltage.

To distinguish a zener diode from an ordinary diode, the symbol shown in Figure 2.15 is used. Zener diode image has a little 'label' connected to the bar of the diode image to recognize its capacity. Zener diodes or reference diodes might be utilized as thyristor devices, or they may be utilized inside integrated circuits. The Simulink model of a zener diode is shown in Figure 2.16.

Figure 2.15 Symbol of a Zener Diode.

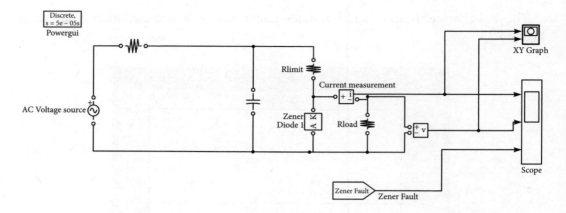

Figure 2.16 Simulink Model of a Zener Diode.

Example 2

To build a closed-loop model of a Zener diode, the required blocks that are chosen from the Simulink Library Browser are as follows:

AC Voltage Source

SimulinkLibraryBrowser>>Simscape>>SimPowerSystems>>ElectricalSources>>ACVoltageSource

Zener Diode

SimulinkLibraryBrowser>>Simscape>>SimPowerSystems>PowerElectronics>>

Diode (Cathode Made Positive w.r.t. Anode)

Voltage and Current Measurement

SimulinkLibraryBrowser>>Simscape>>SimPowerSystems>Measurements>>

Voltage Measurements, Current Measurements

R and C

SimulinkLibraryBrowser>>Simscape>>SimPowerSystems>Elements>RLCBranch

Powergui

SimulinkLibraryBrowser>>Simscape>>SimPowerSystems>>Powergui

Scope

Simulink>>Sinks>>Scope

XY Graph

Simulink>>Sinks>>XYGraph (Optional)

A normal power diode can be used as an alternate. For starters, in the normal power diode, the cathode made positive with respect to the anode would favour voltage regulation.

A Zener diode is a strongly doped diode, made to work in the breakdown area. A diode generally does not conduct when inverse one-sided. However, if the inverse biasing is stretched, at a particular voltage, it starts directing enthusiastically. This voltage is called breakdown voltage. High temperature through the diode can perpetually affect the device. To avoid high temperature, a resistor can be united with a Zener diode. Once the diode starts leading, it endures voltage over the terminals whatever may be the current through it, that is, it has low-dynamic security. Zener diodes are commonly utilized within voltage controllers. The diode is the most straightforward sort of voltage controller and the time when a Zener diode breaks down is known as the 'breakdown voltage'.

Figure 2.17 depicts the switching waveforms of a Zener diode, which is best suited for voltage regulation. The figure clearly pictures the reverse conduction mode and the load voltages and load current values while the diode is operated for R load.

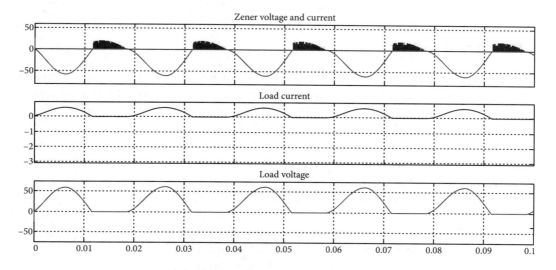

Figure 2.17 Switching Waveforms of a Zener Diode.

2.5.3 Fast recovery diode

Diodes have low recovery time, less than 5 s. The significant field of utilizations is in electrical power change, that is, in free-wheeling AC–DC and DC–AC converter circuits. Their current evaluations are from under 1 A to several amperes with voltage appraisals from 50 V to around 3 kV. Utilization of quick recovery diodes are ideal for free wheeling in thyristor circuits on account of low reverse characteristics, lower intersection temperature and diminished di/dt. For high-voltage appraisals more significant than 400 V, they are fabricated by dispersion process and the recovery time is controlled by platinum or gold diffusion. For under 400 V rating, epitaxial diodes give speedier exchanging rates than diffused diodes. Fast recovery diodes have an exceptionally restricted base width bringing about a quick recovery time of around 50 ns. Table 2.2 represents the various characteristics of different diodes used in power electronics.

Table 2.2 Characteristics of Various Types of Diodes

S.No.	Description	Normal Diode	Fast Recovery Diode	Schottky Diodes
1	Voltage and current ratings	Up to 5000 V and 3500 A	Up to 3000 V and 1000 A	Up to 100 V and 300 A
2	Reverse recovery time	High	Low	Very low
3	T_{rr}	Less than 25 s	$0.1 \geq 5$ s	Nanoseconds
4	Switching frequency	Low	High	Very high

2.5.4 Thyristors

Thyristors are the most important kind of power semiconductor device. They are broadly utilized as part of power electronic circuits. They change from nonleading to a directing state. A thyristor (Figure 2.18) is a four layer semiconductor that is regularly utilized for taking care of a large amount of power. A thyristor can be turned on or off; it can also manage power utilization, which may be referred as phase power control. This permits the measure of power output to be controlled by changing the edge of the current input. A sample of this is the dimmer switch for a light. The thyristor in the dimmer switch is a combination of a thyristor and a thyratron. The thyristor family includes a silicon controlled rectifier (SCR), TRIAC, DIAC and GTO. Among these the SCR has various exceptional attributes. It has three terminals: anode, cathode and gate. The gate is the control terminal, while the primary current streams between the anode and the cathode.

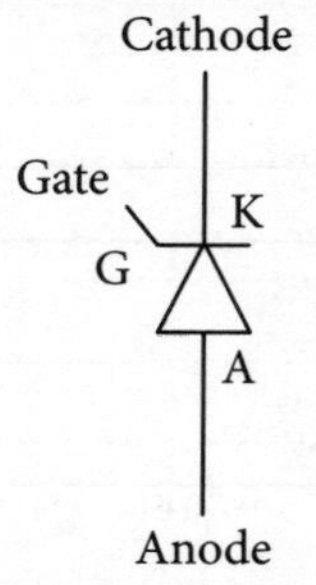

Figure 2.18 Symbol of an SCR.

The SCR is a four layer three terminal device with junctions J_1, J_2, J_3 as shown in Figure 2.19. The development of SCR demonstrates that the gate terminal is kept closer to the cathode. The estimated thickness of every layer and doping density are as shown in the figure. As far as their horizontal measurements are concerned, thyristors are the biggest semiconductor devices made. A complete silicon wafer as big as 10 cm in distance across might be utilized to make a solitary high-power thyristor.

At the point when the anode is made positive with reference, the cathode intersections J_1 and J_3 are forward one-sided and intersection J_2 is opposite one-sided. When the anode to cathode voltage V_{AK} is small, a leakage current courses through the device. The SCR is then said to be in the forward-blocking state. In the event that V_{AK} is further increased, the converse one-sided intersection J_2 will breakdown because of torrential slide impact bringing about an extensive current through the device. The voltage at which this phenomenon happens is known as the forward breakdown voltage V_{BO}. Following alternate intersections J_1 and J_3 are now forward one-sided; there will be free development of transporters over each of the three intersections bringing about a vast forward anode current. Once the SCR is exchanged on, the voltage drop crosswise over is small, usually 1–1.5 V. The anode current is restricted by the outside impedance current in the circuit.

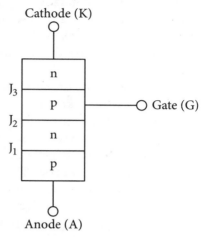

Figure 2.19 Structure of an SCR.

In spite of the fact that an SCR can be turned on by expanding the forward voltage past V_{BO}, usually, the forward voltage is kept up well beneath V_{BO} and the SCR is turned on by applying a positive voltage in between the gate and the cathode. With the positive gate voltage, the leakage current through the junction J_2 increases. This is because subsequent the gate current comprises for the most part the electron stream from the cathode to the gate. The base end layer is intensely doped when compared with the p-layer. Because of the connected voltage, some of these electrons reach intersection J_2 and add to the minority bearer concentration in the p-layer. This raises the reverse leakage current and results in breakdown of intersection J_2 despite the fact that the connected forward voltage is not exactly the breakdown voltage V_{BO}. With increase in gate current, breakdown happens earlier.

The thyristor has three fundamental states:

- Reverse blocking: In this mode, the thyristor hinders the current in the same route as that of a reverse-biased diode.
- Forward blocking: In this mode, the thyristor operation is in such a way that it produces forward current conduction that would typically be conveyed by a forward-biased diode.
- Forward conducting: In this mode, the thyristor has been activated into conduction. It will stay leading until the forward current drops below a limit known as the 'holding current'.

2.5.4.1 V–I Characteristics of SCR

V–I characteristics of a thyristor is explained in Figure 2.20. In the converse bearing, the thyristor behaves like a reverse one-sided diode that leads almost no current until heavy slide breakdown happens. In the forward course, the thyristor has two stable states or methods of operation that are joined together by an insecure mode that shows up as a negative resistance on the V–I characteristics. The low-current high-voltage area is the forward-closing state or the off state and the low-voltage high-current mode is the on state. For the forward-blocking mode, the amount of diversion is the forward blocking voltage V_{BO}, which is characterized for zero gate current. In the event that a positive gate current is connected to a thyristor, the move or break over to the on state will happen at smaller estimations of anode to cathode voltage. Although it has not been demonstrated, the gate current does not need to be a DC current; it can also be a pulse of current having some base length of time. This capacity to switch the SCR for a current pulse is the reason behind the far-reaching uses of the device.

It is not possible to switch off the device once the SCR is in the on state. The best way to turn off is to use an external drive.

To design a thyristor-based MATLAB/Simulink model, choose the following blocks from the Library Simulink Browser:

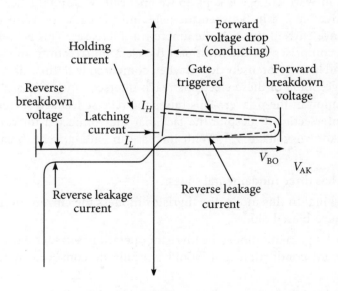

Figure 2.20 V–I characteristics of an SCR.

AC Voltage Source

SimulinkLibraryBrowser>>Simscape>>SimPowerSystems>>ElectricalSources>>ACVoltageSource

Thyristor

SimulinkLibraryBrowser>>Simscape>>SimPowerSystems>PowerElectronics>>

Thyristor

Voltage and Current Measurement

SimulinkLibraryBrowser>>Simscape>>SimPowerSystems>Measurements>>

Voltage Measurements, Current Measurements

R

SimulinkLibraryBrowser>>Simscape>>SimPowerSystems>Elements>RLCBranch

Powergui

SimulinkLibraryBrowser>>Simscape>>SimPowerSystems>>Powergui

Scope

Simulink>>Sinks>>Scope

In operation, the thyristor or SCR will not lead at first. It requires a certain level of current to stream in the gate to trigger. Once terminated, the thyristor will stay in conduction until the voltage over the anode and cathode is reversed. This clearly happens at the end of the half cycle over which the thyristor conducts. The following half cycle will be obstructed as a consequence of the rectifier activity. It will then need current in the gate circuit to trigger the SCR once more.

Example 3

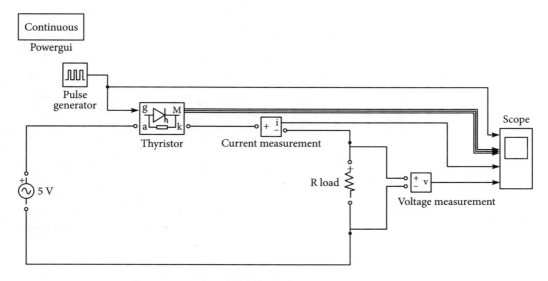

Figure 2.21 Simulink Model of a Thyristor (Forward Bias).

In the forward-biased mode shown in Figure 2.21, no current is separated from the leakage current streams. This is known as the forward-blocking mode. When a gating pulse is connected, the thyristor 'fires' and the forward safety of the device tumbles to a low esteem, permitting huge current and flows to stream in the forward-conduction mode. Thyristors can likewise be made to trigger by applying a huge forward voltage in between the anode and the cathode; yet this is not useful as the device cannot then control conduction. The switching waveforms of a thyristor in the forward-biased condition are shown in Figure 2.22.

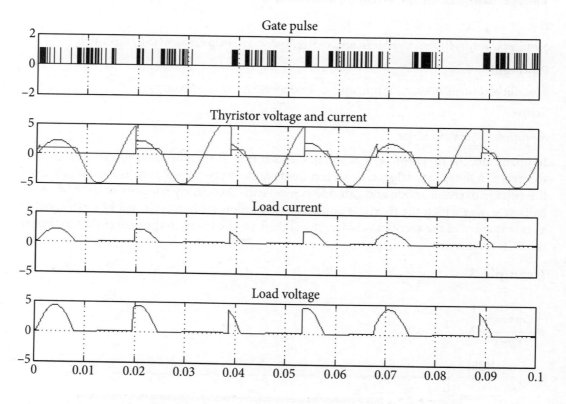

Figure 2.22 Switching Waveform of a Thyristor (Forward Bias).

In the reverse-inclined conditions shown in the Simulink model (Figure 2.23), the pulse is carried in a comparable manner to the thyristor. The current, separated from a small leakage current, is blocked until the opposite breakdown region is arrived at, which shows that the protection due to the consumption layers at the intersections has broken down. The switching waveforms clearly depict the conduction of the thyristor switch in the reverse-biased direction (Figure 2.24).

MATLAB Simulation of Power Semiconductor Devices   81

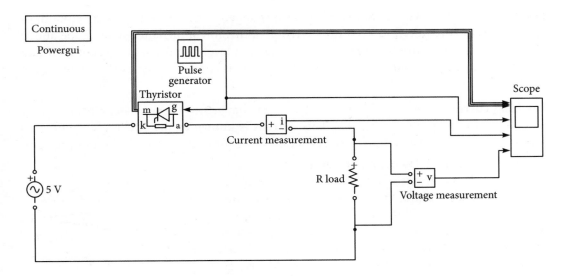

Figure 2.23 Simulink Model of a Thyristor (Reverse Bias).

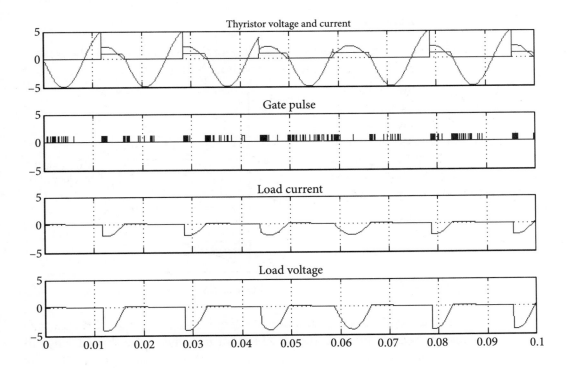

Figure 2.24 Switching Waveform of a Thyristor (Reverse Bias).

Holding current I_H After a thyristor has been changed to the on state, a small amount of anode current is required to keep the thyristor in this low impedance state. If the anode current is reduced beneath the basic holding current level, the thyristor cannot maintain the current through it and returns to its off state, as I is connected with turn the off of the device.

Latching current I_L After the thyristor has been changed to the on state, a base current is required to maintain conduction. This current is known as the latching current. I_L is connected with the turn on state and is generally more prominent than the holding current.

The holding and latching current are depicted in Figure 2.25. While thyristors plays a main role in phase edge control and taking care of a lot of power, they are not as suitable for low power applications. This is because they can only be turned off by changing the flow of current. Hence, a thyristor may take more time to turn on or off compared to other semiconductors. Moreover, thyristors can lead only in one heading, making them unfeasible for applications.

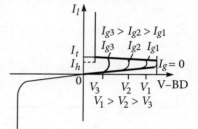

Figure 2.25 Effect of Holding Current and Latching Current.

2.5.4.2 Gate characteristics

Gate voltage can be plotted with respect to gate current for the thyristor. $I_{g(max)}$ is the greatest gate current that can course through the thyristor without harming it.

Similarly, $V_{g(max)}$ is the maximum gate voltage that can be connected. $V_{g(min)}$ and $I_{g(min)}$ are the least gate voltage and current below which the thyristor will not be turned on. Consequently, to turn on the thyristor effectively, the gate current and voltage ought to be

$$I_{g(min)} < I_g < I_{g(max)}$$

$$V_{g(min)} < V_g < V_{g(max)}$$

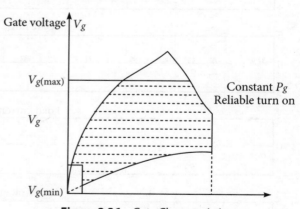

Figure 2.26 Gate Characteristics.

The gate characteristics for Figure 2.25 additionally demonstrate the bend for steady gate power (P_g). Thus, for reliable turn on, the (V_g, I_g) point must lie in the shaded range in Figure 2.26. It turns on thyristor effectively. Any spurious voltage/current spikes at the gate must not be equivalent to $V_{g(min)}$ and $I_{g(min)}$ or the thyristor will be falsely activated. The gate attributes shown here are for DC estimations of gate voltage and current.

2.5.4.3 Protection of thyristor circuits

In general, the thyristor requires some time to spread the current conduction throughout the junctions in a uniform manner. SCRs are sensitive to high voltage, overcurrent and any type of drifters. For ideal and dependable operation, they are required to be ensured against such anomalous working conditions. Therefore, safety is considered in the hardware by selecting devices with evaluations higher than those required for ordinary operation. In actual fact however, it is not economical to use devices of higher evaluations. High forward-voltage security is one of the characteristics in SCRs. The SCR will break down and begin directing before the top forward voltage is achieved so that the high voltage is exchanged to another piece of the circuit. The turn on of SCR causes a substantial current to flow and represents an issue of overcurrent protection.

Overcurrent protection can be given by interfacing an electrical switch and a breaker in arrangement with the SCR, as generally accomplished for the security of any circuit. In fact, there are a few exceptions to their utilization. A semiconductor device is equipped to take overloads for a restricted period, so the breaker utilized ought to have high breaking limit and quick intrusion of current. There must be a similarity between SCR and wire (I^2) × t rating because without proper protection, the SCRs move to the off or unbounded impedance condition. There are conflicting prerequisites for voltage security when quick acting wires are utilized. Wires when utilized, should have their curve voltages kept underneath 1.5 times the crest circuit voltage. For little power applications, it is pointless to utilize rapid breakers for circuit safety due to the fact that it might cost more than the SCR. Current extent location can be used and is utilized as a part of numerous applications. When an overcurrent is identified, the gate circuits are controlled either to switch off the suitable SCRs, or move in phase compensation, to reduce the conduction period for the normal estimation of the current.

If the output to the load from the SCR circuit is exchanging current, LC reverberation gives overcurrent protection and separation. A current restricting device utilizing a saturable reactor will appear. With typical streams, the saturable reactor L offers high impedance and C and L are in resonance arrangement to offer zero impedance to the stream of current of the principal consonant. An overcurrent immerses L, thus giving unimportant impedance. There is LC parallel reverberation and from then on, boundless impedance to the stream of current at high frequency.

There are numerous diadvantages because of voltage surges as SCRs do not generally have a security component incorporated into their appraisals. Outside voltage surges cannot be controlled by the SCR circuit planner. Voltage surges regularly prompt either failing of the circuit by accidental turn on of SCR or changeless harm to the device because of converse breakdown. SCR can be ensured against voltage surges by using shunt joined nondirect resistance devices. Such defensive devices cause a fall in resistance with the increase in voltage thus building up a virtual short out over the SCR when a high voltage is connected.

An overvoltage protection circuit using a thyristor diode, which has low resistance at high voltage and vice versa, includes an inductor L and capacitor C protecting the SCR against huge dV/dt and dI/dt.

2.5.4.3.1 dI/dt protection

If the rate of rise of anode current is very fast compared with the spreading speed of the turn on process, a hot spot may occur due to high current density causing the device to fail. Hence, dI/dt protection is required. See Figure 2.27 for dI/dt protection.

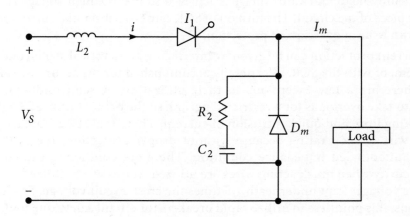

Figure 2.27 *dI/dt* Protection.

2.5.4.3.2 dV/dt protection

The dV/dt across the thyristor is limited using a snubber circuit as shown in Figure 2.28. When the switch S1 is closed at time $t = 0$, the rate of rise of Voltage VAK at the thyristor is limited by the capacitor C. When the thyristor is turned on, the the discharge current of the capacitor can be restricted by adding a resistor if required.

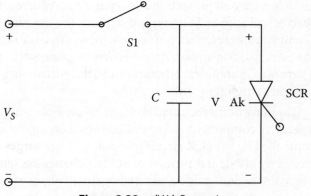

Figure 2.28 *dV/dt* Protection.

2.5.4.4 Thyristor turn on and turn off characteristics

When the thyristor is turned on with the use of the gate trigger pulse, the thyristor does not lead completely. At the outset, there is no obvious increment in the thyristor anode current, because only a little amount of the silicon pellet in the quick region of the gate terminal begins leading. The time period between 90 per cent of the top gate trigger pulse and the moment the forward voltage has reduced to 90 per cent of its beginning quality, is known as the gate-controlled/trigger postponement time T_{gd}. It is additionally characterized as the time period between 90 per cent of the gate trigger pulse and the moment at which the anode current rises to 10 per cent of its maximum. T_{gd} is for the most part in the range of 1 μs. When T_{gd} has passed, the current begins rising towards its maximum. The period during which the anode current rise from 10 per cent to 90 per cent of its crest quality is known as the rise time. It is likewise characterized as the ideal opportunity when the anode voltage reduces from 90 per cent to 10 per cent of its maximum. The summation of T_{gd} and T_r gives the turn on time t_{on} of the thyristor.

When a thyristor is turned on by the gate terminal (Figure 2.29), the gate loses control over the device and the power device can be taken back to the blocking state by diminishing the forward current to a level beneath that of the holding current. In AC circuits, the current experiences a characteristic zero worth and the device will consequently switch off. In DC circuits, where no impartial zero estimation of current exists, the forward current is diminished by applying an opposite voltage crosswise over the anode and the cathode and accordingly, compelling the current through the thyristor to zero.

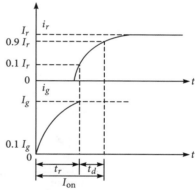

Figure 2.29 Turn on of Thyristor.

The delay time is measured from the moment at which the gate current achieves 90 per cent of its last value to the moment at which the anode current achieves 10 per cent of its last value. It can likewise be characterized as the time between which the anode voltage falls from the introductory anode voltage V_a to 0.9 V_a. Consider Figure 2.29. Until the time t_d, the SCR is in the forward-blocking mode so the anode current is the little leakage current. When the gate sign is connected at 90 per cent of I_g, the gate current becomes 0.1 I_a and correspondingly, the anode to cathode voltage tumbles to 0.9Va.

With the gate signal connected, there will be nonuniform appropriation of current over the cathode surface so the current value is much higher at the gate terminal. Moreover, it quickly diminishes as the separation from the gate increases. Subsequently, the postponement time t_d is the time during which anode current streams in a restricted area at which current density gate current is most noteworthy. This is the time taken by the anode current to rise from 10 per cent to 90 per cent of its last value. It is also the time required for the forward-blocking voltage to reduce from $0.9V_a$ to $0.1V_a$. This rising time is converse compared to the gate current and its rate of change. Thus, high and steep current pulses are associated with a gate that reduces the rising time t_r. Furthermore, if the load is inductive, this rising time will inevitably be higher; for resistive and capacitive load, it will be lower. In the midst of this time, turn on problems in the SCR are high due to the broad anode current; high anode voltage happens in the meantime.

t_s is the time taken by the anode current to ascend from $0.9I_a$ to I_a. It is also the time required for the forward-blocking voltage to tumble from $0.1V_a$ to its on state voltage drop, which is in the range 1–1.5 V. During this time, anode current spreads over the whole leading area of an SCR from a thin directing area. After the spreading time, a full anode current courses through the device with little on state voltage drop.

In this way, the aggregate turn-on time,

$$T_{on} = t_r + t_d + t_s$$

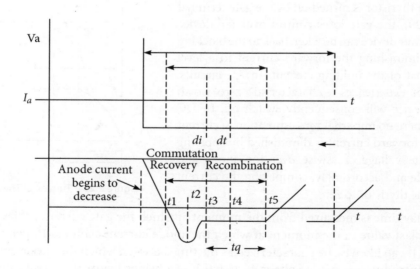

Figure 2.30 Turn off of Thyristor.

Turn on time is in the range 1–4 micro seconds; it relies on the gate signal wave shapes and anode circuit parameters. To decrease the turn on time of the SCR, the gate pulse should be 3–5 times the base gate current of the SCR. The thyristor has reverse recovery attributes t_{rr}, which is because of charge storing in the intersections of the thyristor. Some time has to be set aside for recombination – the gate recovery time or switch recovery time t_{gr}. Consequently, the turn off time t_q is the entirety of the terms for which reverse recovery current streams after the use of opposite voltage and the time required for the recombination of every overabundance current bearer. Towards the end of the turn off time, a consumption layer is created crosswise over J_2 and the intersection can now withstand the forward voltage. The turn off time is dependent on the anode current; the extent of opposite V_g connected shows the importance and rate of use of the forward voltage. The turn off time for converter evaluation thyristor is 50–100 s and that for the inverter review thyristor is 10–20 μs.

To ensure that a thyristor has successfully turned off, it is required that the circuit off time t_c be greater than thyristor turn off time t_q.

2.5.4.5 Series and parallel operation of thyristors

To expand the off state or blocking voltage, thyristors can be associated with different arrangements. Voltage dispersion must be guaranteed – utilizing parallel resistors for closing and off state and parallel *RC* components for replacement. Parallel resistors must be evaluated so that the current that moves through the resistors is 5–10 times the thyristor off state current in the hot state. To guarantee that the thyristor termination is as concurrent as would be prudent, adequately high, soak-rising trigger pulses are required. The voltage load on each thyristor has to be no less than 10 per cent lower than for individual operation. In parallel thyristor circuit courses of action, homogenous current circulation is required from the bit terminating and all through the whole current stream time. For this reason, steeply rising trigger pulse of adequate value and also symmetrical line impedances in the fundamental circuit are required. While joining thyristors in parallel, they must be organized by small conceivable forward voltage contrasts. To take care of any remaining asymmetry, parallel thyristors must be set without any distortion for mean forward current. It might be important to confine the rate of change of current in individual thyristors.

2.5.4.5.1 Series operation of thyristors

Thyristors are arranged specifically to enhance their general voltage rating. The qualities of thyristors of the same sort are not the same, as shown in Figure 2.31 and hence, assistant segments must be added to thyristors associated with an arrangement to guarantee legitimate operation. In Figure 2.31, it can be seen that for the same off state current, the off state voltages vary. Voltage sharing systems are required for both switch and off state conditions. Resistors set in parallel with thyristors are utilized to achieve voltage sharing between thyristors set in arrangement. The voltage sharing resistors for *n* thyristors in arrangement are shown in Figure 2.32. Refer to Figure 2.33 for equal voltage sharing during forward leakage currents.

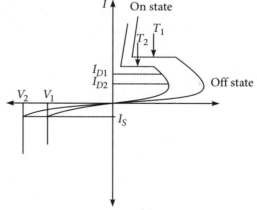

Figure 2.31 Off State Characteristics of Two Thyristors of the Same Type.

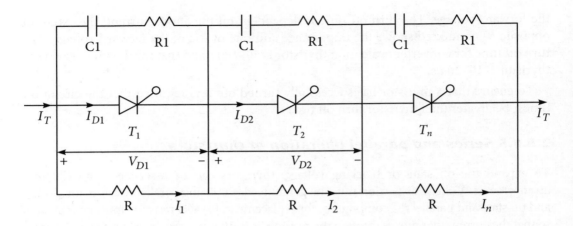

Figure 2.32 Series Operation of Thyristors.

Let the off state current of thyristor T_1 be represented by I_{D1}. For n_s thyristors in the arrangement, and alternate $n_s - 1$ thyristors having the same off state current,

$$I_{D2} = I_{D3} = I_{Dn}$$

Moreover,

$$I_{D1} < I_{D2}$$

Since thyristor T_1 has the minimum off state current, this thyristor will share the most elevated voltage in light of the fact that the voltage drop over the resistor R in parallel with this thyristor will be bigger than that over the resistor R for alternate thyristors. If I_1 is the current through resistor R, which is associated crosswise over thyristor T_1, then the current through alternate resistors are equivalent and given by

$$I_{D2} = I_{D3} = I_{Dn}$$

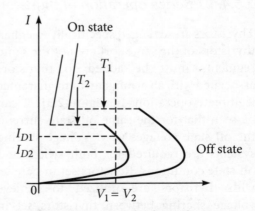

Figure 2.33 Equal Voltage Sharing.

The off state current spread is given by DI_D where

$$DI_D = I_{D2} - I_{D1}$$

$$DI_D = (I_T - I_2) - (I_T - I_1)$$

$$DI_D = I_1 - I_2$$

Thus,

$$I_2 = I_1 - DI_D$$

The voltage drop crosswise over thyristor T_1 is given by V_{D1}, where

$$V_{D1} = RI_1$$

Using Kirchhoff's voltage law where supply voltage is given by V_s yields

$$V_s = V_{D1} + (n_s - 1)I_2R = V_{D1} + (n_s - 1)(I_1 - DI_D)R$$

$$V_s = V_{D1} + (n_s - 1)I_1R - (n_s - 1)DI_DR$$

$$V_s = n_sV_{D1} - (n_s - 1)DI_DR$$

$$V_{D1} = (V_s + (n_s - 1)R\Delta I_D)/n_s$$

Solving this mathematical statement for V_{D1}, the voltage crosswise over thyristor T_1 yields $I_{D1} = 0$.

The voltage over the thyristor T1 will be the greatest when the off state current spread DI_D is most extreme. The worst case scenario of voltage over the thyristor T1 is seen when no current courses through thyristor T1. Under these conditions,

$$I_{D1} = 0$$

Also,

$$DI_D = I_{D2}$$

The voltage crosswise over thyristor T1 is currently given as $V_{DS(max)}$, where

$$V_{DS(max)} = (V_s + (n_s - 1)RI_{D2})/n_s$$

During turn off, the distinctions in forward leakage current streams causes contrasts in storage charge, which thus causes contrasts in backward voltage sharing. Thyristors with the slightest recovery have the most minimal converse recovery, resulting in the most astounding transient voltage. The intersection capacitances that control the transient voltage conveyance will be deficient for this procedure and an outside capacitor C1 has to be utilized as seen in Figure 2.32. Resistor R1 limits the capacitor release current. The same segments R1 and C1 are utilized for both transient voltage sharing and dV/dt protection.

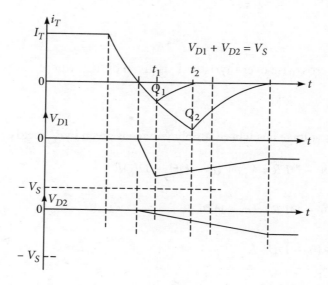

Figure 2.34 Voltage Sharing and Reverse Recovery Time.

The voltage difference is given by

$$\Delta V = R\Delta ID = (Q_2 - Q_1)/C_1 = \Delta Q/C_1$$

where Q_1 is the charge in thyristor T1 and Q_2 is the charge in the other thyristor.

Here $Q_2 = Q_3 = Q_n$, also $Q_1 < Q_n$.

The transient voltage across the thyristor is given by

$$V_{D1} = 1/n_s(V_s + ((n_s - 1)\Delta Q)/C_1)$$

The voltage factor is normally used to increase the reliability, and hence the string factor is given by

$$\mathrm{DRF} = 1 - V_s/(n_s V_{ds(max)})$$

2.5.4.5.2 Parallel operation of thyristors

When the load current surpasses the rating of a solitary thyristor, thyristors are placed in parallel to build the general current capacity. As opposed to what may be normal, the load current is not shared similarly between the thyristors, as thyristors are not flawlessly coordinated. Figure 2.35 demonstrates the V–I attributes of two thyristors T1 and T2, associated in parallel.

The thyristor conveying the higher current would disseminate more power, which thus will increase the intersection temperature and diminish the interior resistance. This will increase its current sharing limit and possibly harm the thyristor. This procedure is termed thermal runaway and is not normal to thyristors. Thermal runaway may be counteracted by utilizing one normal warm sink to guarantee that both thyristors are working at the same temperature. Measuring up to the current needed can be refined by the utilization of a resistor or inductor in arrangement with each thyristor as shown in Figure 2.36.

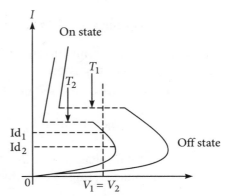

Figure 2.35 Characteristics of Two Parallel Thyristors.

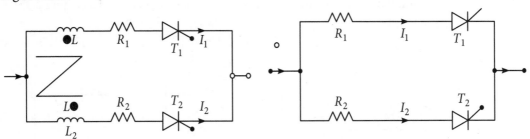

Figure 2.36 Parallel Operation of Thyristors.

If a resistor is used to measure up to current sharing, there are many issues in the arrangement resistor are high and the result might be unsatisfactory. Appropriately coupled inductors can also be utilized for current sharing. In this case, if thyristor T1 current increases, a voltage of maximum inverse to that of the loop in arrangement with T1 will be actuated in the curl in the arrangement with thyristor T2. The maximum of this voltage increases the anode capability of thyristor T2, thus increasing the current moving through this thyristor.

String Efficiency When thyristors are connected in parallel, they do not share the total current equally. When they are connected in series, they do not share the total voltage equally. String efficiency is defined as follows.

$$\text{String efficiency} = \text{Number of thyristors} \times \frac{(\text{Voltage/current rating of string})}{(\text{Voltage/current rating of one thyristor})}$$

Derating If thyristors share the voltage/current equally, string efficiency will be one and the utilization of thyristors will be maximum. When thyristors are connected in parallel, the current derating is given as follows

$$\text{Current derating} = 1 - I_m/n_p I_T$$

where, I_m is the total circuit current, n_p is the number of thyristors in parallel and I_T is the current rating of each thyristor.

When thyristors are connected in series, the voltage derating is as follows.

$$\text{Voltage derating} = 1 - V_s/n_s V_D$$

where, V_s is the total voltage across the string, n_s is the number of thyristors in series and V_D is the forward voltage rating of each thyristor.

2.5.4.6 Two-transistor model of a thyristor

The two-transistor model shows the latching activity due to the feedback in the thyristor. The thyristor is a combination of two transistors of *pnp* model and *npn* model (Figure 2.37).

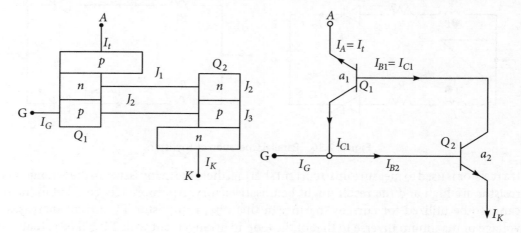

Figure 2.37 Two-transistor Model of Thyristor.

The relation between the collector current, emitter current, and the leakage current of the collector–base junction, I_{CBO} is given by

$$I_C = \alpha I_E + I_{CBO} \tag{2.1}$$

For transistor Q_1 with anode current I_A, collector current I_{C1} is given by

$$I_{C1} = \alpha_1 I_A + I_{CBO1} \qquad (2.2)$$

For transistor Q_2 with cathode current I_k, collector current I_{C2} is given by

$$I_{C2} = \alpha_2 I_K + I_{CBO2} \qquad (2.3)$$

where α_1, and α_2 are the current gain values, respectively, and I_{CBO1} and I_{CBO2} are the leakage current values of the two transistors, respectively.

The sum of both collector current yields:

$$I_A = I_{C1} + I_{C2} \qquad (2.4)$$

$$I_A = \alpha_1 I_A + I_{CBO1} + \alpha_2 I_K + I_{CBO2} \qquad (2.5)$$

When a gate current I_G is applied to the thyristor

$$I_K = I_A + I_G \qquad (2.6)$$

Solving for anode current I_A in Equation (2.6) yields

$$I_A = (\alpha_2 I_K + I_{CBO2} + I_{CBO1})/1 - (\alpha_1 + \alpha_2) \qquad (2.7)$$

The current gain α_1 varies with emitter current I_{E1}, which is equal to I_A; and α_2 varies with emitter current I_{E2}, which is equal to I_K.

2.5.4.7 Turn on methods of SCR

With a voltage connected to the SCR, if the anode is made positive with respect to the cathode, the SCR gets to be forward one-sided. Accordingly, the SCR enters the forward-blocking state. The SCR is made direct or changed into conduction mode by any of the accompanying strategies.

- Forward-voltage triggering
- Temperature triggering
- dV/dt triggering
- Light triggering
- Gate triggering

2.5.4.7.1 Forward-voltage triggering

By increasing the forward anode to cathode voltage, the consumption layer width is additionally increased at intersection J_2. This causes the minority charge transporters to increase, thus also increasing the voltage at intersection J_2. This further prompts a torrential slide breakdown of the intersection J_2 at a forward breakover voltage V_{BO}. At this stage, SCR transforms into conduction mode and consequently a huge current streams through it with a low-voltage drop crosswise over it. During the turn on state, the forward-voltage drop over the SCR is in the range 1–1.5 V and this might be increased with the load current. This technique is not utilized as it needs an increasing anode to cathode voltage. Furthermore, once the voltage is more than V_{BO}, it creates high streams, which may harm the SCR. Hence in most cases, this kind of method is not preferred.

2.5.4.7.2 Temperature triggering

The converse leakage current relies on temperature. If the temperature is increased to a specific value, the quantity of opening combines also increases. This causes a build up of leakage current and further increases the current in the SCR. Thus, the regenerative activity inside the SCR begins, since the $(\alpha_1 + \alpha_2)$ value deals with solidarity (as the present additions increases). Increasing the temperature at intersection J_2 causes the breakdown of the intersection and consequently, it conducts. This activation happens in a few circumstances especially when the device temperature is also false activating. Temperature triggering is also not utilized often because it causes warm runaway and subsequently, the device or SCR might be harmed.

2.5.4.7.3 dV/dt triggering

In forward-blocking state, intersections J_1 and J_3 are forward one-sided and J_2 is converse one-sided. Hence, the intersection J_2 carries on as a capacitor of two leading plates J_1 and J_3 with a dielectric J_2 because of the space charges in the consumption area. The charging current of the capacitor is given as

$$I = C\, dV/dt$$

where dV/dt is the rate of progress of connected voltage and C is the intersection capacitance. From the aforementioned mathematical statement, if the rate of progress of the connected voltage is substantial, it prompts increase in the charging current, which is sufficient to build the estimation of α. Hence, the SCR gets to be turned on without a gate signal. Nonetheless, this technique is for all intents and purposes not useful as it is a false turn on procedure, and can create high-voltage spikes over the SCR resulting in significant harm.

2.5.4.7.4 Light triggering

An SCR turned on by light radiation is also called a light activated SCR (LASCR). This kind of activation is utilized for stage-controlled converters in high-voltage, direct current (HVDC) transmission systems. In this strategy, light beams with suitable wavelength and power are permitted to strike the intersection J_2. The intersection has a special inward p-layer. When the light strikes this layer, electron-gap sets are produced at the intersection J_2, which generates extra charge transporters at the intersection, prompting turn on of the SCR.

2.5.4.7.5 Gate triggering

This is one of the most basic and common strategy to turn on the SCR. When the SCR is forward one-sided, an adequate voltage at the gate terminal infuses a few electrons into the intersection J_2. This leads to build up of reverse leakage current and consequently, the breakdown of intersection J_2 even at a voltage lower than the V_{BO}. Relying on the value of the SCR, the gate current fluctuates from a couple of mA to 200 mA or more. If the gate current connected is increased, then more electrons are infused into the intersection J_2 and results in the SCR entering the conduction state at much lower connected voltage. In gate activating strategy, a positive voltage is connected between the gate and the cathode terminals. We can utilize three sorts of signals to turn on the SCR – DC signal, AC flag and pulse signal.

In case of DC signal, an adequate DC voltage is connected between the anode and the cathode terminals so that the gate is made positive with respect to the cathode. The gate current drives the SCR into conduction mode. In this, a consistent gate sign is connected at the terminal and consequently causes internal power scattering.

AC signal is the most commonly utilized technique for AC applications where the SCR is used as an exchanging device. With the best possible confinement between the power and the control circuit, the SCR is activated by the phase shift AC voltage got from the principal supply. The terminating point is controlled by changing the phase edge of the gate signal. This is one of the constraints of AC activating – isolated venture down or pulse transformer is expected to supply the voltage to the gate drive from the principal supply.

The most common technique for setting off the SCR is pulse activating. In this strategy, the gate is supplied with a single pulse or continuous pulses. The fundamental advantage of this technique is that given that the gate drive is irregular or does not require persistent pulse to turn the SCR, problems with the gate are decreased by applying single or occasional pulses. For disconnecting the gate drive from the primary supply, a pulse transformer is utilized.

2.5.4.8 Commutation techniques of a thyristor

The procedure of commutation of a thyristor includes (a) the reduction of current through the leading thyristor to zero: the current may get to be zero either because the load current itself is made to decrease to zero or shifted to an optional way comprising another thyristor, a diode, a capacitor, or a capacitor and inductor blend; (b) subjecting the thyristor to an opposite predisposition sufficiently long for it to recapture its forward-voltage blocking ability.

A thyristor can be turned on by activating the gate terminal with a low-voltage brief length of time pulse. Subsequent to turning on, it will direct constant until the thyristor is converse one-sided or the load current falls to zero. This persistent conduction of thyristors causes some issues in a few applications. The procedure utilized for stopping a thyristor is called commutation. By the comutation process, the thyristor working mode is changed from forward-directing mode to forward-blocking mode. Commutation techniques are categorized into natural commutation and forced commutation.

2.5.4.8.1 Natural commutation

By and large, if we consider AC supply, the current will course through the zero intersection line while going from the positive maximum to the negative crest. In this way, an opposite voltage will show up over the device at the same time, which will turn off the thyristor promptly. This procedure is called the normal compensation or natural commutation (Figure 2.38) as the thyristor is turned off actually without utilizing any outer parts or circuit or supply for replacement. Characteristic compensation can be seen in AC voltage controllers, phase-controlled rectifiers, and cycloconverters.

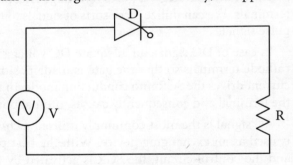

Figure 2.38 Natural Commutation.

2.5.4.8.2 Forced commutation

The thyristor can be turned off by biasing the SCR or dynamic or latent segments. Thyristor current can be diminished to a value below the holding current. Since the thyristor is turned off coercively, it is termed as a forced commutation process. Fundamental hardware and electrical parts, for example, inductance and capacitance are utilized as commutating components for compensation. Constrained substitution can also be seen while utilizing DC supply; consequently, it is also called as DC replacement. The outer circuit utilized for forced compensation procedure is called the substitution circuit and the components utilized as part of this circuit are called commutating components.

2.5.4.8.2.1 Characterization of forced commutation methods

Forced commutation can be characterized into distinctive techniques:

- Class A: Self commutated by a resonating load
- Class B: Self commutated by an *LC* circuit
- Class C: *C* or *LC* exchanged by another load conveying SCR
- Class D: *C* or *LC* exchanged by an assistant SCR
- Class E: An outer pulse hot spot for commutation
- Class F: AC line commutation

2.5.4.8.2.1.1 Class A: Self commutated by a resonating load

Class A is one of the most commonly utilized thyristor recompense strategies. When the thyristor is activated or turned on, anode current will stream by making the capacitor *C* positive. The underdamped circuit is also shaped by the inductor or AC resistor, capacitor and resistor. When current develops through an SCR and finishes its half cycle, then the inductor current will move through the SCR in the converse bearing, which will turn off the thyristor. Refer to Figure 2.39 for class A commutation.

After the thyristor is replaces or turned off, the capacitor will begin releasing current from its maximum through the resistor in an exponential way. The thyristor will be in a backward predisposition condition until the capacitor voltage comes back to the supply voltage level.

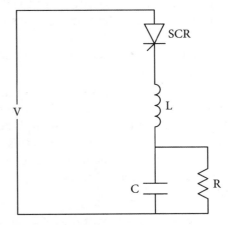

Figure 2.39 Class A Commutation.

2.5.4.8.2.1.2 Class B: Self commutated by an LC circuit

The significant difference between the class A and class B thyristor compensation procedures is that the *LC* is associated in series with the thyristor in class A, while in parallel with the thyristor in class B. Before activating the SCR, the capacitor is energized. If the SCR is activated or given

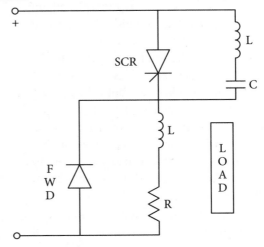

Figure 2.40 Class B Commutation.

activating pulse, the subsequent current has two segments. A steady load current moving through the RL load is guaranteed by the huge reactance associated in series with the load, which is clipped with a freewheeling diode. If the sinusoidal current moves through the LC circuit, then the capacitor C is negative towards the end of the half cycle.

See Figure 2.40 for class B Commutation. If the aggregate current coursing through the circuit or inverse current turns out to be more than the load current, then the SCR will be turned off.

2.5.4.8.2.1.3 Class C: C or LC switched by another load-carrying SCR

In the thyristor recompense methods mentioned earlier, there is only one SCR; yet in class C compensation systems of thyristors, there will be two SCRs. One SCR is considered as the primary thyristor and the other as the assistant thyristor. Both may be considered as the principle SCRs conveying load current; they can also be outlined with four SCRs with the load burden over the capacitor by utilizing a current hot spot for supplying a necessary converter.

Refer to Figure 2.41 for class C commutation systems. If the thyristor SCR_2 is activated, then the capacitor will be energized. If the thyristor SCR_1 is activated, then the capacitor will release current and this release current of C will restrict the stream of load current in SCR_2 as the capacitor is exchanged crosswise over SCR_2 by means of SCR_1.

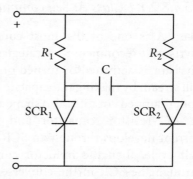

Figure 2.41 Class C Commutation.

2.5.4.8.2.1.4 Class D: LC or C switched by an auxiliary SCR

Class C and class D thyristor replacement methods can be distinguished by the load current: in class D, one of the SCRs will convey the load current while the alternate goes about as an assistant thyristor; in class C, both SCRs will convey load current. The helper thyristor comprises a resistor in its anode, which has a resistance roughly 10 times the load resistance. Refer to Figure 2.42 for a class D commutation system.

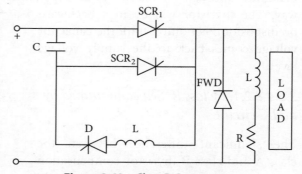

Figure 2.42 Class D Commutation.

By setting off the SCR_2 (helper thyristor), the capacitor is energized to supply voltage. Afterward, the SCR_2 power lines will be released through the diode–inductor–load circuit. If SCR$_1$ (primary thyristor) is activated, then the current will stream in two ways: commutating current will move through the C–SCR$_1$–L–D way and load current will course through the load. If the charge on the capacitor is turned around and held at that level utilizing the diode and if SCR$_2$ is reactivated, then the voltage over the capacitor will show up over the SCR$_1$ through SCR$_2$. In this way, the fundamental thyristor SCR$_1$ will be turned off.

2.5.4.8.2.1.5 Class E: External pulse source for commutation

In class E thyristor substitution systems, a transformer cannot act as a pulse transformer. Refer to Figure 2.43 for a class E commutation system.

An external pulse generator is used to make a positive pulse, which is supplied to the cathode of the thyristor through a pulse transformer. The capacitor C is charged to around 1 V and it is considered to have zero impedance for the turn off pulse time allotment. The voltage over the thyristor is exchanged by the pulse from the electrical transformer, which supplies the reverse recovery current; for the required turn off time, it holds the negative voltage.

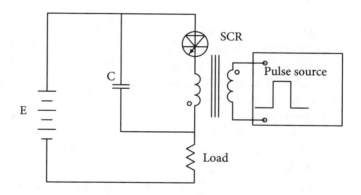

Figure 2.43 Class E Commutation.

2.5.3.8.2.1.6 Class F: AC line commutation

In class F thyristor recompense methods, a substituting voltage is utilized for supply. During the positive half cycle of this supply, stack current will flow. If the load is exceedingly inductive, then the current will stay until the energy stored in the inductive load is scattered. During the negative half cycle, as the load current goes to zero, the thyristor will be turned off. If voltage exists during turn off time of the device, then the negative maximum of the voltage over the active thyristor will turn it off. Here, the time period of the half cycle must be more than the turn off time of the thyristor. This substitution procedure is similar to the concept of a three-phase converter. Thyristors can be considered to be directing with the activating point of the converter, which is equivalent to 60°, and working in constant conduction mode with an exceptionally inductive load.

If the thyristors are activated, then the current through the approaching device will not ascend to the load current level. If the current through the approaching thyristors achieves the load current level, then the replacement procedure of active thyristors will be started. This converse biasing voltage of a thyristor will continue until the forward-blocking state is reached. Thyristors can be essentially called as controlled rectifiers. There are diverse sorts of thyristors, which are utilized for outlining power device-based creative electrical undertakings.

2.5.5 Power MOSFET

A MOSFET is produced by consolidating the ranges of field-impact ideas and MOS engineering. A power MOSFET is a three-terminal (gate, drain and source), four-layer unipolar semiconductor device.

It is a greater branch carrier device, and as the share bearers have no recombination activity, the MOSFET has amazingly high data transfer capacities and exchange times. The gate is electrically detached from the source, giving the MOSFET its high input impedance; it also has a good capacitor. MOSFETs do not have optional breakdown range; their channel to source has a positive temperature coefficient, so they have a tendency to act naturally resistive. It has low on safety and no intersection voltage drop when forward biased. These peculiarities make MOSFET a good device for power supply exchanging in power electronics application. As Figure 2.44 shows, for the *n*-channel MOSFET, the direction is inwards. For a *p*-channel MOSFET, the direction will be outwards.

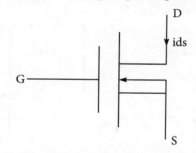

Figure 2.44 Symbol of MOSFET.

There is no option for the current to stream between the source terminal and the drain terminal, because none of the terminals are reverse biased for maximum operation of the connected voltage between the source and the drain terminal. Also, there is no option for current infusion of the triggering gate terminal in case of MOSFET switch. The application of a positive voltage at the gate terminal from the source will change over. To design a MOSFET-based MATLAB/Simulink model, choose the following blocks from the Library Simulink Browser:

DC Voltage Source

SimulinkLibraryBrowser>>Simscape>>SimPowerSystems>>ElectricalSources>>DCVoltageSource

DC Current Source

SimulinkLibraryBrowser>>Simscape>>SimPowerSystems>>ElectricalSources>>DCCurrentSource

MOSFET

SimulinkLibraryBrowser>>Simscape>>SimPowerSystems>PowerElectronics>>

MOSFET

Diode

SimulinkLibraryBrowser>>Simscape>>SimPowerSystems>PowerElectronics>>

Diode

Voltage and Current Measurement

SimulinkLibraryBrowser>>Simscape>>SimPowerSystems>Measurements>>

Voltage Measurements, Current Measurements

RLC

SimulinkLibraryBrowser>>Simscape>>SimPowerSystems>Elements>RLCBranch

Powergui

SimulinkLibraryBrowser>>Simscape>>SimPowerSystems>>Powergui

Scope

Simulink>>Sinks>>Scope

Example 4

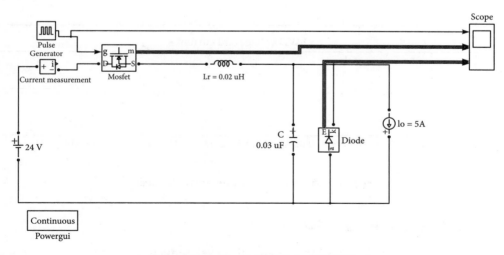

Figure 2.45 Simulink Model of MOSFET.

The Simulink model of MOSFET is shown in Figure 2.45. While the input state of the voltage level of the gate is zero, the MOSFET switch conducts no current; therefore the output voltage would equal the supply voltage. Thus, the MOSFET switch can be set in the cutoff region. At the second stage, while a suitable pulse is triggered by the gate, the switch turns to the on state to pass high current. Thus, the ability to turn on the switch is a reasonable advantage of the MOSFET switch. Refer to the switching waveforms of MOSFET in Figure 2.46.

Since the MOSFET is a larger part bearer device, its on state safety has a positive temperature coefficient. There is no difference between forward biasing and reverse biasing for the MOSFET. They are indistinguishable. The MOSFET will work appropriately and the device will not get corrupted inside the safe operating limits (SOAs) of the device. From these determination parameters, we can get the cutoff points of maximum channel to source voltage and channel present for the device.

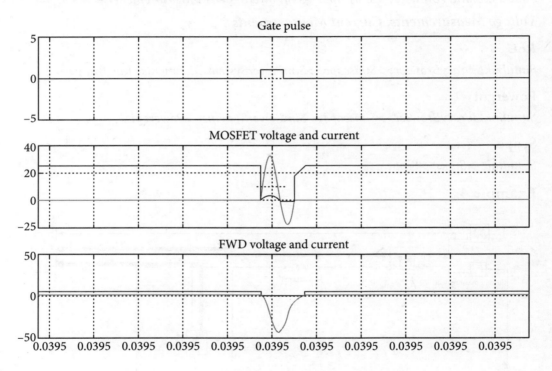

Figure 2.46 Switching Waveforms of MOSFET.

2.5.6 Gate turn off thyristors

Figure 2.47 shows four-layered three-terminal power semiconductor devices. GTOs are exchanging devices that can be turned on by a gate pulse and set off by applying a reverse gate pulse.

As shown in Figure 2.47, the GTO switch comprises an anode, a cathode, and a gate. The ideal GTO has a two way gate terminal. Hence, GTO switches are the most well-known sort available. They are ordinarily used as antiparallel diodes. They do not have high invert-blocking capacity. They are utilized as part of voltage fed converters. The symmetrical sort of GTO has an equivalent forward and converse blocking ability. They are utilized as part of current fed converters.

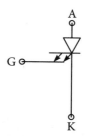

Figure 2.47 Symbol of GTO.

The opposing gate current depends more on the anode current, which has to be turned off. There is no prerequisite for an outside circuit to turn it off. Hence, the usage of these devices is reduced to a greater extent. The circuit can be turned on by applying the gate current and turned off by applying the gate cathode voltage.

To build a Simulink model of a power GTO, the various blocks that are chosen from the Simulink Library Browser are as follows:

DC Voltage Source

SimulinkLibraryBrowser>>Simscape>>SimPowerSystems>>ElectricalSources>>DCVoltageSource

GTO

SimulinkLibraryBrowser>>Simscape>>SimPowerSystems>PowerElectronics>>

GTO

Voltage and Current Measurement

SimulinkLibraryBrowser>>Simscape>>SimPowerSystems>Measurements>>

Voltage Measurements, Current Measurements

R

SimulinkLibraryBrowser>>Simscape>>SimPowerSystems>Elements>RLCBranch

Powergui

SimulinkLibraryBrowser>>Simscape>>SimPowerSystems>>Powergui

Scope

Simulink>>Sinks>>Scope

The Simulink model of GTO is shown in Figure 2.48. The gate turn off compatibility is very worthwhile as it gives increased adaptability in circuit application. The power loss due to triggering of the gate pulse is negligible and can be supplied even using low-voltage power MOSFETs.

Example 5

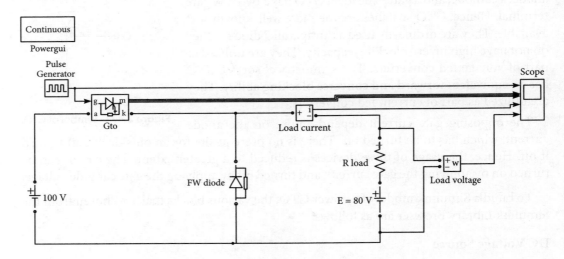

Figure 2.48 Simulink Model of a GTO.

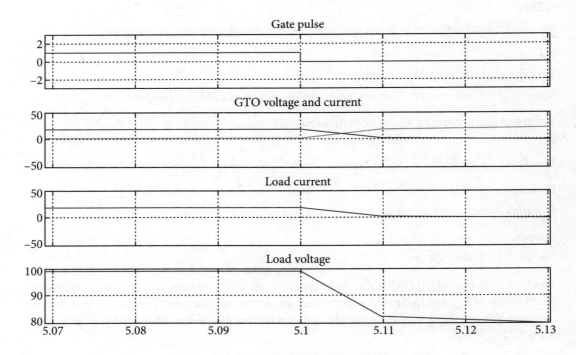

Figure 2.49 Switching Waveform of GTO.

The switching waveforms shown in Figure 2.49 depict the responses of GTO. GTOs are adaptable to control in DC circuits without using substitution hardware. The prime configuration objective of GTO devices are to accomplish quick turn off time and high current turn off ability and to upgrade the safe working range during turn off.

2.5.7 Insulated-gate bipolar transistor (IGBT)

A combination of the bipolar junction transistor (BJT) and MOSFET results in another device called the insulated-gate bipolar transistor. This is particularly valid in high frequency circuits where the IGBT is especially important because of its naturally high exchanging speed. It has prevalent on state attributes, great exchanging speed and astounding safe working zone. The symbol is shown in Figure 2.50 with the three terminals – emitter, collector and gate.

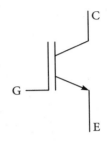

Figure 2.50 Symbol of IGBT.

There is a bit of ambiguity regarding the proper symbol and nomenclature of the IGBT. Some like to consider the IGBT as essentially a BJT with MOSFET gate input information and therefore use the altered BJT symbol for the IGBT as shown in Figure 2.50. Some prefer to consider the terminals as drain and source rather than collector and emitter. The MATLAB/Simulink model of an IGBT is shown in Figure 2.51. To build a closed-loop model of a power IGBT, the various blocks that are chosen from the Simulink Library Browser are as follows:

DC Voltage Source

SimulinkLibraryBrowser>>Simscape>>SimPowerSystems>ElectricalSources

IGBT

SimulinkLibraryBrowser>>Simscape>>SimPowerSystems>PowerElectronics>>IGBT

Voltage and Current Measurement

SimulinkLibraryBrowser>>Simscape>>SimPowerSystems>Measurements

R Load

SimulinkLibraryBrowser>>Simscape>>SimPowerSystems>Elements>RLCBranch

Pulse Generator

Simulink>>Sources>>PulseGenerator

Powergui

SimulinkLibraryBrowser>>Simscape>>SimPowerSystems>>Powergui

Scope

Simulink>>Sinks>>Scope

Example 6

The Simulink model of an IGBT is shown in Figure 2.51. When the gate pulse is connected, a little leakage current moves through the device while the current–emitter voltage for all intents and purposes measures up to the supply voltage. The device in this condition is said to be working in the cutoff range. The maximum forward voltage the device can withstand in this mode is controlled by the torrential slide breakdown voltage of the collector.

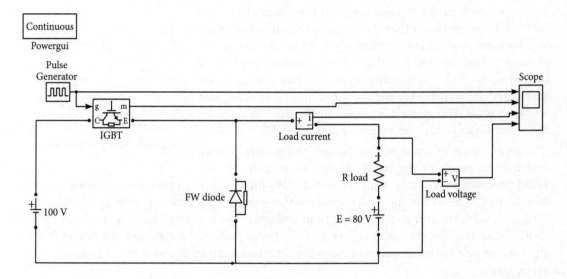

Figure 2.51 Simulink Model of an IGBT.

In contrast to a BJT, the breakdown voltage is autonomous of the collector current. IGBTs can obstruct the most extreme opposite voltage equivalent to V_{CE} in the cutoff mode. As the gate emitter voltage increases past the edge voltage, the IGBT goes into dynamic operation. In this mode, the collector current I_c is controlled and linear over most of the collector current scope.

The exchanging waveforms of an IGBT (Figure 2.52) are, in various respects, similar to that of a power MOSFET. This is expected, since the introductory period of an IGBT is a MOSFET. Similarly, in the late period of an IGBT, a real part of the total device current courses through the MOSFET. Hence, the trading voltage and current waveforms are similar to those of a MOSFET. The output has an effect on the trading characteristics of the device, particularly while turn off. To keep the components separate from each other, the gate emitter voltage of an IGBT is kept at a negative quantity when the device is off.

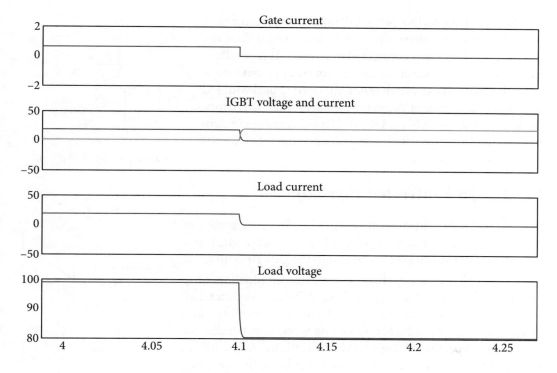

Figure 2.52 Switching Waveforms of IGBT.

2.6 Other Semiconductor Devices

2.6.1 DIAC

A DIAC is a type of switch used to switch AC voltage. The DIAC shown in Figure 2.53 is a two-terminal four-layer semiconductor device that can direct current in either side, when the maximum is dynamic.

It fits with the class of switches known as thyristors. It is similar to an intersection transistor without a base lead, that is, a two-lead device and finishes its exchanging activity by breaking down at a certain voltage.

Figure 2.53 Symbol of DIAC.

2.6.2 TRIAC

The TRIAC switch is a five-layered device with three terminals, which holds a pair of controlled thyristors, and is associated with the equivalent approach on the similar chip. The TRIAC is said to be a bidirectional device (refer to Figure 2.54), which can conduct current in both directions.

The TRIAC looks like two *PNP* transistors placed a back to back. However, physically, the TRIAC switch does not embody two thyristors connected in parallel. It is parallelly associated with substituting current on two opposite sides. Laterally, the TRIAC switch is not intended to deal with DC and unlike a couple of converse parallel joined thyristors, does not work steadily on DC. A TRIAC is more efficient than a couple of thyristors in parallel arrangement and its control is less complex.

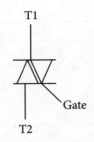

Figure 2.54 Symbol of TRIAC.

2.6.3 MOS controlled thyristor

An MCT (MOS controlled thyristor) is one of a number of newly developed devices (Figure 2.55). Basically, MCT is a thyristor with two MOSFETs built into the gate structure.

It is a high frequency, high power, low conduction drop exchanging device. The MCTs have different qualities like less state troubles, substantial current conveyability, and quick exchanging paces. MCTs can be worked in parallel for re current imparting at high-level applications. A few MCTs can be paralleled to bigger structure modules with individual devices. It has a few points of interest contrasted with different advanced devices like power MOSFET and IGBT. MCTs have low forward-voltage drop and high current thickness. Hence, it is utilized as a part of high power applications.

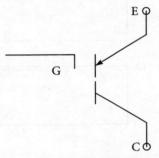

Figure 2.55 Symbol of MCT.

2.6.4 Integrated gate-commutated thyristors

IGCT shown in Figure 2.56 is also a newly developed power semiconductor. It is a state of the art high-control semiconductor switch utilized for power change, which sets new execution standards as to power, uniform quality, speed, effectiveness, cost, weight, and volume.

The IGCT can be turned on and off by triggering the gate signal. When compared with GTO thyristors, the IGCTs have lesser conducting problems; they can also withstand higher voltage ascent ratio. Hence, snubber circuits are not necessary for a majority of the applications.

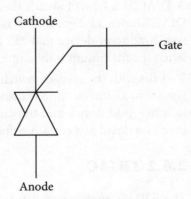

Figure 2.56 Symbol of IGCT.

Compared to a GTO thyristor, it is smaller in cell size, has more significant gate association, bringing about a much lower inductance in the gate drive circuit. The IGCT device can be with or without the converse blocking ability. Current source inverters use IGCTs.

Note The switches mentioned in Section 2.6 are not available in MATLAB/Simulink model.

2.7 MATLAB/Simulink Model of Semiconductor Devices in Electronics

In the field of electronics study, a few common semiconductor switches are used. These devices are not of much importance in power electronics sector. However, a general outline is given in this section.

2.7.1 Schottky diode

The Schottky diode is a hardware device that is generally utilized in radio frequency, *RF* applications as a detector diode. The diode is also used in power applications as a rectifier, in view of its low forward-voltage drop prompting lower levels of power loss compared with customary diodes. It is also sometimes called the surface boundary diode, the hot bearer diode and even the hot electron diode.

The Schottky diode depicted in Figure 2.57 is used as a part of numerous circuit schematic outlines. It could be seen from the figure that it is focused around a typical diode with extra components to the bar over the triangle shape.

Figure 2.57 Symbol of a Schottky Diode.

Example 7

Both forward and reverse qualities of a Schottky diode demonstrate a superior level of execution. The structure is fused to enhance the opposite breakdown ability.

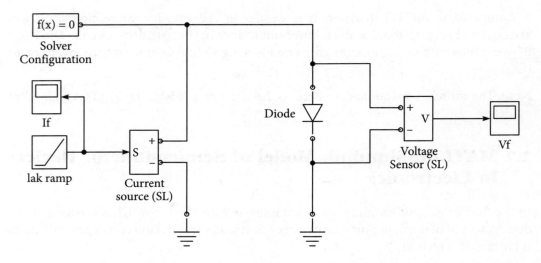

Figure 2.58 Model of a Schottky Diode.

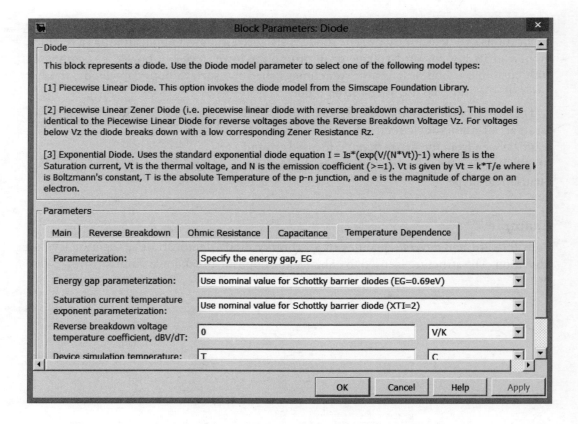

Figure 2.59 Block Parameter of a Schottky Diode.

The Simulink model shown in Figure 2.58 generates the current versus voltage curve for a Schottky barrier diode. The dataset for this device gives $V_f = 0.4$ V when $I_f = 10$ mA, and $V_f = 0.65$ V when $I_f = 100$ mA. The ohmic resistance is set to one over the gradient of the I–V curve at higher voltages. The temperature dependence is then modelled by selecting the default energy gap and saturation current temperature exponent values for a Schottky barrier diode in the block parameter (Figure 2.59). The plot produced by this illustration can be used to validate the I–V plots (Figure 2.60).

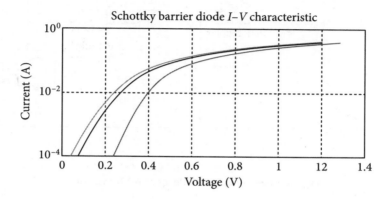

Figure 2.60 *I–V* Plot (Schottky Diode).

2.7.2 Bipolar junction transistors

The BJT constitutes three terminals, which includes the emitter, base and collector terminals. The symbol of the power BJT is the same as the signal level transistor (refer to Figure 2.61).

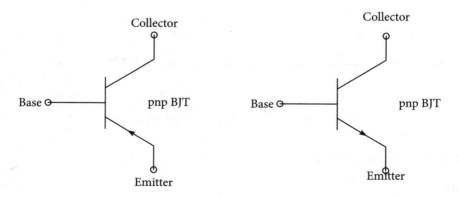

Figure 2.61 Symbol of BJT.

2.7.2.1 *PNP* Transistors

Example 8

Figures 2.62 and 2.63 depict the *PNP* and *NPN* transistors.

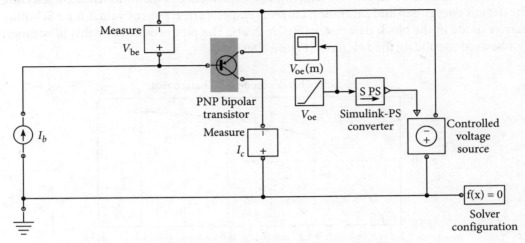

Figure 2.62 Simulink Model of a *PNP* Transistor.

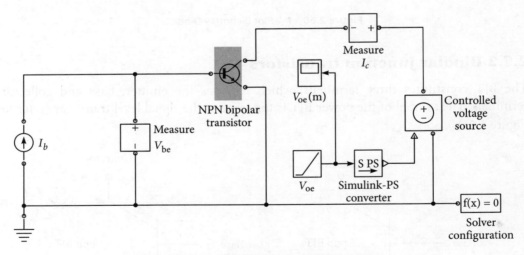

Figure 2.63 Simulink Model of an *NPN* Transistor.

2.7.2.2 *NPN* Transistors

Example 9

This model creates the I_c versus V_{ce} bend for an *NPN* bipolar transistor. Characterize the vector of the base momentums and the minimum and maximum collector–emitter voltages.

MATLAB Simulation of Power Semiconductor Devices 113

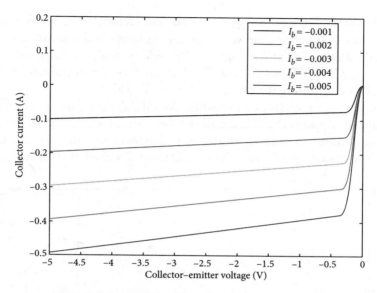

Figure 2.64 Switching Characteristics of a Transistor.

The plot in Figure 2.64 can be compared to a signal dataset that executes the transistor parameters. We can also determine the transistor qualities by the range of negative V_{ce} values – the increase in negative values is characterized by the reverse current exchange proportion.

2.7.3 MOSFET

Example 10

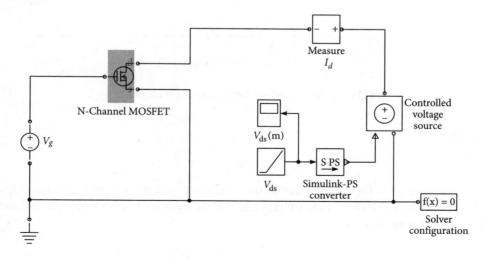

Figure 2.65 Simulink Model of *N* Channel MOSFET.

The model shown in Figure 2.65 produces the characteristic curves for an *N*-channel MOSFET. We can characterize the vector of gate voltages and minimum and maximum channel source voltages. The plot in Figure 2.66 could be compared against a signal dataset that executes the MOSFET parameters. We can analyze the MOSFET attributes in the converse field by determining the range of negative V_{ds} values.

To investigate the properties of a *P*-channel MOSFET, duplicate a *P*-channel MOSFET from the library to supplant the *N*-channel device, taking the precaution to swap over the two output associations so that the source is still associated with ground. To determine typical operation, the vector of gate voltages and the V_{ds} scope has to be negative.

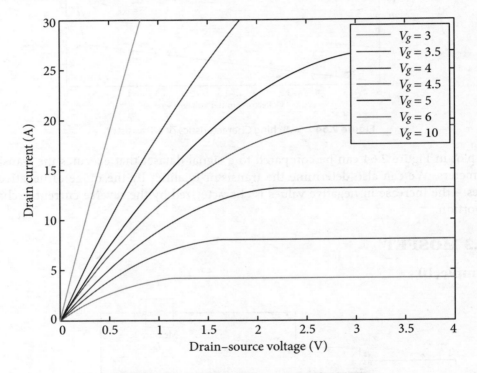

Figure 2.66 Switching Characteristics of MOSFET.

2.7.4 IGBT

Example 11

The Simulink model in Figure 2.67 creates the I_c versus V_{ce} curve for a protected gate bipolar transistor. Characterize the vector of gate emitter voltages and minimum and maximum collector–emitter voltages.

MATLAB Simulation of Power Semiconductor Devices 115

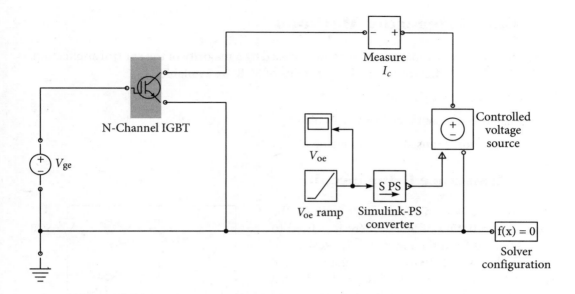

Figure 2.67 Simulink Model of *N* Channel IGBT.

The plot in Figure 2.68 can be compared with a signal input executing the transistor parameters.

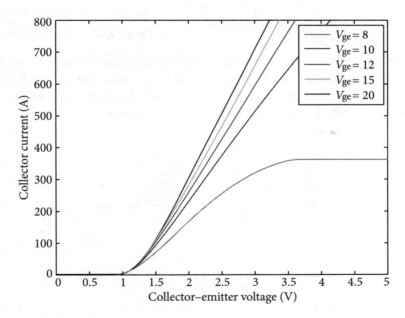

Figure 2.68 Switching Characteristics of IGBT.

2.8 Gate Triggering Methods

Gate activating is the most proficient strategy used by a majority of the control applications for SCR turning. Different terminating circuits of SCR are available.

- Resistance Firing Circuit
- Resistance–Capacitance Firing Circuit
- Uni Junction Transistor (UJT) Firing Circuit

2.8.1 Resistance firing circuit

The circuit shown in Figure 2.69 exhibits the resistance initiating of SCR, where it is used to drive the load from the input supply. Resistance and diode circuit works as a gate control equipment to switch the SCR to the desired condition. As the positive voltage increases, the SCR is forward biased and does not lead until its gate current is more than the minimum gate current of the SCR. When the by varying the resistance $R2$, the gate current is more than the base estimation of gate current, the SCR is turned on. In this manner, the load current starts coursing through the SCR.

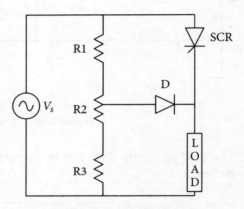

Figure 2.69 Resistance Firing Circuits.

The SCR stays on until the anode current is equivalent to the holding current of the SCR. It will also switch off when the voltage connected to it is zero. Hence, the load current is zero as the SCR behaves like an open switch. The diode shields the gate drive circuit from converse gate voltage during the negative half cycle of the input.

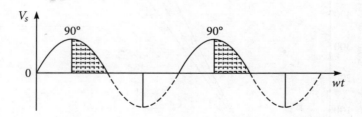

Figure 2.70 Waveform of Resistance Firing Circuits.

The resistance $R1$ limits the current coursing through the gate terminal and its quality is such that the gate current will not surpass the maximum gate current. It is the least complex and conservative sort of activating yet restricted in use because of its inconveniences.

The waveform shown in Figure 2.70 represents the activating edge, which is restricted to 90°. Since the connected voltage is maximum at 90°, the gate current needs to achieve the least gate current value some place between 0° and 90°.

2.8.2 Resistance–capacitance firing circuit

The confinement of a resistance terminating circuit can be overcome by the *RC* activating circuit, which gives the terminating point control from 0° to 180°. By changing the stage and sufficiency of the gate current, a wide range in terminating points is acquired utilizing this circuit. Figure 2.71 shows the *RC* activating circuit comprising two diodes with an *RC* system for turning the SCR. By fluctuating the variable resistance, activating or terminating edge is controlled to a full positive half cycle of the data signal.

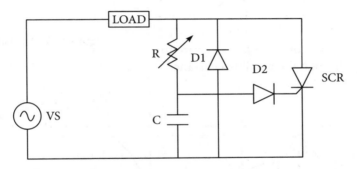

Figure 2.71 Resistance–Capacitance Firing Circuit.

During the negative half cycle of the input signal, the capacitor is positive through diode D2 up to the maximum supply voltage V_{max}. This voltage stays at V_{max} over the capacitor till the supply voltage achieves zero crossing. During the positive half cycle of the data, the SCR gets to be forward one-sided and the capacitor begins charging through variable imperviousness up to the activating voltage value of the SCR. When the capacitor charging voltage is equivalent to the gate trigger voltage, SCR is turned on and the capacitor holds a little voltage. Accordingly, the capacitor voltage is useful for setting off the SCR even after 90° of the input waveform. The diode D1 keeps the negative voltage between the gate and the cathode amid the negative half cycle of the input through diode D2. Refer to Figure 2.72 for the waveform of a resistance–capacitance firing circuit.

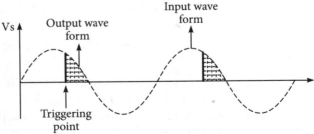

Figure 2.72 Waveform of a Resistance–Capacitance Firing Circuit.

2.8.3 UJT firing circuit

This is the most common technique for setting off the SCR in light of the fact that the drawn pulse at the gate utilizing R and RC activating strategies causes more power dissemination at the gates than by utilizing UJT as the activating device; and the power loss is restricted as it delivers continuous pulses.

The RC system is associated with the emitter terminal of the UJT, which frames the timing circuit. The capacitor is altered while the resistance is variable; thus, the charging rate of the capacitor relies on the variable resistance implying that the RC time constant is controlled. When the voltage is connected, the capacitor begins charging through the variable resistance.

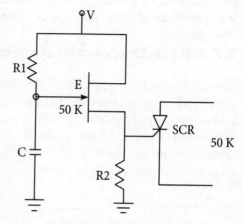

Figure 2.73 UJT Firing Circuit.

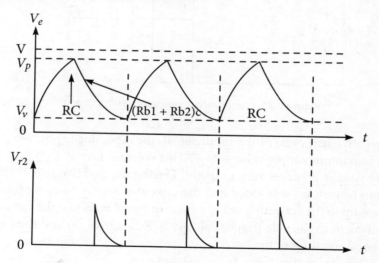

Figure 2.74 Waveform of UJT Firing Circuit.

By changing the resistance value, voltage over the capacitor gets shifted. Once the capacitor voltage is equivalent to the crest value of the UJT, it begins directing and subsequently delivers a pulse output till the voltage over the capacitor is equivalent to the voltage V_v of the UJT. This procedure repeats itself and creates continuous pulses at the base terminal 1. The output pulse at the base terminal 1 is utilized to turn on the SCR at regular time interims.

2.8.4 Pulse transformers

Pulse transformers shown in Figure 2.75 have one essential winding and one or more auxiliary windings. Different auxiliary windings permit concurrent gating signs for series and parallel associated transistors. The transformer should have minimum leakage inductance and ascent time of output. The transformer should soak at low exchanging frequency and the output will be misshaped.

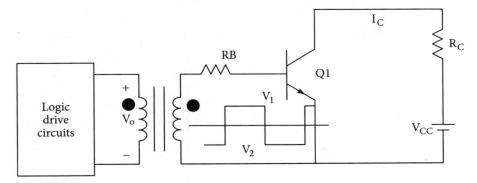

Figure 2.75 Pulse Transformer.

2.8.5 Optocoupler

Optocouplers (Figure 2.76) comprise an infrared LED and a silicon photo transistor. The input sign is connected to the LED and the output is taken from the photo transistor. The rise and fall times of the photo transistor are small: turn on time = 2.5 s and turn off = 300 ns. Thus, the optocoupler cannot be used for high-frequency applications. The photo transistor could be a Darlington pair. It requires separate power supply and adds to the multifaceted nature, cost and weight of driver circuits.

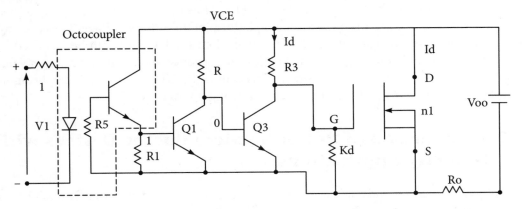

Figure 2.76 Optocoupler.

2.8.6 Ramp-pedestrial triggering

The circuit demonstrated in Figure 2.77 uses a UJT to trigger an SCR. The UJT is utilized to precisely trigger the SCR. When the source voltage surpasses 20 V, the Zener diode D_Z will start to direct, applying a DC voltage over the base associations of the UJT. In the meantime, diode D1 will be forward one-sided, and the capacitor will rapidly charge through $R1$ and $R2$.

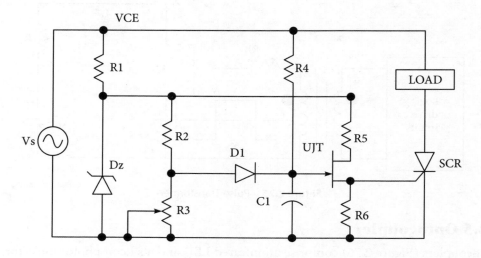

Figure 2.77 Ramp-Pedestrial Triggering Circuit.

This shows the one-side hand platform segment of the emitter voltage. Once the capacitor charges to the voltage crosswise over $R3$, D1 will get to be converse one-sided and the capacitor will slowly charge through $R4$. This causes the slope bit of the emitter voltage. The capacitor keeps on charging until the UJT fires. Now the capacitor will rapidly release through $R6$; this speaks to the one-side hand platform of the emitter voltage. The capacitor release is adequate to trigger the SCR. The time when the UJT fires can be balanced by shifting the plot $R3$. With a bigger setting on $R3$, the capacitor must charge to a bigger value before D2 gets to be converse one-sided. This causes the UJT to flash quicker, bringing about a greater amount of the source voltage showing up over the SCR.

2.9 Comparison of Power Semiconductor Devices with Industry Applications

The physical processes and applications performance of the different power semiconductor devices are listed in Table 2.3.

Table 2.3 Applications of Power Semiconductor Devices

S.No.	Device	Description	Application
1	DIAC	Bi-directional switch to improve operation of alternating current power switching systems. Voltage rating: 18–20 V	Dimmers Florescent lamps
2	Diode	A device that allows an electric current to pass in one direction. Voltage rating: 10 V–10 kV	Radio demodulation Power conversion Over-voltage protection Temperature measurements Clamper
3	Schottky diode	A device to exhibit high-voltage, high competence, and power conversion capability. Voltage rating: up to 100 V	Smart grid power Traction Ship and vessel power systems Wind turbine
4	Zener diode	A switch that allows the current to flow in the forward conduction mode similar to an ideal diode. Voltage rating: up to 0.6 V	Voltage regulators Surge protectors
5	Light-emitting diode	A two-lead semiconductor light source. Voltage rating: 1.8–2.2 V	Visual signals Light sensors
6	TRIAC	A three-terminal device that conducts either in the positive mode or in the negative mode when triggered by a gate signal. Voltage rating: 200–800 V	Light dimmers Electric fans Electric motors
7	Thyristors	A semiconductor switch with four layers where the current flows between two electrodes is triggered by the third electrode. Voltage rating: 200–6 kV	Television Motion pictures Theatre

8	Bipolar Junction Transistor	A transistor that relies on the contact of two types of semiconductor for its operation. Voltage rating: 25–100 V	Amplifiers Logarthamic converters
9	Static induction transistor	A high-power, high-frequency device. Voltage rating: 125 V	Audio power amplifiers
10	MOS controlled thyristor	Voltage-controlled fully controllable thyristor. Voltage rating: 500 V	Amplifiers Television
11	Gate turn off thyristors	Special type of thyristor, which is a high-power semiconductor device. Voltage rating: 100 kV	Motor drives High-power inverters Traction
12	Power MOSFET	A specific type of metal – oxide semiconductor field-effect transistor (MOSFET) designed to handle significant power levels. Voltage rating: <200 V	Amplifiers Television
13	Insulated-gate bipolar transistor (IGBT)	An electronic switch with high effectiveness and fast switching. Voltage rating: 600–1200 V	Variable speed refrigerators Electric cars Trains Air-conditioners Stereo systems
14	Integrated gate-commutated thyristor (IGCT)	A fully controllable power switch. Voltage rating: 4500–6500 V	Variable frequency inverters drives Traction

2.9.1 Other devices

Asymmetrical SCR

Asymmetrical SCRs are unidirectional devices that can lead current in one direction and also they do not block reverse voltage. Hence, ASCRS are generally used where reverse blocking methods are not much important. The reverse voltage rating is 20 to 30V and forward voltage rating is 400-2000V.

Reverse Control Thyristors

A reverse control thyristor (RCT) has an incorporated opposite diode; so, it is not suitable for converse blocking. RCTs can be used where an opposite or freewheel diode is needed. Since the SCR and diode do not direct in the meantime, they do not create heat all the time and can undoubtedly be incorporated and cooled together. RCTs are regularly utilized as part of frequency changers and inverters.

Silicon Unilateral Switch

The silicon unilateral switch (SUS) is a device from the thyristor family; it is utilized as a trigger component in circuits utilizing SCRs. The SUS is an SCR with an anode gate and an implicit Zener diode. The critical part of an SUS is the activating voltage or the interior zener voltage. The maximum current and the voltage is bolstered by the device. Silicon one-sided switches are not rare nowadays; they can be found in circuits utilizing SCRs as a trigger component.

Silicon Bilateral Switch

Silicon bilateral switches are specifically used for applications where exchanging voltage over a wide temperature range and coordinated two-sided qualities are a benefit. They are perfectly suited for half wave and full wave activation in low-voltage SCR and TRIAC stage control circuits.

Silicon Controlled Switch

A semiconductor device with an extra terminal in the fundamental SCR is said to be a silicon controlled switch (SCS). In an SCS, the cathode gate and anode terminals are the control lead terminals, where the load current is passed on by the anode gate and cathode terminals. The SCS switch can be switched on by applying a positive voltage in between the cathode gate and the gate terminals. The switch can be turned off either by applying negative voltage between the anode and cathode terminals or by short circuit, where the anode terminal is kept positive.

Light Activated SCR

Light activated SCRs or LASCRs have been fabricated using an additional terminal of gate that can be activated by electrical signals. LASCRs are mostly activated using gate terminals that perform like normal SCRs. When the light falls on the exhaustive solid layer, the valence electrons become free electrons and stream out to the transistor.

Emitter Turn-off Thyristor

The emitter turn-off thyristor (ETO) is another popular high-power semiconductor switch, which combines the benefits of GTO thyristor's high voltage and current capacity and MOS's simple gate control. Its unrivalled control qualities consolidated with its rapid, more extensive reverse bias safe operating area (RBSOA), higher controllable current, forward current immersion ability, its on-device, current detection and minimal effort make the

ETO the best power device in high power and control applications. Future ETO switches (a work in progress) will be likewise packed with extra components that no competing device can offer, incorporating worked in voltage, current and temperature detecting ability, control-power self-era capacity and high-voltage current immersion ability. These abilities help ETO to lessen the expense of converter-based transmission controllers while enhancing the controller output power, dynamic execution, and uniform working quality.

Silicon Carbide Devices

Silicon carbide (SiC) control devices can be an alternate option to Si-based power devices in high-productivity and high-control thickness applications. SiC gadgets have many advantages such as with high breakdown voltage, high working electric field, high working temperature, high exchanging frequency, and low power loss. Despite the fact that the cost of SiC devices is still higher than its Si partners, SiC devices are finding more applications where the advantages are sufficiently noteworthy to counterbalance the higher device cost.

Experts estimate that nowadays, the electrical energy consumed in a developed country entirely depends on the ability of its electronic equipments and its growing proportion. These equipments have engulfed the complete chain of production, transport, distribution, and consumption of voltage. There is a major 'electronification' of electrical equipments through that the latter's gain of new functions and options. Table 2.4 summarizes this international development of how electrical equipments incorporate power electronic systems.

Table 2.4 Power Electronic Systems

S.No.	Field	Description
1	Commercial	Drives with electrical motors, machine tools, and robots, electrical furnaces, heat treatment
2	Transport	Electrical trains, electrical underground railways, tramways, trolleys, hybrid or electrical cars, elevators, cranes with electrical drives
3	Energetics	Equipments for traditional power plants, renewable energy resources, microgrids, high-voltage DC power transmission, energy storage systems
4	Home appliances	Vacuum cleaners, laundry machines, electrical plates, microwaves ovens, food processors, lamps with adjustable light-weight, air conditioners, computers, and audio–video players
5	Business	Escalators, elevators, heat installations, ventilation systems, air conditioners, decorative lighting, code readers, computers, UPS
6	Telecommunications	DC power sources, uninterruptible power sources, target-hunting radio antennas

7	Medicine	Laboratory and surgery robots, computerized axial tomography and nuclear resonance, air conditioners, elevators, medical equipments
8	Aerospace	Aero planes, satellites and communication installations, launch bases
9	Military	Measuring devices for tracking installations, rockets, target-hunting launch ramps, military vehicles, target-hunting antiaircraft batteries

It is to be noted that from all the applications that embrace power electronic systems, a high proportion include applications in which the energy is generated or the movement is decided by the electrical motors. These electrical drives are the biggest power users of the generated energy.

Thus, several power electronic applications were found within the field of electrical drives. For this reason, the study of power electronics typically intersect with different domains like electrical machines, electrical drives.

Similarly, different disciplines are tied to the management of power electronic systems such as electronics, digital systems with chip (microcontrollers – μC, digital signal processors – DSP), automatics, management theory, electronic measurements, and so on.

Summary

This chapter will possibly be a clear conception of power electronics and the role of power semiconductor devices in the electrical and electronics field. Readers/readers will be able to understand how to choose semiconductor switches depending on the requirement. A detailed study of these concepts would effectively help learners design and simulate their own circuits that could show the pathway for creating big applications in the power electronics field, from the lower to the higher end.

Solved Examples

1. A BJT is specified to have a range 8–40. The load resistance in $R_c = 11\ \Omega$. The DC supply voltage is $V_{CC} = 200$ V and the input voltage to the base circuit is $V_B = 10$ V. If $V_{CE(sat)} = 1.0$ V and $V_{BE(sat)} = 1.5$ V, find

    (a) The value of R_B that results in saturation with an overdrive factor of 5.

    (b) The forced β_f.

    (c) The power loss P_T in the transistor.

    Solution

    (a) $I_{CS} = \dfrac{V_{CC} - V_{CC(sat)}}{R_C} = \dfrac{200 - 1.0}{11\ \Omega} = 18.1$ A

Therefore, $I_{BS} = \dfrac{I_{CS}}{\beta_{min}} = \dfrac{18.1}{8} = 2.2625$ A

Therefore, $I_B = ODF \times I_{BS} = 11.3125$ A

$$I_B = \dfrac{V_B - V_{BE(sat)}}{R_B}$$

Therefore, $R_B = \dfrac{V_B - V_{BE(sat)}}{I_B} = \dfrac{10 - 1.5}{11.3125} = 0.715\ \Omega$

(b) Therefore $\beta_f = \dfrac{I_{CS}}{I_B} = \dfrac{18.1}{11.3125} = 1.6$

(c) $P_T = V_{BE}I_B + V_{CE}I_C$

$P_T = 1.5 \times 11.3125 + 1.0 \times 18.1$

$P_T = 16.97 + 18.1 = 35.07$ W

2. The β of a bipolar transistor varies from 12 to 75. The load resistance is $R_c = 1.5\ \Omega$. The DC supply voltage is $V_{CC} = 40$ V and the input voltage of the base circuit is $V_B = 6$ V. If $V_{CE(sat)} = 1.2$ V, $V_{BE(sat)} = 1.6$ V and $R_B = 0.7\ \Omega$, determine

(a) The overdrive factor (ODF)
(b) The forced β_f.
(c) Power loss in transistor P_T

Solution

$$I_{CS} = \dfrac{V_{CC} - V_{CE(sat)}}{R_C} = \dfrac{40 - 1.2}{1.5} = 25.86\ A$$

$$I_{BS} = \dfrac{I_{CS}}{\beta_{min}} = \dfrac{25.86}{12} = 2.15\ A$$

Also $I_B = \dfrac{V_B - V_{BE(sat)}}{R_B} = \dfrac{6 - 1.6}{0.7} = 6.28\ A$

(a) Therefore, $ODF = \dfrac{I_B}{I_{BS}} = \dfrac{6.28}{2.15} = 2.92$

(b) Forced $\beta_f = \dfrac{I_{CS}}{I_B} = \dfrac{25.86}{6.28} = 4.11$

(c) $P_T = V_{BE}I_B + V_{CE}I_C$

$P_T = 1.6 \times 6.25 + 1.2 \times 25.86$

$P_T = 41.032$ W

3. For the transistor switch shown in the figure,
   (a) Calculate forced beta, β_f of transistor
   (b) If the manufacturer's specified β is in the range of 8–40, calculate the minimum overdrive factor (ODF).
   (c) Obtain power loss P_T in the transistor.

   $V_B = 10$ V, $\quad R_B = 0.75\ \Omega$

   $V_{BE\ sat} = 1.5$ V, $\quad R_C = 11\ \Omega$

   $V_{CE\ sat} = 1$ V, $\quad V_{CC} = 200$ V

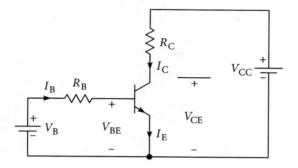

Solution

(a) $I_B = \dfrac{V_B - V_{BEsat}}{R_B} = \dfrac{10 - 1.5}{0.75} = 11.33$ A

$I_{CS} = \dfrac{V_{CC} - V_{CEsat}}{R_C} = \dfrac{200 - 1.0}{11} = 18.09$ A

Therefore, $I_{BS} = \dfrac{I_{CS}}{\beta_{min}} = \dfrac{18.09}{8} = 2.26$ A

$\beta_f = \dfrac{I_{CS}}{I_B} = \dfrac{18.09}{11.33} = 1.6$

(b) $\text{ODF} = \dfrac{I_B}{I_{BS}} = \dfrac{11.33}{2.26} = 5.01$

(c) $P_T = V_{BE}I_B + V_{CE}I_C = 1.5 \times 11.33 + 1.0 \times 18.09 = 35.08$ W

4. A simple transistor switch is used to connect a 24 V DC supply across a relay coil, which has a DC resistance of 200 Ω. An input pulse of 0–5 V amplitude is applied through a series base resistor R_B at the base so as to turn on the transistor switch. Sketch the device current waveform with reference to the input pulse. Calculate
   (a) I_{CS}
   (b) Value of resistor R_B required to obtain an overdrive factor of two.
   (c) Total power dissipation in the transistor that occurs during the saturation state.

128 Power Electronics with MATLAB

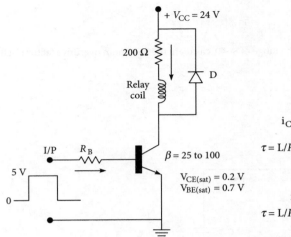
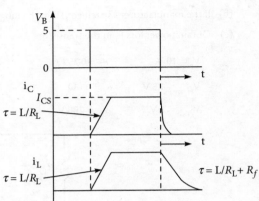

Solution

(I) $I_{CS} = \dfrac{V_{CC} - V_{CEsat}}{R_C} = \dfrac{24 - 0.2}{200} = 0.119\,A$

(ii) Value of R_B

$I_{BS} = \dfrac{I_{CS}}{\beta_{min}} = \dfrac{0.119}{25} = 4.76\,mA$

Therefore, $I_B = ODF \times I_{BS} = 2 \times 4.76 = 9.52\,mA$

Therefore, $R_B = \dfrac{V_B - V_{BEsat}}{I_B} = \dfrac{5 - 0.7}{9.52} = 450\,\Omega$

(iiI) $P_T = V_{BEsat} \times I_B + V_{CEsat} \times I_{CS} = 0.7 \times 9.52 + 0.2 \times 0.119 = 6.68\,W$

5. The various parameters of the transistor circuit are as follows: $V_{CC} = 220$ V, $V_{CE(sat)} = 2$ V, $I_{CS} = 80$ A, $t_d = 0.4$ μs, $t_r = 1$ μs, $t_n = 50$ μs, $t_s = 3$ μs, $t_f = 2$ μs, $t_o = 40$ μs, $f = 5$ kHz, $I_{CEO} = 2$ mA. Determine the average power loss due to collector current during t_{on} and t_n. Find also the peak instantaneous power loss due to collector current during turn on time.

Solution

During delay time, the time limits are $0 \leq t \leq t_d$. Also, $i_c t = I_{CEO}$ and $V_{CE} t = V_{CC}$. Therefore, instantaneous power loss during delay time is

$P_d t = i_c V_{CE} = I_{CEO} V_{CC} = 2 \times 10^{-3} \times 220 = 0.44\,W$

Average power loss during delay time $0 \leq t \leq t_d$ is given by

$P_d = \dfrac{1}{T} \int_0^{t_d} i_c t V_{CE} t\, dt$

$$P_d = \frac{1}{T}\int_0^{t_d} I_{CEO} V_{CC} dt$$

$$P_d = f I_{CEO} V_{CC} t_d$$

$$P_d = 5 \times 10^3 \times 2 \times 10^{-3} \times 220 \times 0.4 \times 10^{-6} = 0.88 \text{ mW}$$

During rise time $0 \leq t \leq t_r$

$$i_c t = \frac{I_{cs}}{t_r} t$$

$$v_{CE} t = \left[V_{CC} - \left(\frac{V_{CC} - V_{CE(sat)}}{t_r} \right) t \right]$$

$$v_{CE} t = V_{CC} + \left[V_{CE(sat)} - V_{CC} \right] \frac{t}{t_r}$$

Therefore, average power loss during rise time is

$$P_r = \frac{1}{T} \int_0^{t_r} \frac{I_{cs}}{t_r} t \left[V_{CC} + V_{CEsat} - V_{CC} \frac{t}{t_r} \right] dt$$

$$P_r = f \cdot I_{cs} t_r \left[\frac{V_{CC}}{2} - \frac{V_{CC} - V_{CES}}{3} \right]$$

$$P_r = 5 \times 10^3 \times 80 \times 1 \times 10^{-6} \left[\frac{220}{2} - \frac{220 - 2}{3} \right] = 14.933 \text{ W}$$

Instantaneous power loss during rise time is

$$P_r t = \frac{I_{cs}}{t_r} t \left[V_{CC} - \frac{V_{CC} - V_{CEsat}}{t_r} t \right]$$

$$P_r t = \frac{I_{cs}}{t_r} t V_{CC} - \frac{I_{cs}^2 t}{t_r^2} \left[V_{CC} - V_{CEsat} \right]$$

Differentiating these equations,

$$\frac{dP_r t}{dt} = \frac{I_{cs} V_{CC}}{t_r} - \frac{I_{cs}^2 t}{t_r^2} V_{CC} - V_{CEsat}$$

At $t = t_m$, $\dfrac{dP_r t}{dt} = 0$

Therefore, $0 = \dfrac{I_{cs}}{t_r} V_{CC} - \dfrac{2 I_{cs} t_m}{t_r^2} \left[V_{CC} - V_{CEsat} \right]$

$$\frac{t_r V_{CC}}{2} = t_m [V_{CC} - V_{CEsat}]$$

Therefore, $t_m = \dfrac{t_r V_{CC}}{2[V_{CC} - V_{CEsat}]}$

$$t_m = \frac{V_{CC} t_r}{2[V_{CC} - V_{CEsat}]} = \frac{220 \times 1 \times 10^{-6}}{2[200-2]} = 0.5046\ \mu s$$

Peak instantaneous power loss P_m during rise time is obtained by substituting the value of $t = t_m$ in the below equation

$$P_{rm} = \frac{I_{CS}}{t_r} \frac{V_{CC}^2 t_r}{2[V_{CC} - V_{CEsat}]} - \frac{I_{CS}}{t_r^2} \frac{V_{CC} t_r^2 [V_{CC} - V_{CEsat}]}{4[V_{CC} - V_{CEsat}]^2}$$

$$P_{rm} = \frac{80 \times 220^2}{4[220-2]} = 4440.4\ W$$

Total average power loss during turn on

$$P_{on} = P_d + P_r = 0.00088 + 14.933 = 14.9339\ W$$

During conduction time $0 \le t \le t_n$

$$i_c t = I_{CS}\ \text{and}\ v_{CE} t = V_{CEsat}$$

Instantaneous power loss during t_n is

$$P_n t = i_c v_{CE} = I_{CS} V_{CEsat} = 80 \times 2 = 160\ W$$

Average power loss during the conduction period is

$$P_n = \frac{1}{T} \int_0^{t_n} i_c v_{CE} dt = f I_{CS} V_{CES} t_n = 5 \times 10^3 \times 80 \times 2 \times 50 \times 10^{-6} = 40\ W$$

6. Design a suitable RC triggering circuit for a thyristorized network operation on a 220 V, 50 Hz supply. The specifications of SCR are $V_{gtmin} = 5\ V$, $I_{gtmax} = 30\ mA$.

Solution

$$R = \frac{V_s - V_{gt} - V_D}{I_g} = 7143.3\ \Omega$$

Therefore, $RC \ge 0.013$

$R \le 7.143\ k\Omega$

$C \ge 1.8199\ \mu F$

7. A UJT is used to trigger a thyristor whose minimum gate triggering voltage is 6.2 V. The UJT ratings are $\eta = 0.66$, $I_p = 0.5$ mA, $I_v = 3$ mA, $R_{B1} + R_{B2} = 5$ kΩ, leakage current = 3.2 mA, $V_p = 14$ V and $V_v = 1$ V. Oscillator frequency is 2 kHz and capacitor $C = 0.04$ µF. Design the complete circuit.

Solution

$$T = R_C C \ln\left[\frac{1}{1-\eta}\right]$$

Here, $T = \dfrac{1}{f} = \dfrac{1}{2 \times 10^3}$. Since $f = 2$ kHz and substituting other values,

$$\frac{1}{2 \times 10^3} = R_C \times 0.04 \times 10^{-6} \ln\left(\frac{1}{1-0.66}\right) = 11.6 \text{ k}\Omega$$

The peak voltage is given as $V_p = V_{BB} + V_D$

Let $V_D = 0.8$, then substituting other values,

$$14 = 0.66 V_{BB} + 0.8$$

$$V_{BB} = 20 \text{ V}$$

The value of R_2 is given by

$$R_2 = \frac{0.7 R_{B2} + R_{B1}}{\eta V_{BB}}$$

$$R_2 = \frac{0.7 \times 5 \times 10^3}{0.66 \times 20}$$

$\therefore$ $R_2 = 265 \ \Omega$

Value of R_1 can be calculated by the equation

$$V_{BB} = I_{leakage} (R_1 + R_2 + R_{B1} + R_{B2})$$

$$20 = 3.2 \times 10^{-3} (R_1 + 265 + 5000)$$

$$R_1 = 985 \ \Omega$$

The value of R_{cmax} is given by the equation

$$R_{cmax} = \frac{V_{BB} - V_p}{I_p}$$

$$R_{cmax} = \frac{20 - 14}{0.5 \times 10^{-3}}$$

$$R_{cmax} = 12 \text{ k}\Omega$$

Similarly, the value of R_{cmin} is given by the equation

$$R_{cmin} = \frac{V_{BB} - V_v}{I_v}$$

$$R_{cmin} = \frac{20 - 1}{3 \times 10^{-3}}$$

$$R_{cmin} = 6.33 \text{ k}\Omega$$

8. Design the UJT triggering circuit for SCR. Given $-V_{BB} = 20$ V, $\eta = 0.6$, $I_p = 10$ μA, $V_v = 2$ V, $I_v = 10$ mA. The frequency of oscillation is 100 Hz. The triggering pulse width should be 50 μs.

Solution

The frequency $f = 100$ Hz. Therefore, $T = \dfrac{1}{f} = \dfrac{1}{100}$

$$T = R_c C \ln\left(\frac{1}{1-\eta}\right)$$

Substituting values in this equation,

$$\frac{1}{100} = R_c C \ln\left(\frac{1}{1-0.6}\right)$$

$$\therefore \quad R_c C = 0.0109135$$

Let us select $C = 1$ μF. Then, R_c will be

$$R_{cmin} = \frac{0.0109135}{1 \times 10^{-6}}$$

$$R_{cmin} = 1091 \text{ k}\Omega$$

The peak voltage is given as

$$V_p = V_{BB} + V_D$$

Let $V_D = 0.8$ and substituting other values,

$$V_p = 0.6 \times 20 + 0.8 = 12.8 \text{ V}$$

The minimum value of R_c can be calculated from

$$R_{cmin} = \frac{V_{BB} - V_v}{I_v}$$

$$R_{cmin} = \frac{20 - 2}{10 \times 10^{-3}} = 1.8 \text{ k}\Omega$$

Value of R_2 can be calculated from

$$R_2 = \frac{10^4}{\eta V_{BB}}$$

$$R_2 = \frac{10^4}{0.6 \times 20} = 833.33\,\Omega$$

Here the pulse width is given, that is 50 μs.

Hence, value of R_1 will be.

$$\tau_2 = R_1 C$$

The width τ_2 = 50 μs and C = 1 μF, hence this equation becomes,

$$50 \times 10^{-6} = R_1 \times 1 \times 10^{-6}$$

$$\therefore\ R_1 = 50\,\Omega$$

Thus, we obtain the values of the components in the UJT triggering circuit as

$$R_1 = 50\,\Omega,\ R_2 = 833.33\,\Omega,$$

9. A bipolar transistor is specified to have β_F in the range 8–40. The load resistance is R_c = 11 Ω. The DC supply voltage is V_{CC} = 200 V and the input voltage to the base circuit is V_B = 10 V. If $V_{CE(sat)}$ = 10 V and $V_{BE(sat)}$ = 1.5 V, find (a) the value of R_B that results in saturation with an ODF of 5, (b) the β_{forced} and (c) the power loss P_T in the transistor.

Solution

V_{CC} = 200 V, β_{min} = 8, β_{max} = 40, R_c = 11 Ω, ODE = 5, V_B = 10 V, $V_{CE(sat)}$ = 1.0 V, $V_{BE(sat)}$ = 1.5 V

$$I_{CS} = \frac{V_{CC} - V_{CE(sat)}}{R_C} = \frac{(200 - 1.0)}{11} = 18.1\,A$$

$$I_{BS} = \frac{I_{CS}}{\beta_F} = \frac{18.1}{\beta_{min}} = \frac{18.1}{8} = 2.2625\,A$$

$$ODF = \frac{I_B}{I_{BS}}$$

$$I_B = 5 \times 2.2625 = 11.3125\,A$$

$$I_B = \frac{V_B - V_{BE}}{R_B}$$

$$R_B = \frac{V_B - V_{BE(sat)}}{I_B} = \frac{10 - 1.5}{11.3125} = 0.7514\,\Omega$$

$$\beta_{forced} = \frac{I_{CS}}{I_B}$$

$$\beta_{forced} = \frac{18.1}{11.3125} = 1.6$$

(a) Total power loss, $P_T = V_{BE} I_B + V_{CE} I_C$

$P_T = (1.5 \times 11.3125) + (1.0 \times 18.1)$

$P_T = 16.97 + 18.1 = 35.07$ W

10. The parameters of a transistor are given as follows $V_{CC} = 250$ V, $V_{BE(sat)} = 3$ V, $I_B = 8$ A, $V_{CS(sat)} = 2$ V, $I_{CS} = 100$ A, $t_d = 0.5$ μs, $t_s = 5$ μs, $t_f = 3$ μs and $f_s = 10$ kHz. The duty cycle is k = 50 per cent. The collector-to-emitter leakage current is $I_{CEO} = 3$ mA. Determine the power loss due to collector current (a) during turn on $t_{on} = t_d + t_r$, (b) during conduction period t_n.

Solution

$$T = \frac{1}{f_s} = 100 \ \mu s$$

$k = 0.5$, $kT = t_d + t_r + t_n = 50 \ \mu s$,

$t_n = 50 - 0.5 - 1 = 48.5 \ \mu s$,

$(1 - k)T = t_s + t_f + t0 = 50 \ \mu s$

$t_0 = 50 - 5 - 3 = 42 \ \mu s$

(a) During delay time, $0 \leq t \leq t_d$:

$i_c(t) = I_{CEO}$

$v_{CE}(t) = V_{CC}$

The instantaneous power due to the collector current is

$P_c(t) = i_c v_{CE} = I_{CEO} V_{CC}$

$= 3 \times 10^{-3} \times 250 = 0.75$ W

The average power loss during the delay time is

$$P_d = \frac{1}{T} \int_0^{t_d} P_c(t) dt = I_{CEO} V_{CC} t_d f_s$$

$= 3 \times 10^{-3} \times 250 \times 0.5 \times 10^{-6} \times 10 \times 10^3 = 3.75$ mW

During rise time, $0 \leq t \leq t_r$:

$i_c(t) = \frac{I_{CS}}{t_r} t$

$v_{CE}(t) = V_{CC} + \left(V_{CE(sat)} - V_{CC}\right)\frac{t}{t_r}$

$P_c(t) = i_c v_{CE} = I_{CS} \frac{t}{t_r}\left[V_{CC} + \left(V_{CE(sat)} - V_{CC}\right)\frac{t}{t_r}\right]$

The power $P_c(t)$ is maximum when $t = t_m$, where

$$t_m = \frac{t_r V_{CC}}{2[V_{CC} - V_{CE(sat)}]}$$

$$= 1 \times \frac{250}{2(250-2)} = 0.504 \ \mu s$$

$$P_p = \frac{V_{CC}^2 I_{CS}}{4[V_{CC} - V_{CE(sat)}]}$$

$$= 250^2 \times \frac{100}{4(250-2)} = 6300 \ W$$

$$P_r = \frac{1}{T}\int_0^{t_r} P_c(t)\,dt = f_s I_{CS} t_r \left[\frac{V_{CC}}{2} + \frac{V_{CE(sat)} - V_{CC}}{3}\right]$$

$$= 10 \times 10^{-3} \times 100 \times 1 \times 10^{-6} \left[\frac{250}{2} + \frac{2-250}{3}\right] = 42.33 \ W$$

The total power loss during the turn on is

$$P_{on} = P_d + P_r$$
$$= 0.00375 + 42.33 = 42.33 \ W$$

(a) The conduction period, $0 \le t \le t_n$:

$$i_c(t) = I_{CS}$$
$$v_{CE}(t) = V_{CE(sat)}$$
$$P_c(t) = i_c v_{CE(sat)} I_{CS}$$
$$= 2 \times 100 = 200 \ W$$

$$P_n = \frac{1}{T}\int_0^{t_n} P_c(t)\,dt = V_{CE(sat)} I_{CS} t_n f_s$$

$$= 2 \times 100 \times 48.5 \times 10^{-6} \times 10 \times 10^3 = 97 \ W$$

11. For the parameters in Problem 2, calculate the average power loss due to the base current.

Solution

$V_{BE(sat)} = 3 \ V, I_B = 8 \ A, t = 1/f_s = 100 \ \mu s, k = 0.5, kT = 50 \ \mu s, t_d = 0.5 \ \mu s, t_r = 1 \ \mu s, t_n = 50 - 1.5 = 48.5 \ \mu s,$

$t_s = 5 \ \mu s, t_f = 3 \ \mu s, t_{on} = t_d + t_r = 1.5 \ \mu s, t_{off} = t_s + t_f = 5 + 3 = 8 \ \mu s.$

During the period, $0 \le t \le (t_{on} + t_n)$:

$$i_b(t) = I_{BS}$$
$$v_{BE}(t) = V_{BE(sat)}$$

The instantaneous power due to the base current is

$$P_b(t) = i_b v_{BE} = I_{BS} V_{BS(sat)} = 8 \times 3 = 24 \text{ W}$$

During the period, $0 \leq t \leq t_o = (T - t_{on} - t_n - t_s - t_f)$:

$$P_B(t) = 0.$$

The average power loss is

$$P_B = I_{BS} V_{BE(sat)} (t_{on} + t_n + t_s + t_f) f_s$$

$$= 8 \times 3 \times (1.5 + 48.5 + 5 + 3) \times 10^{-6} \times 10 \times 10^3 = 13.92 \text{ W}$$

12. The maximum junction temperature of a transistor is $T_J = 150°$ C and the ambient temperature is $T_A = 25°$ C. If the thermal impedances are $R_{JC} = 0.4°$C/W, and $R_{SA} = 0.5°$C/W, calculate (a) the maximum power dissipation and (b) the case temperature.

Solution

$$T_J = T_A = P_T (R_{JC} + R_{CS} + R_{SA}) = P_T R_{JA}$$

$$R_{JA} = 0.4 + 0.1 + 0.5 = 1.0$$

$150 - 25 = 1.0 P_T$, which gives the maximum power dissipation as $P_T = 125$ W.

(a) $T_C = T_J - P_T R_{JC} = 150 - 125 \times 0.4 = 100°$C.

13. A bipolar transistor is operated as a chopper switch at a frequency of $f_s = 10$ kHz. The DC voltage of the chopper is $V_s = 220$ V and the load current is $I_L = 100$ A. $V_{CE(sat)} = 0$ V. The switching times are $t_d = 0$, $t_r = 3$ μs and $t_f = 1.2$ μs. Determine the values of (a) L_s, (b) C_s, (c) R_s, for a critically damped condition, (d) R_s, if the discharge time is limited to one-third of switching period, (e) R_s, if the peak discharge current is limited to 10 per cent of load current and (f) power loss due to RC snubber P_s, neglecting the effect of inductor L_s on the voltage of the snubber capacitor C_s.

Solution

$I_L = 100$ A, $V_s = 220$ V, $f_s = 10$ kHz, $t_r = $ μs, $t_f = 1.2$ μs.

$$L_s = \frac{V_s t_r}{I_L} = 220 \times \frac{3}{100} = 6.6 \text{ μH}$$

$$C_s = \frac{I_L t_f}{V_s} = 100 \times \frac{1.2}{220} = 0.55 \text{ μF}$$

(a) $R_s = 2\sqrt{\frac{L_s}{C_s}} = 2\sqrt{\frac{6.6}{0.55}} = 6.93 \; \Omega$.

$$R_s = \frac{1}{3 f_s C_s} = \frac{10^3}{(3 \times 10 \times 0.55)} = 60.66 \; \Omega$$

$$\frac{V_s}{R_s} = 0.1 \times I_L$$

$$\frac{220}{R_s} = 0.1 \times 100$$

$$R_s = 22\,\Omega$$

(a) The snubber loss, neglecting the loss in diode D_s, is

$$P_s \cong 0.5\, C_s\, V_s^2 f_s = 0.5 \times 0.55 \times 10^{-6} \times 220^2 \times 10 \times 10^3 = 133.1\text{ W}$$

14. MOSFETs that are connected in parallel carry a total current of $I_T = 20$ A. The drain-to-source voltage of MOSFET M_1 is $V_{DS1} = 2.5$ V and that of MOSFET M_2 is $V_{DS2} = 3$ V. Determine the drain current of each transistor and difference in current sharing if the current sharing series resistances are (a) $R_{s1} = 0.3\,\Omega$ and $R_{s2} = 0.2\,\Omega$ and (b) $R_{s1} = R_{s2} = 0.5\,\Omega$.

Solution

(a) $I_{D1} + I_{D2} = I_T$

$$V_{DS1} + I_{D1} R_{s1} = V_{DS2} + I_{D2} R_{s2} = D_{s2} = R_{s2}(I_T - I_{D1}).$$

$$SI_{D1} = \frac{V_{DS2} - V_{DS1} + I_T R_{s2}}{R_{s1} + R_{s2}}$$

$$= \frac{3 - 2.5 + (20 \times 0.2)}{0.3 + 0.2} = 9\text{ A or } 45\text{ per cent}$$

$I_{D2} = 20 - 9 = 11$ A   or   55 per cent

$\forall I = 55 - 45 = 10$ per cent

$$I_{D1} = \frac{3 - 2.5 + (20 \times 0.5)}{0.5 + 0.5} = 10.5\text{ A} \quad \text{or} \quad 52.5 \text{ per cent}$$

$I_{D2} = 20 - 10.5 = 9.5$ A   or   47.5 per cent

$\forall I = 52.5 - 47.5 = 5$ per cent

15. A bipolar transistor has current gain $\beta = 40$. The load resistance $R_C = 10\,\Omega$, DC supply voltage $V_{CC} = 130$ V and input voltage to base circuit, $V_B = 10$ V. For $V_{CES} = 1.0$ V and $V_{BES} = 1.5$ V, calculate:

(a) the value of R_B for operation in the saturated state,

(b) the value of R_B for an overdrive factor 5,

(c) forced-current gain, and

(d) power loss in the transistor for both parts (a) and (b).

Solution

Here $\beta = 40$, $R_C = 10\,\Omega$, $V_{CC} = 130$ V, $V_B = 10$ V, $V_{CES} = 1.0$ V, $V_{BES} = 1.5$ V

(a) For operation in saturated state, $I_{CS} = \dfrac{V_{CC} - V_{CES}}{R_C} = \dfrac{130 - 1.0}{10} = 12.90$ A

Base current that produces saturation

$$I_{BS} = \frac{I_{CS}}{\beta} = \frac{12.90}{40} = 0.3225 \text{ A}$$

Value of R_B for $I_{BS} = 0.3225$ A is given by

$$R_B = \frac{V_B - V_{BES}}{I_B} = \frac{10 - 1.5}{0.3225} = 26.357 \text{ }\Omega$$

(b) Base current with overdrive,

$$I_B = \text{ODF} \times I_{BS} = 5 \times 0.3225 = 1.6125 \text{ A}$$

$$R_B = \frac{10 - 1.5}{1.6125} = 5.27 \text{ }\Omega$$

(c) Forced current gain,

$$\beta_f = \frac{I_{CS}}{I_B} = \frac{12.90}{1.6125} = 8 \text{, which is less than the natural current gain } \beta = 40.$$

(d) Power loss in transistor,

$$P_T = V_{BES} I_{BE} + V_{CES} I_{CS}$$

For normal base drive, $P_T = 1.5 \times 0.3225 + 1.0 \times 12.9 = 13.384$ W

With overdrive, $P_T = 1.5 \times 1.6125 + 1.0 \times 12.9 = 15.32$ W

It seen that power loss with hard drive of transistor is more.

16. A transistor has a current gain of 0.99 when used in common base (CB) configuration. How much will be the current gain of this transistor in common emitter (CE) configuration?

Solution

The current gain in common base circuit is written as α, and it has been given equal to 0.99.

α and current gain in common emitter configuration β are related as

$$\beta = \frac{\alpha}{1-\alpha}$$

Therefore,

$$\beta = \frac{0.99}{1-0.99} = 99$$

or $\beta = 99$

17. Find the operating point current I_{CQ} and voltage V_{CBQ} in the circuit shown. (VBE = 0.7 V, Beta of transistor is 200).

Solution

$$I_E \approx I_C = \frac{V_{EE} - V_{BE}}{R_E} = \frac{9 - 0.7}{4.3 \times 10^3}$$

or $\quad I_C = I_{CQ} = 1.93$ mA

Voltage summation in the collector circuit leads to

$$V_{CC} = R_C I_C + V_{CE}$$

or $\quad V_{CE} = V_{CC} - R_C I_C$

$$= 9 - 2.5 \times 10^3 \times 1.93 \times 10^{-3}$$

$$= 9 - 4.83$$

or $\quad V_{CE} = V_{CEQ} = 4.17$ V

18. For a reasonably linear small operation of a BJT, V_{BE} must be limited to no longer than 10 mV. To what percentage change of bias current does this correspond? For a design in which the required output voltage signal is 10 mA peak, what bias current is required?

Solution

The total (AC and DC) collector current i_C is,

$$i_C = I_C e^{V_{BE}/V_T}$$

$$= I_C + i_c$$

$$\frac{i_c}{I_C} = e^{V_{BE}/V_T} - 1$$

Taking $V_{BE} = 10$ mV and $V_T = 25$ mV, we get:

$$\frac{i_c}{I_C} = e^{10/25} - 1$$

$$= 0.492$$

$$= 49.2 \text{ per cent}$$

For $i_c = 10$ mV while $V_{BE} = 10$ mV, we get:

$$I_C = \frac{10}{0.492} = 20.3 \text{ mA}$$

19. For the class C commutation, $R_1 = R_2 = 5$ W, $C = 10$ mF, supply voltage $E_{DC} = 100$ V. Determine the turns off time.

Solution

Capacitor charging equation is

$$V_c = E_{DC} * \left(1 - e^{\frac{-t_c}{R_1 C}}\right)$$

$$100 = 200 * \left(1 - e^{\frac{-t_c}{R_1 C}}\right)$$

$$t_c = 0.693 \, R_1 C = 0.693 * 5 * 10 * 10^{-6} = 34.65 \, \mu s.$$

20. In a class C commutations circuit, determine the value of R, R_L and C for commutating the main SCR when it is conducting a full current of 15 A. The minimum time for which this SCR is to be reverse biased for proper commutation is 30 ms. It is given that the complementary SCR will undergo natural commutation when its forward current falls below the holding current of 3 mA. Assume supply voltage to be 100 V.

Solution

$$R_L = \frac{E_{DC}}{I_L} = \frac{100}{15} = 6.66 \, \Omega$$

$$V_c = E_{DC} \times \left(1 - e^{\frac{-t}{R_c}}\right)$$

$t = 30 \, \mu s$, $E_{CD} = 200$ V, $V_C = 100$ V

$$100 = 200 \times \left(1 - e^{\frac{-t}{R_L C}}\right)$$

$$\frac{-t}{R_c C} = \log_e 0.5 = -0.693$$

$$C = \frac{t}{0.693 \, R_L} = \frac{30 \times 10^{-6}}{0.693 \times 6.66}$$

$$C = 6.49 \, \mu F$$

Select R such that

$$R > \frac{E_{DC}}{3 \, mA}$$

$$R > \frac{100 \, V}{3 * 10^{-3}}$$

R > 33.33 kΩ

21. An SCR with $\int i^2 \, dt$ rating of 20 A² s is used to act as a rectifier to feed a load. If an earth fault occurs at the output of the rectifier when the input AC voltage (100 sin ωt) is at its positive peak, find the fault current and the safe time that the SCR can withstand the fault without damage. Assume the wire resistance and thyristor resistance to be 1 Ω ohm.

Solution

$$i^2 dt = 20$$

$$\therefore \int_0^{t_f} 100^2 \, dt = 20$$

$$t_f = \frac{20}{100^2} = 2 \, ms$$

Thus, the safe time the fault of 100 A can be withstood without damage of SCR is 2 m s.

22. The latching current for a thyristor inserted between a DC source voltage of 100 V and a load is 75 mA. Calculate the minimum width of the gate pulse required to turn on the thruster when the load is purely inductive having an inductance of 100 mH and consisting of resistance and inductance of 10 Ω ohm and 100 mH, respectively.

Solution

$$V = L \frac{di}{dt} \quad \text{i.e.,} \quad di = \frac{V}{L} dt$$

$$i = \frac{V}{L} t$$

$$t = \frac{L \cdot i}{V} = \frac{100 \times 10^{-3} \times 75 \times 10^{-3}}{100}$$

$$t = 75 \, \mu s$$

$$V = R \cdot i + L \frac{di}{dt}$$

$$i = \frac{V}{R}\left(1 - e^{\frac{R}{L}t}\right)$$

$$100 \times 10^{-3} = \frac{100}{10}\left(1 - e^{\frac{10}{100 \times 10^{-3}}t}\right)$$

$$t = 100 \, \mu s$$

Practice Questions

1. With a diagram, explain the on and off characteristics of an SCR.
2. Explain any two operating mechanisms of SCR under on and off mode.

3. Explain the transfer, output and switching characteristics of IGBT.
4. Explain the switching characteristics of BJT in detail with relevant diagrams.
5. (i) Explain the two-transistor model of SCR along with its forward characteristics

   (ii) Compare IGBT and MOSFET
6. (i) Compare the performance characteristics of MOSFET with BJT.

   (ii) Briefly discuss the RC triggering of SCR.
7. Compare the various features of the different commutation methods.
8. Explain the operation of protection circuits in power electronics.
9. Explain with a diagram, the various modes of working of TRIAC.
10. Investigate the Simulink model of a P channel MOSFET and obtain the characteristics.
11. Model the P channel IGBT and generate its characteristics using MATLAB.
12. Design a simple MATLAB model of any electrical machine to calculate its efficiency.
13. Design a Simulink model with RL load, using
    - Thyristor
    - MOSFET
    - IGBT
    - GTO

    Compare the performance levels. Take $R = 2\ \Omega$ and $L = 3$ mH. Justify your results.

Review Questions

1. Explain the role of the PN diode in electronics era.
2. Why is IGBT very popular nowadays?
3. What are the different methods to turn on the thyristor?
4. What is the difference between a power diode and a signal diode?
5. IGBT is a voltage-controlled device. Why?
6. Power MOSFET is a voltage-controlled device. Why?
7. Power BJT is a current-controlled device. Why?
8. What are the different types of power MOSFET?
9. How can a thyristor be turned off?
10. Define latching current.
11. Define holding current.
12. What is a snubber circuit?
13. What losses occur in a thyristor during working conditions?
14. Define hard driving or overdriving.
15. Define circuit turn off time.
16. Why should the circuit turn off time should be greater than the thyristor turn off time?

17. What is the turn off time for converter grade SCRs and inverter grade SCRs?
18. What are the advantages of GTO over SCR?
19. What is meant by phase-controlled rectifier?
20. Mention some applications of controlled rectifier.
21. Compare the various types and the characteristics of two-terminal and three-terminal devices.
22. Explain the switching modes of a thyristor in detail.
23. Explain the role of a Zener diode as a voltage regulator.
24. Describe the benefits of MOSFET being used in the field of power electronics.
25. Investigate the role of IGBT in industrial applications and elucidate with a suitable sample model.
26. List some of the recent trends in power electronics engineering.
27. Classify the various types of modern semiconductor devices used in power electronics system.

Multiple Choice Questions

1. The working principle of a Zener diode is
   (a) Tunnelling of charge carriers across the junction
   (b) Thermionic emission
   (c) Diffusion of charge carriers across the junction
   (d) Hopping of charge carriers across the junction
2. During the negative bias of the thyristor
   (a) All the three junctions are negatively biased
   (b) Outer junctions are positively biased and the inner junction is negatively biased
   (c) Outer junctions are negatively biased and the inner junctions is positively biased
   (d) The junctions near the anode is negatively biased and the one near the cathode is positively biased
3. SITs possess
   (a) High dV/dt and low dI/dt
   (b) Low dV/dt and high dI/dt
   (c) Low dV/dt and low dI/dt
   (d) High dV/dt and high dI/dt
4. In general, MOSFET is a
   (a) Voltage-controlled device
   (b) Current-controlled device
   (c) Frequency-controlled device
   (d) None of the above
5. The switch that is used for VHF and UHF applications is
   (a) BJT
   (b) MOSFET
   (c) SIT
   (d) IGBT
6. For a conventional type reverse blocking thyristor
   (a) External layers are lightly doped and internal layers are heavily doped
   (b) External layers are heavily doped and internal layers are lightly doped

(c) The p layers are heavily doped and the n layers are lightly doped

(d) The p layers are lightly doped and the n layers are heavily doped

7. MCTs possess

   (a) Low forward-voltage drop during conduction
   (a) Fast turn on and turn off time
   (c) Low switching losses
   (d) High reverse voltage blocking capability

8. MOS controlled thyristors have

   1. low forward voltage drop during conduction
   2. fast turn-on and turn-off time
   3. low-switching losses
   4. high reverse voltage blocking capability
   5. low gate input impedance Of these statements

   (a) 1, 2, and 3 are correct
   (b) 3, 4, and 5 are correct
   (c) 2, 3, and 4 are correct
   (d) 1, 3, and 5 are correct

9. The MOSFET time is of the order of

   (a) Seconds
   (b) Milliseconds
   (c) Microseconds
   (d) Nanoseconds

10. Identify the correct statement.

    1. Thyristor is a current-driven device
    2. GTO is a current-driven device
    3. Giant transistor (GTR) is a current-driven device
    4. SCR is a pulse triggered device

    (a) 1 and 2
    (b) 1, 2, 3
    (c) All
    (d) 4 only

11. Identify the correct statement.

    1. GTO is a pulse triggered device
    2. MOSFET is a unipolar device
    3. SCR is a bipolar device
    4. Continuous gate signal is not required to maintain the SCR in the on state

    (a) 1, 2, 4 only
    (b) 1, 2 only
    (c) 4 only
    (d) All

12. Identify which is not a fully controlled semiconductor switch.

    (a) MOSFET
    (b) IGBT
    (c) IGCT
    (d) SCR

13. Which of the following is incorrect related to PN junction?

    (a) Junction capacitance
    (b) Charge storage capacitance
    (c) Depletion capacitance
    (d) Channel length modulation

14. Under revere biased condition, the magnitude of electric field at maximum condition is
    (a) The edge of the depletion region on the P side
    (b) The edge of the depletion region on the N side
    (c) The P–N junction
    (d) The centre of the depletion region on the N side
15. During on state, the MOSFET switch is equivalent to
    (a) Resistor
    (b) Inductor
    (c) Capacitor
    (d) Battery
16. The MOSFETs effective channel length in saturation decreases with increase in
    (a) Gate voltage
    (b) Drain voltage
    (c) Source voltage
    (d) Body voltage
17. The BJTs early effect is due to
    (a) Fast turn on
    (b) Fast turn off
    (c) Large collector—base reverse bias
    (d) Large emitter–base forward bias
18. MOSFET can also be used as
    (a) Current-controlled capacitor
    (b) Voltage-controlled capacitor
    (c) Current-controlled inductor
    (d) Voltage-controlled inductors
19. The number of PN junctions in an SCR is
    (a) 1
    (b) 2
    (c) 3
    (d) 4
20. In the SCR, when the anode terminal is positive with respect to the cathode terminal, the number of blocked PN junctions is
    (a) 1
    (b) 2
    (c) 3
    (d) 4
21. In the SCR, when the cathode terminal is positive with respect to the anode terminal, the number of blocked PN junctions is
    (a) 1
    (b) 2
    (c) 3
    (d) 4
22. In the SCR, the anode current is made up of
    (a) Electrons only
    (b) Electrons or holes
    (c) Electrons and holes
    (d) Holes only
23. During the triggering of SCR, the forward-blocking mode shifts to conduction state. At this state, the anode cathode voltage is equal to
    (a) Peak repetitive off state forward voltage
    (b) Peak working off state forward voltage
    (c) Peak working off state reverse voltage
    (d) Peak nonrepetitive off state forward voltage

3
Phase-Controlled Rectifiers Using MATLAB (AC–DC Converters)

Learning Objectives

- ✓ To understand the operation of commutated rectifiers
- ✓ To examine the concepts of single-phase and three-phase converters and their various modes of operation
- ✓ To develop the characteristic equations describing the behaviour of rectifiers
- ✓ To list the various applications of rectifiers

3.1 Introduction

Single-phase uncontrolled rectifiers are comprehensively used in different converters in the power electronics field. Power electronic converters are frequently used to give a moderate unregulated DC voltage source, which can be further changed to get a DC or AC output. Rectifiers are capable of changing power phase. The only real demerit of rectifiers could be a disability to govern the DC load voltage or current increase after the load parameters are settled. Converters are also unidirectional since they permit electrical power to flow from the AC to the DC side.

Generally, power diodes are used inside these converters. If these power diodes are replaced by thyristors, the resultant converters are referred to as fully-controlled converters. Fully controlled converters cannot be turned off from the gate terminals. Thus, they continue to display output voltage or current waveforms similar to their uncontrolled counterparts. However, in fully-controlled converters, the thyristor can create forward-biased voltage, the load voltage or current waveforms can be measured by governing its turn on of the thyristors. The working principle of thyristors is based on single-phase fully controlled converters with assorted loads supplying an *R* or *RL* load. Half-wave converters are also used inside rectifiers, although single-phase fully controlled rectifiers are the most standard setup. In this chapter, we will examine and execute rectifiers using MATLAB/Simulink.

3.2 Rectification and Its Classification

A circuit that changes over from AC to DC can be referred to as rectifiers or converters. A converter circuit that utilizes diodes is known as an uncontrolled converter circuit. Thyristors can also be utilized to change AC to DC current; they also govern output voltage. Unlike power diodes, the thyristor can conduct even after its voltage has ended up positive. It needs to be activated by a triggering method. Hence, the thyristor can conduct anytime, which puts a positive supply to its anode. Therefore, controlled output voltage can be obtained. AC to DC converters have many applications, including as segments for supply of power and indicators of radio waves. Rectifiers can be classified on the basis of the following characteristics.

- Control qualities
- The time of conduction in the middle of each cycle of AC input voltage
- The number of phases in the supply side
- The count of firing on the load side in a single period

3.2.1 Based on control characteristics

Line frequency diode rectifiers transform line frequency AC to DC in an uncontrolled way. Most power device applications utilize, for example, exchanging DC power supplies, AC engine drives, DC servo drives, etc. The circuits consist of switches like diodes and give DC load voltage with respect to AC voltage input. Based on these characteristics, converters can be further classified into half-wave rectifiers, full-wave rectifiers with centre-tapped transformers and bridge type of rectifiers.

3.2.1.1 Uncontrolled rectifiers

Most uncontrolled rectifiers works at a comparatively large voltage level; the voltage drop tends to be small compared to this high voltage. The perfect diode model can be used for this. Diode circuits are generally termed uncontrolled circuits.

3.2.1.1.1 Half-wave rectifier

A perfect diode takes no conduction drops while the power is forward biased and no current when it is reverse biased. An explanation along with an analysis is given here using the diode model with R load as shown in Figure 3.1.

Example 1

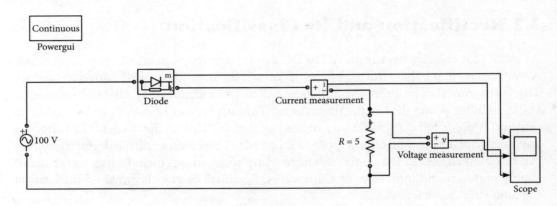

Figure 3.1 Simulation Model of a Half-Wave Diode Rectifier with *R* Load.

Refer to Figure 3.1 that shows the single phase of a half-wave diode rectifier with *R* load. Consider the voltage source to be a sinusoidal voltage source V_s. The diode starts to lead when the anode voltage is made more positive with respect to its cathode voltage; the load current flows through the circuit as a result. The power diode continues to be in the on state during the positive voltage half-cycle and switches over to be in the off state during its negative voltage half-cycle. Figure 3.2 shows different current and voltage waveforms of a half-wave diode rectifier with *R* load. These waveforms demonstrate that both the load voltage and current have high swells. Hence, a single-phase half-wave diode rectifier cannot be used as such in power electronic devices.

The normal or DC output voltage can be obtained by considering the waveforms demonstrated in Figure 3.2.

$$V_{DC} = \frac{V_m}{\pi} \qquad (3.1)$$

where V_m is the greatest value of the supply voltage.

Since the load is a resistor, the normal or DC segment of load current is:

$$I_{DC} = \frac{V_m}{\pi R} \qquad (3.2)$$

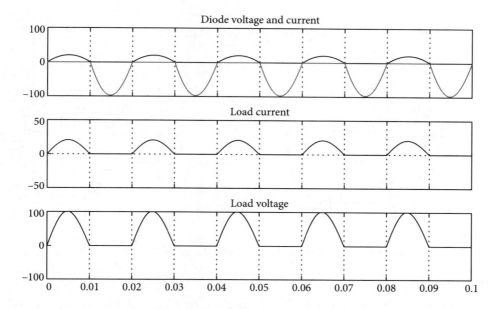

Figure 3.2 Output Waveforms for a Half-Wave Diode Rectifier with R Load.

The root mean square (rms) estimation of voltage at the load is given as:

$$V_{rms} = \frac{V_m}{2} \tag{3.3}$$

The rms estimation of the load current is characterized as:

$$I_{DC} = \frac{V_m}{2R} \tag{3.4}$$

3.2.1.1.2 Full-wave rectifier

In a full-wave rectifier circuit, the current passes over the load in the same course for equal number of half-cycles of the input supply. Such a type of circuit can be attained with dual power diodes. There are two types of full-wave rectifiers.

3.2.1.1.2.1 Center taped rectifier

A simple model of a center taped rectifier is shown in Figure 3.3. The circuit comprises two diodes. An inside tapped step down transformer is used to connect the two diodes. The three terminals, to be specific, are 1, 2 and 3. The aggregate auxiliary voltage is isolated into two equivalent sections. It can be accessible over the terminals 1 and 2 and 3 and 2. Hence, every power diode utilizes a unique half-cycle of input AC voltage.

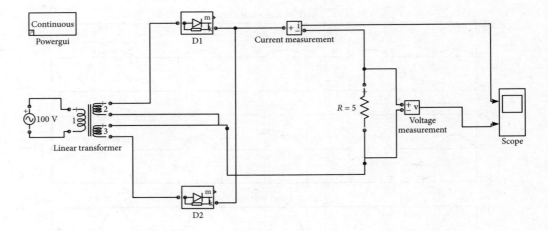

Figure 3.3 Simulation Model of a Center Taped and Bridge Rectifier.

Example 2

The current streams over the load in the identical bearing for both half-cycles of input AC voltage. The diode D1 is forward biased and set on and diode D2 is reverse biased and set off in the positive half-cycle of the input voltage. While in the negative half-cycle, the diode D1 is reverse biased and set to off and diode D2 is forward biased and set on. Therefore, DC output is acquired over the load.

The simulation results of a full-wave rectifier using a centre-tapped transformer forming a bridge rectifier is shown in Figure 3.4.

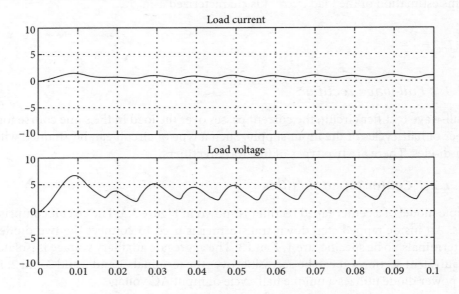

Figure 3.4 Simulation Results of a Center Taped Rectifier.

3.2.1.1.2.2 Full-wave bridge rectifier

An alternate option to obtain a single-phase full-wave rectifier is by utilizing four diodes as shown in Figure 3.5. This is known as a single-phase full-wave bridge diode rectifier. It holds four diodes namely D1, D2, D3 and D4 placed in a bridge circuit.

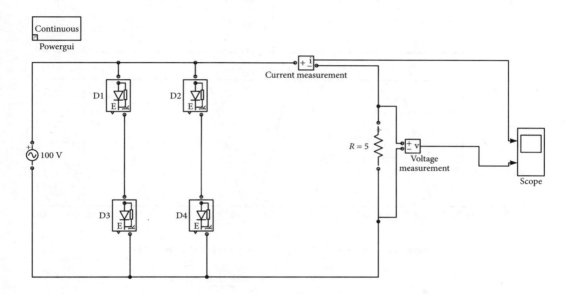

Figure 3.5 Simulation Model of a Single-Phase Full-Wave Bridge Rectifier with R Load.

Example 3

When the sinusoidal voltage supply is said to be the input, the current starts flowing through the power diodes D1 and D4 all the way through the positive half-cycle of the input voltage; the power diodes D2 and D3 are set to be in the off condition. Similarly, during the negative half-cycle, the power diodes D2 and D3 are said to be in conduction, whereas power diodes D1 and D4 are set to be in the off state.

In the positive half-cycle of the supply voltage, the current moves crosswise over D1. In the negative half-cycle of the supply voltage, the current moves crosswise over D3. In this way, there is response in both the positive and negative half-cycles of the supply voltage. Thus, a DC output current can be acquired at the load in both the cycles of the supply voltage. The switching waveform for this rectifier is shown in Figure 3.6.

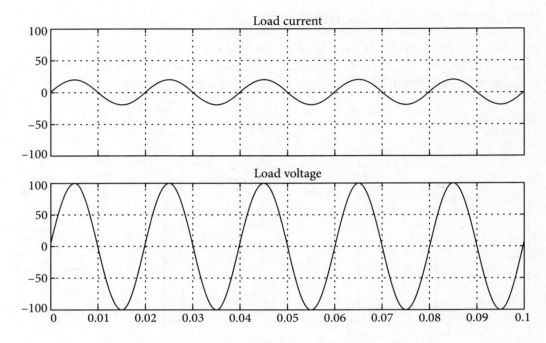

Figure 3.6 Simulation Results of a Full-Wave Bridge Rectifier.

3.2.1.2 Half-controlled rectifiers

Half-controlled rectifier circuits contain a blend of thyristors and diodes, which inverse the load voltage. These circuits permit control of normal estimation of the DC side load voltage. Half-controlled rectifiers are explained in more detail in Section 3.4.

3.2.1.3 Fully-controlled rectifiers

As explained earlier, in fully-controlled circuits, the rectifying components used are thyristors. These circuits are bidirectional; they allow the power to stream in whichever direction the middle of the supply and DC load is. Fully-controlled rectifiers are also explained in detail in Section 3.4.

3.2.2 Based on period of conduction

Based on the time of conduction period per series of AC input, rectifiers are grouped into the following classes.

(i) Half-wave rectifiers

(ii) Full-wave rectifiers

- Full-wave rectifiers using focus-tapped transformers
- Full-wave rectifiers using scaffold design

3.2.3 Based on number of phases

Based on the number of phases, converters are divided into the following groups.
- Single-phase rectifiers
- Three-phase rectifiers

3.2.4 Based on number of pulses

Based on the number of pulses at the DC side, rectifiers are divided into two classes.
- Single pulses
- Two pulses
- Three pulses
- Six pulses

In a few applications, for example, battery chargers and a class of AC and DC engine drives, it may be necessary for the DC output to be in control. This can be done by line frequency phase controlled converters using thyristors. These thyristor converters are now used often in three-phase high power applications.

Thyristors can be triggered at their gate terminal instantly; they cannot be turned off. Thus, as mentioned before, the fully controlled converter continues exhibiting load on account of their uncontrolled parts.

Note In the same way as done for *R* load in the examples, MATLAB/Simulink models can be designed for *RL*, *LC*, and *RLC* load – these are left as an exercise to the reader.

3.3 Selection of Components from the Simulink Library Browser

Before we introduce MATLAB simulation, the choice of components is very important. Listed are the components and their locations:

New File

>>*Type Simulink on Matlab Command Window>>File>>New>>Model*

AC Voltage Source

>>*Libraries>>Simscape>>SimPowerSystems>>Electrical Sources>>AC Voltage Source*

DC Source

>>Libraries>>Simscape>>SimPowerSystems>>ElectricalSources>>DC Voltage Source

Thyristor

>>Libraries>>Simscape>>SimPowerSystems>>PowerElectronics>>Thyristor

Diode

>>Libraries>>Simscape>>SimPowerSystems>>PowerElectronics>>Diode

Series *RLC* Branch

>>Libraries>>Simscape>>SimPowerSystems>>Elements>>Series RLC Branch

Pulse Generator

>>Libraries>>Sources>>PulseGenerator

Three-Phase *V–I* Measurement

>>Libraries>>Simscape>>SimPowerSystems>>ExtraLibrary>>Measurements>>Three-Phase V–I Measurement

Voltage Measurement

>>Libraries>>Simscape>>SimPowerSystems>>Measurements

Current Measurement

>>Libraries>>Simscape>>SimPowerSystems>>Measurements

Mean

>>Libraries>>Simscape>>SimPowerSystems>>Extra Library>>Measurement>>Mean Value

RMS

>>Libraries>>Simscape>>SimPowerSystems>>Extra Library>>Measurement>>RMS Value

Subsystem

>>Simulink>>Commonly Used Blocks>>Subsystem

THD

>>Libraries>>Simscape>>SimPowerSystems>>Extra Library>>Measurement>>Total Harmonic Distortion

From

>>Simulink>>Signal Routing>>From

Goto

>>Simulink>>Signal Routing>>Goto

Display

>>Simulink>>Sinks>>Display

Scope

>>Simulink>>Sinks>>Scope

Ground

>>Libraries>>Simscape>>SimPowerSystems>>Application Libraries>>Elements>>Ground

3.4 One Pulse Converters

Rectifier circuits are used to transform AC input into DC output. If a variable voltage is needed at the output, a controlled converter is utilized, phase-controlled thyristor being a part of the diode. The least complex of these circuits is the half-wave-controlled converter with a single thyristor. The triggering pulse is phase deferred to get reduced voltage. One pulse converter is a device that gets activated with one pulse for each cycle of the supply. Models of single-phase half-wave-controlled converters with diverse load qualities will be explained in this section.

3.4.1 Single-phase half-wave-controlled rectifiers

In the single-phase half-wave-controlled rectifier, the thyristor is utilized as part of the diode. Most of the applications work at a high voltage; in such cases, the drop over the thyristor is comparatively small. Frequently, the transmission drop over the thyristor is zero. Sometimes, the current through the thyristor is zero, and when it is not leading, the thyristor can lead to transmission in either heading.

3.4.1.1 Single-phase half-wave-controlled rectifier with R load

The phase control in the device can be depicted by considering the half-wave-controlled thyristor circuit with R load. The Simulink model of the circuit is shown in Figure 3.7. The circuit is energized by a line voltage $V_s = V_m \sin \omega t$, where V_s is positive, when $0 < \omega t < \pi$, and negative when $\pi < \omega t < 2\pi$. When V_s begins to be positive, the thyristor T is forward-biased yet stays in the blocking state till it is activated. Besides the positive half-cycle of the supply voltage, the thyristor T is activated at $wt = \alpha$, and starts conducting. During the period $\alpha < wt < \pi$, the output voltage and current levels can be measured. During the negative half-cycle of the supply voltage, the thyristor obstructs the flow of current and no voltage is connected over the load R. By fluctuating the triggering angle α, the output voltage can be controlled all through the time of conduction. The simulation waveforms of a single-phase half-wave-controlled rectifier with R load are shown in Figure 3.8.

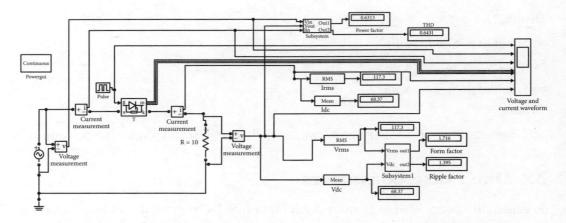

Figure 3.7 Simulation Model of a Single-Phase Half-Wave-Controlled Rectifier with *R* Load.

In this simulation model, the peak amplitude is set for V_p = 230 V operated at 50 Hz. The phase delay is calculated at α = 30° for thyristors and the load is considered as R = 10 Ω. After simulation, the voltage and current relations obtained from the circuit are shown in Example 4.

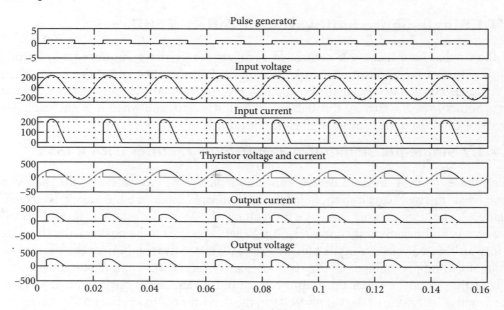

Figure 3.8 Simulation Waveform of a Single-Phase Half-Wave-Controlled Rectifier with *R* Load.

Example 4

Considering the waveforms mentioned earlier, the performance parameters of rectifier circuits are as follows:

The average load voltage is given by

$$V_{DC} = \frac{V_m}{2\pi}(1+\cos\alpha) \qquad (3.5)$$

The maximum output voltage is obtained when $\alpha = 0$.

The average load current is given by

$$I_{DC} = \frac{V_m(1+\cos\alpha)}{2\pi R} \qquad (3.6)$$

The rms load voltage is given by

$$V_{rms} = \frac{V_m}{2}\left[\frac{1}{\pi}\left(\pi - \alpha + \frac{\sin 2\alpha}{2}\right)\right]^{1/2} \qquad (3.7)$$

The rms load current is given by

$$I_{rms} = V_{rms}/R \qquad (3.8)$$

The form factor (FF) is given by

$$FF = V_{rms}/V_{DC} \qquad (3.9)$$

Ripple factor (RF) is given by

$$RF = [(FF)^2 - 1]^{1/2} \qquad (3.10)$$

The input power factor is given by

$$PF = V_{rms}/V_s \qquad (3.11)$$

The output power is the product of voltage and current, i.e., DC power ($V_{dc} \times I_{dc}$) and $P_{ac} = V_{ac} \times I_{ac}$ or $P_{rms} = V_{rms} \times I_{rms}$. The efficiency is given by P_{dc}/P_{rms}.

Using the aforementioned equations, the MATLAB circuit is simulated for different values of firing angle and the values of V_{DC}, V_{rms}, FF, RF and the input power factor.

3.4.1.2 Single-phase half-wave-controlled rectifier with RL load

The source V_s represents a sinusoidal voltage source with, $V_s = V_m \sin(\omega t)$, where V_s is positive for the time period $0 < \omega t < \pi$, and negative for the time period $\pi < \omega t < 2\pi$. When V_s begins to be positive, the thyristor is forward biased; however, the blocking state is still activated. The simulation model of a single-phase half-wave-controlled rectifier with RL load is shown in Figure 3.9.

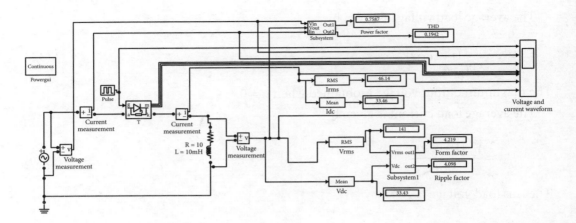

Figure 3.9 Simulation Model of a Single-Phase Half-Wave-Controlled Rectifier with *RL* Load.

Example 5

The thyristor is activated the instant when $\omega t = \alpha$. It starts leading when the source of the supply is positive; the thyristor is in the conducting state until π radians. When the current passing through the circuit is not said to be zero, a certain amount of energy is put away in the inductor at $\omega t = \pi$; the magnitude of voltage through the inductor leads to be positive while the current tends increase and when the current falls, the current leads to be negative. At this instant, the thyristor is said to be forward biased; it keeps on leading till the energy put away in the inductor becomes zero. After that, the current has a tendency to stream in the reverse bias and the thyristor is turned off. Figure 3.10 represents the input–output voltage and current levels along with the thyristor switching waveforms.

In this simulation model, the peak amplitude is set for V_p = 230 V operated at 50 Hz. The phase delay is calculated at $\alpha = 30°$ for the thyristors and the load is considered to be $R = 10\ \Omega$ and $L = 10$ mH. After simulation, the voltage and current relations obtained from the circuit are as follows.

Considering the waveforms, the average load voltage is given by

$$V_{DC} = \frac{V_m}{2\pi}(\cos\alpha - \cos\beta) \qquad (3.12)$$

The maximum output voltage can be obtained when $\alpha = 0$.

The average load current is given by

$$I_{DC} = \frac{V_m}{2\pi R}(\cos\alpha - \cos\beta) \qquad (3.13)$$

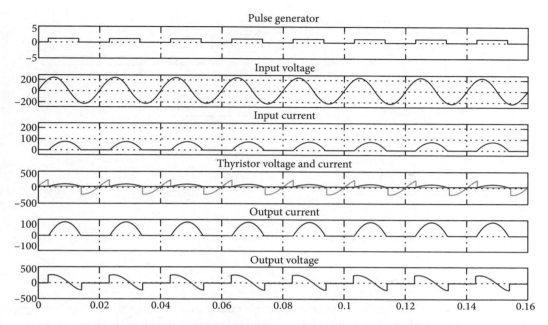

Figure 3.10 Single-Phase Half-Wave-Controlled Rectifier with *RL* Load.

The rms load voltage is given by

$$V_{rms} = \frac{V_m}{2\sqrt{\pi}}\left[(\beta-\alpha)-0.5(\sin 2\beta - \sin 2\alpha)\right]1/2 \tag{3.14}$$

The rms load current is given by

$$I_{rms} = V_{rms}/R \tag{3.15}$$

Using these equations, the MATLAB circuit is simulated for different values of firing angle and the values of V_{DC}, V_{rms}, FF, RF and the input power factor can be calculated.

3.4.1.3 Single-phase half-wave-controlled rectifier with RL load and a freewheeling diode

Consider a single-phase half-wave-controlled rectifier with *RL* load along with a freewheeling diode marked FWD. The voltage source V_s is a rotating sinusoidal source. $V_s = V_m \sin(\omega t)$; V_s is set to be positive during the time period $0 < \omega t < \pi$ and negative during the time period $\pi < \omega t < 2\pi$. The simulation model of a single-phase half-wave-controlled rectifier with *RL* load and FWD is shown in Figure 3.11.

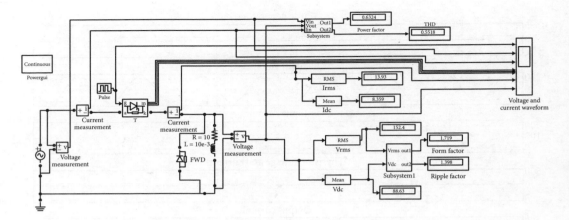

Figure 3.11 Single-Phase Half-Wave-Controlled Rectifier with *RL* Load and Freewheeling Diode.

Example 6

When V_s is forward biased, the thyristor is also forward biased and stays in the blocking state till it is activated. At the point when the thyristor is activated in the forward-biased state, it starts driving and the positive source keeps the thyristor in conduction till ωt touches π radians. During this time, some amount of energy is stored in the inductor, where the current throughout the circuit is not zero at $\omega t = \pi$ radians.

The inductor in the circuit would keep the thyristor switch in regular conduction mode for a period of negative half-cycle, even without the FWD, until the energy is released. If FWD is available as shown in the circuit of Figure 3.11, the source voltage will force the FWD to get into conduction in such a way that the anode voltage gets to be more positive than the cathode voltage. The inductor also releases its energy during the time period $\pi < \omega t < (2\pi + \alpha)$ all the way through the load. In the presence of FWD, the load is said to be consistent under perfect conditions; hence at the point when the diode is in conduction, the thyristor is said to be reverse biased. The switching waveforms of the circuit are shown in Figure 3.12.

In this simulation model, the peak amplitude is set for $V_p = 230$ V operated at 50 Hz. The phase delay is calculated at $\alpha = 30°$ for the thyristors and the load is considered as $R = 10\ \Omega$ and $L = 10$ mH. After simulation, the voltage and current relations obtained from the circuit are as follows.

Considering the waveforms, the average load voltage is given by

$$V_{DC} = \frac{V_m}{2\pi}(1+\cos\alpha) \tag{3.16}$$

The maximum output voltage can be obtained when $\alpha = 0$.

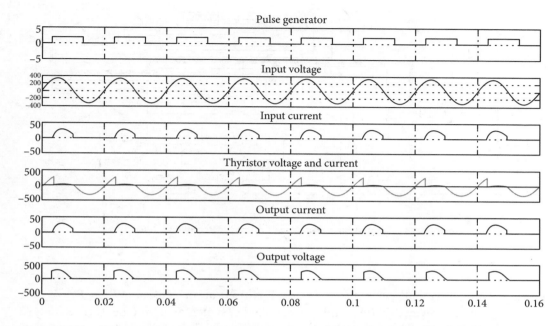

Figure 3.12 Switching Waveforms of a Single-Phase Half-Wave-Controlled Rectifier with RL Load and Freewheeling Diode.

The average load current is given by

$$I_{DC} = \frac{V_m(1+\cos\alpha)}{2\pi R} \tag{3.17}$$

The rms load voltage is given by

$$V_{rms} = V_m/2\pi\left[(\pi-\alpha)+(\sin2\alpha/2)/)\right]^{1/2} \tag{3.18}$$

The rms load current is given by

$$I_{rms} = V_{rms}/R \tag{3.19}$$

Using these equations, the MATLAB circuit is simulated for different values of firing angle and the values of V_{DC}, V_{rms}, FF, RF and the input power factor can be calculated.

3.4.1.4 Single-phase half-wave-controlled rectifier with RLE load

Consider a single-phase half-wave-controlled rectifier where the load is *RLE* as shown in Figure 3.13. Here, *E* is the emf in the load that may be a battery or back emf. The voltage source V_s is an alternating sinusoidal source of $V_s = V_m \sin(\omega t)$. The simulation model of the single-phase half-wave-controlled rectifier with *RLE* load is shown in Figure 3.13.

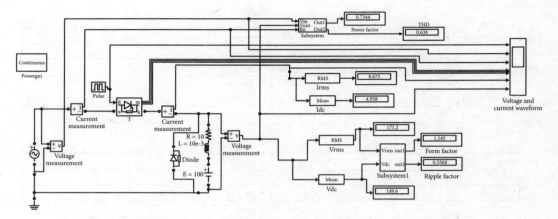

Figure 3.13 Single-Phase Half-Wave-Controlled Rectifier with *RLE* Load.

Example 7

During the positive half-cycle, the thyristor T is fired at an angle α. When $E > V_s$, the thyristor will be reverse biased and therefore, the switch will not be turned on. At ωt = α, the thyristor will be in the on state, where the load voltage follows the source voltage and the load current increases. At ωt = π, the load current will not be equal to zero. Owing to the presence of inductance *L* in the load, the load current gets reduced and finally reaches zero at ωt = β; then the load voltage becomes negative. At this condition, the load current reaches zero where the thyristor automatically becomes zero. During the negative half-cycle, the thyristor *T* is reverse biased. During the period from β to 2π + α, the load current reaches zero and the load voltage $V_0 = E$. The switching waveforms are shown in Figure 3.14. In this simulation model, the peak amplitude is set for V_p = 230 V operated at 50 Hz. The phase delay is calculated at α = 30° for the thyristors and the load is considered as R = 10 Ω, L = 10 mH and E = 100 V. After simulation, the voltage and current relations obtained from the circuit are as follows.

Considering the waveforms, the average load voltage is given by

$$V_{DC} = \frac{1}{2\pi}\left[V_m(\cos\alpha - \cos\beta) + E(2\pi + \alpha - \beta)\right] \tag{3.20}$$

The average load current is given by

$$I_{DC} = \frac{V_{DC} - E}{R} \tag{3.21}$$

The rms load voltage is given by

$$V_{rms} = V_s \tag{3.22}$$

The rms load current is given by

$$I_{rms} = V_{rms}/R \tag{3.23}$$

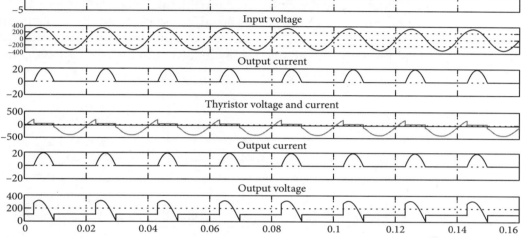

Figure 3.14 Switching Waveforms of a Single-Phase Half-Wave-Controlled Rectifier with *RLE* Load.

Using the aforementioned equations, the MATLAB circuit is simulated for different values of firing angle and the values of V_{DC}, V_{rms}, FF, RF and the input power factor can be calculated.

3.5 Two Pulse Converters

The two pulse converter is a prearrangement comprising thyristors in a bridge setup, which implies that either the polarization of expected output voltage is one or that there is extra polarization of input voltage, which is most distant connected in common applications like changing AC into DC output. Separated from other rectifier circuits, this arrangement does not turn into an auto transformer. The heading of the current is chosen by two thyristors leading at any given time. The heading of the current through the load is dependably the same.

3.5.1 Single-phase full-wave bridge rectifiers

A bridge rectifier gives full-wave correction from a two-wire AC circuit, bringing about lower cost and weight as contrasted with an inside tapped transformer arrangement. This full-wave-controlled extension rectifier that utilizes thyristors has a more extensive control over the level of DC output voltage.

3.5.1.1 Single-phase full-wave bridge rectifier with R load

The ultimate role of the full-wave bridge circuit is to produce a variable DC output. Figure 3.15 represents the single-phase completely-controlled rectifier. The circuit holds four thyristors, where V_s acts as the supply voltage source. When the voltage supply is positive, thyristors T1 and T2 are initiated. The current streams from V_s through thyristor T1, load resistor R, thyristor T2 and go into the source. At this stage, the load current is zero. In the negative half-cycle, the other pair of thyristors T3 and T4 are used and the load current streams through thyristor T4 and goes into the source. Although the flow of current through the source shifts, starting with one half-cycle then onto the next half-cycle, the current through the load stays unidirectional.

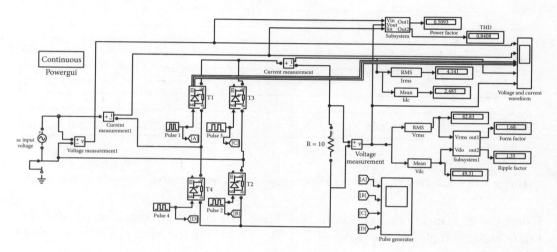

Figure 3.15 Simulink Model of Single-Phase Full-Wave Bridge Rectifier with R Load.

Example 8

In the simulation model shown in Figure 3.16, the peak amplitude is set for V_p = 230 V operated at 50 Hz. The phase delay is calculated at $\alpha = 30°$ for the thyristors and the load is considered as R = 10 Ω. After simulation, the voltage and current relations obtained from the circuit are as follows.

Considering the waveforms, the average load voltage is given by

$$V_{DC} = \frac{V_m}{\pi}(1+\cos\alpha) \tag{3.24}$$

The maximum output voltage can be obtained when $\alpha = 0$.

The average load current is given by

$$I_{DC} = \frac{V_m(1+\cos\alpha)}{\pi R} \tag{3.25}$$

The rms load voltage is given by

$$V_{rms} = V_m[(\pi - \alpha/2\pi) + (\sin 2\alpha/4\pi)]^{1/2} \tag{3.26}$$

The rms load current is given by

$$I_{rms} = V_{rms}/R \tag{3.27}$$

Using the aforementioned equations, the MATLAB circuit is simulated for different values of firing angle and the values of V_{DC}, V_{rms}, FF, RF and the input power factor can be calculated.

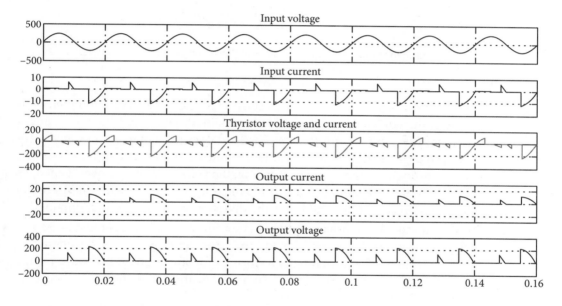

Figure 3.16 Switching Waveforms of a Single-Phase Full-Wave Bridge Rectifier with R Load.

3.5.1.2 Single-phase full-wave bridge rectifier with RL load

Figure 3.17 shows a single-phase-controlled bridge-controlled rectifier where the circuit produces a variable DC output voltage that fluctuates the ending edge. Similar to the R load circuit, the inductive circuit has four thyristors. For this circuit, V_s is a sinusoidal voltage source. During the positive half-cycle of the voltage source, the thyristors T1 and T2 are actuated and the current streams from the +ve side of the voltage source, V_s through T1–L–R–T2 and then to the −ve side of the voltage source. During the negative half-cycle, the current starts streaming from the −ve side of the voltage source, the supply voltage V_s through T3–R–L–T4 and returns to the +ve side of the source voltage.

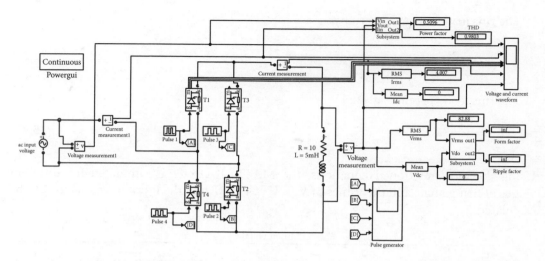

Figure 3.17 Simulink Model of a Single-Phase Full-Wave Bridge Rectifier with *RL* Load.

Example 9

Although the current through the source shifts, starting with one half-cycle and then onto the next half-cycle, the current through the load stays unidirectional. When V_s changes from positive to negative, the current through the load does not fall to zero immediately at $\omega t = \pi$ radians, since the load contains an inductor and the thyristors continue heading. When the current through the inductor starts falling, the voltage crosswise over it changes sign in such a way that the inductor permits load to the resistor. It feeds some energy to the AC source under certain conditions and keeps the thyristor in forward conduction mode.

In this simulation model, the peak amplitude is set for $V_p = 230$ V operated at 50 Hz. The phase delay is calculated at $\alpha = 30°$ for the thyristors and the load is considered as $R = 10$ Ω and $L = 10$ mH. The switching waveforms of a single-phase full-wave bridge rectifier with *RL* load is shown in Figure 3.18. After simulation, the voltage and current relations obtained from the circuit are as follows.

Considering the waveforms, the average load voltage is given by

$$V_{DC} = \frac{2V_m}{\pi}(\cos\alpha). \tag{3.28}$$

The maximum output voltage can be obtained when $\alpha = 0$.

The average load current is given by

$$I_{DC} = \frac{2V_m(\cos\alpha)}{\pi R}. \tag{3.29}$$

The rms load voltage is given by

$$V_{rms} = V_s. \tag{3.30}$$

The rms load current is given by

$$I_{rms} = V_{rms}/R. \tag{3.31}$$

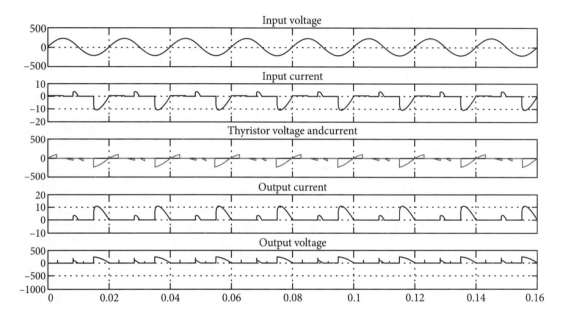

Figure 3.18 Switching Waveforms of a Single-Phase Full-Wave Bridge Rectifier with *RL* Load.

Using the aforementioned equations, the MATLAB circuit is simulated for different values of firing angle and the values of V_{DC}, V_{rms}, FF, RF and the input power factor can be calculated.

3.5.1.3 Single-phase full-wave bridge rectifier with RLE load

The power circuit consists of four thyristors along with the *RLE* load. Assuming the load inductance to be high enough, the load current is said to be continuous and free from ripples. When thyristors T1 and *T2* are triggered at $\omega t = \alpha$, they conduct till $\pi + \alpha$. The load voltage follows the source voltage. When $\omega t = \pi + \alpha$, the thyristors *T3* and *T4* are forward biased. At this condition, T1 and T2 gets turned off and the load current is transferred to T3 and T4. The Simulink model of a single-phase full-wave bridge rectifier with *RLE* load is shown in Figure 3.19.

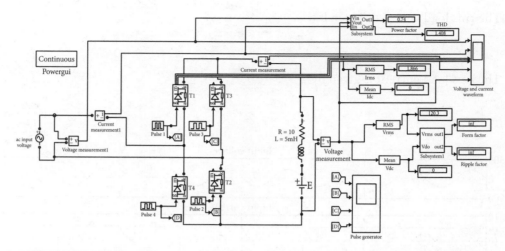

Figure 3.19 Simulink Model of a Single-Phase Full-Wave Bridge Rectifier with *RLE* Load.

Example 10

In this simulation model, the peak amplitude is set for $V_p = 230$ V operated at 50 Hz. The phase delay is calculated at $\alpha = 30°$ for the thyristors and the load is considered as $R = 10\ \Omega$, $L = 10$ mH and $E = 100$ V. The switching waveforms of a single-phase full-wave bridge rectifier with *RLE* load are shown in Figure 3.20. After simulation, the voltage and current relations obtained from the circuit are as follows.

The voltage and current equations are the same as for the *RL* load circuit. Using those equations, the MATLAB circuit is simulated for different values of firing angle and the values of V_{DC}, V_{rms}, FF, RF and the input power factor can be calculated.

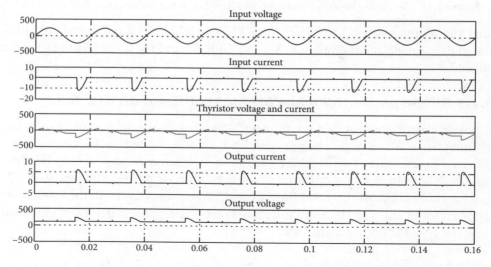

Figure 3.20 Switching Waveforms of a Single-Phase Full-Wave Bridge Rectifier with *RLE* Load.

3.5.2 Single-phase midpoint bridge rectifiers

The single-phase midpoint bridge rectifier holds two thyristors T1 and T2 along with a centre-tapped transformer. The midpoint rectifiers are generally used for rectification of low power ratings.

3.5.2.1 Single-phase midpoint bridge rectifier with R load

The Simulink model of a single-phase midpoint bridge rectifier with R load is shown in Figure 3.21. When the upper portion of the transformer is made positive, thyristor T1 is made to lead and the current courses through the load from left to right. At the point when the lower half of the transformer is positive and thyristor T2 is activated, T2 will lead and the current moves through the load from the left end to the right end. Along these lines, every half of the input wave and unidirectional voltage is connected over the load.

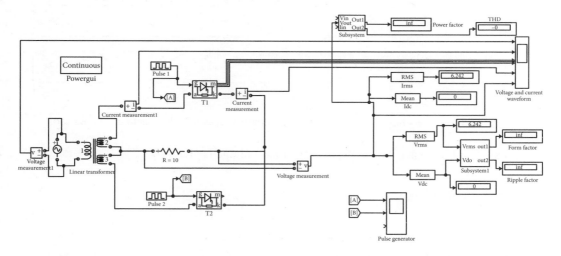

Figure 3.21 Simulink Model of a Single-Phase Midpoint Bridge Rectifier with R Load.

Example 11

During the positive half-cycle, the thyristor T1 is triggered at a firing angle α, where the load current flows through the resistive load R and gets back to the centre tap of the transformer. This is continuous until the α value reaches π, when the line voltage is reverse polarized and the thyristor T1 is turned off.

During the negative half-cycle, the thyristor T2 is forward biased. When $\omega t = \pi + \alpha$, the current flows through the thyristor T2 and resistive load R and gets back to the centre tap transformer. This is continuous till the angle 2π, and then the thyristor T2 gets turned off. In this circuit operation, the thyristors T1 and T2 are triggered with the same firing angle.

In the simulation model, the peak amplitude is set for V_p = 230 V operated at 50 Hz. The phase delay is calculated at α = 30° for the thyristors and the load is considered as R = 10 Ω. After simulation, the voltage and current relations obtained from the waveforms shown in Figure 3.22 are as follows.

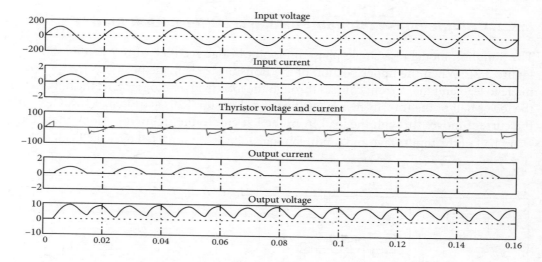

Figure 3.22 Switching Waveform of a Single-Phase Midpoint Bridge Rectifier with R Load.

Considering the waveforms, the average load voltage is given by

$$V_{DC} = \frac{V_m}{\pi}(1+\cos\alpha) \tag{3.32}$$

The maximum output voltage can be obtained when α = 0.
The average load current is given by

$$I_{DC} = \frac{V_m(\cos\alpha)}{\pi R} \tag{3.33}$$

The rms load voltage is given by

$$V_{rms} = V_m[(\pi-\alpha/2\pi) + (\sin2\alpha/4\pi)]^{1/2} \tag{3.34}$$

The rms load current is given by

$$I_{rms} = V_{rms}/R \tag{3.35}$$

Using these, the MATLAB circuit is simulated for different values of firing angle and the values of V_{DC}, V_{rms}, FF, RF and the input power factor can be calculated.

3.5.2.2 Single-phase midpoint bridge rectifier with RL load

The single-phase midpoint bridge rectifier with *RL* load is shown in Figure 3.23. When the thyristor T1 is turned on, current starts building up in the inductive load, which maintains the thyristor T1 in the on state during the positive half-cycle of the source voltage. During the negative half-cycle, the thyristor T2 takes on the load current. The load current may be continuous or discontinuous depending on the inductive load.

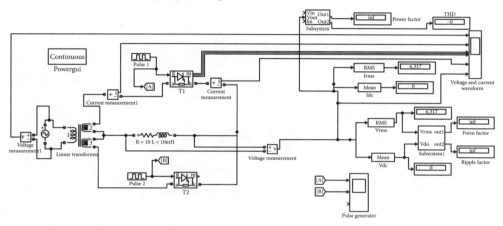

Figure 3.23 Simulink Model of a Single-Phase Midpoint Bridge Rectifier with *RL* Load.

Example 12

In this simulation model, the peak amplitude is set for V_p = 230 V operated at 50 Hz. The phase delay is calculated at $\alpha = 30°$ for the thyristors and the load is considered as $R = 10\ \Omega$ and $L = 10$ mH. After simulation, the voltage and current relations obtained from the waveforms shown in Figure 3.24 are as follows.

Considering the waveforms, the average load voltage is given by

$$V_{DC} = \frac{V_m}{\pi}(1+\cos\alpha) \tag{3.36}$$

The maximum output voltage can be obtained when $\alpha = 0$.

The average load current is given by

$$I_{DC} = \frac{V_m(1+\cos\alpha)}{2\pi R} \tag{3.37}$$

The rms load voltage is given by

$$V_{rms} = V_m[(\pi-\alpha/2\pi) + (\sin 2\alpha/4\pi)]^{1/2} \tag{3.38}$$

The rms load current is given by

$$I_{rms} = V_{rms}/R \tag{3.39}$$

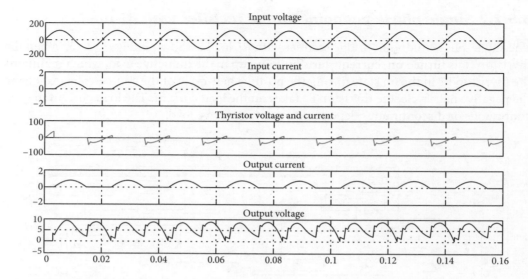

Figure 3.24 Switching Waveforms of a Single-Phase Midpoint Bridge Rectifier with *RL* Load.

Using the aforementioned equations, the MATLAB circuit is simulated for different values of firing angle and the values of V_{DC}, V_{rms}, FF, RF and the input power factor can be calculated.

3.5.2.3 Single-phase midpoint bridge rectifier with freewheeling diode

When a FWD is connected across the load, the thyristors are triggered at an angle α. As the input voltage reaches zero, the load voltage will not become negative since the FWD starts to conduct and clamps the load voltage. The main purpose of the FWD is to maintain a constant load current while the inductance load circulates the load current through the FWD. The Simulink model of a single-phase midpoint bridge rectifier with FWD is shown in Figure 3.25.

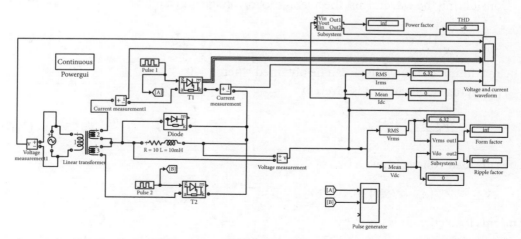

Figure 3.25 Simulink Model of a Single-Phase Midpoint Bridge Rectifier with Freewheeling Diode.

Example 13

The voltage and current equations are the same as that of the RL load circuit. In this simulation model, the peak amplitude is set for V_p = 230 V operated at 50 Hz. The phase delay is calculated at $\alpha = 30°$ for the thyristors and the load is considered as $R = 10\ \Omega$ and $L = 10$ mH. After simulation, the voltage and current relations obtained from the waveforms shown in Figure 3.26 are as follows.

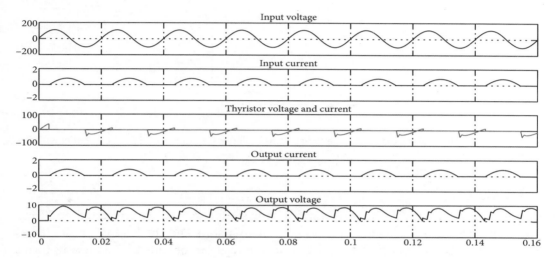

Figure 3.26 Switching Waveforms of a Single-Phase Midpoint Bridge Rectifier with Freewheeling Diode.

3.5.3 Single-phase semiconverter half-controlled bridge rectifiers

The semiconverter circuit is a mixture of diodes and thyristors, where there is a limited control over the level of DC output voltage. The voltage and current values will not be negative.

3.5.3.1 Single-phase semiconverter half-controlled bridge rectifier with R load

The single-phase semiconverter half-controlled bridge rectifier with R load shown in Figure 3.27 consists of two thyristors T1 and T2 along with two diodes $D1$ and $D2$. During the positive half-cycle, the thyristor T1 and diode $D1$ are set to be forward biased when triggered at an angle α, thus switching on T1 and $D1$. The load current flows through $T1$ and $D1$ through the resistive load. When the triggering angle reaches π, the load current reaches zero, thus bringing T1 and $D1$ to the off state.

During the negative half-cycle, T2 and D2 are set to be forward biased, when the thyristor $T2$ is triggered at an angle $\pi + \alpha$, which makes $T2$ and $D2$ to be in the on state. When the load current reaches zero, the thyristor $T2$ and $D2$ are set to the off state.

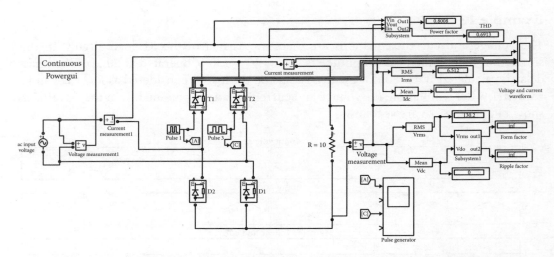

Figure 3.27 Simulink Model of a Single-Phase Semiconverter Half-Controlled Bridge Rectifier with R Load.

Example 14

In this simulation model, the peak amplitude is set for $V_p = 230$ V operated at 50 Hz. The phase delay is calculated at $\alpha = 30°$ for the thyristors and the load is considered as $R = 10\ \Omega$. After simulation, the voltage and current relations obtained from the waveforms shown in Figure 3.28 are as follows.

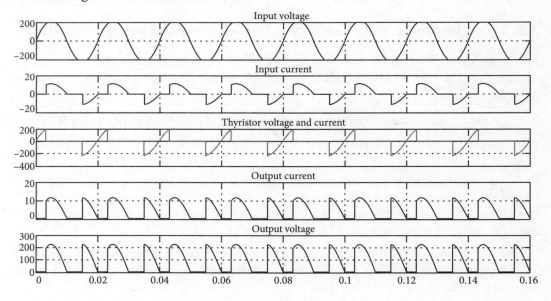

Figure 3.28 Switching Waveforms of a Single-Phase Semiconverter Half-Controlled Bridge Rectifier with R Load.

Considering the waveforms, the average load voltage is given by

$$V_{DC} = \frac{V_m}{2\pi}(1+\cos\alpha) \qquad (3.40)$$

The maximum output voltage can be obtained when $\alpha = 0$.

The average load current is given by

$$I_{DC} = \frac{V_m(1+\cos\alpha)}{2\pi R} \qquad (3.41)$$

The rms load voltage is given by

$$V_{rms} = V_m[(\pi - \alpha/2\pi) + (\sin 2\alpha/4\pi)]^{1/2} \qquad (3.42)$$

The rms load current is given by

$$I_{rms} = V_{rms}/R \qquad (3.43)$$

Using these equations, the MATLAB circuit is simulated for different values of firing angle and the values of V_{DC}, V_{rms}, FF, RF and the input power factor can be calculated.

3.5.3.2 Single-phase semiconverter half-controlled bridge rectifier with RL load

The single-phase semiconverter half-controlled bridge rectifier with *RL* load shown in Figure 3.28 consists of two thyristors T1 and T2 along with two diodes D1 and D2. During the positive half-cycle, the thyristor T1 and diode D1 are set to be forward biased when triggered at an angle α, thus switching on T1 and D1. The load current flows through T1 and D1 to the *RL* load. When the triggering angle reaches π, diode D2 is forward biased and the load current passes through D2 and T1, thus making D1 reverse biased and turning it off.

Example 15

During the negative half-cycle, T2 is set to forward bias, where it is triggered at an angle π + α, which makes T1 reverse biased and turns it off. When ω*t* = 2π, the thyristor T2 and diode D2 conduct. Then when the supply is reversed, the diode D1 is forward biased and the load current passes through D1 and T2. Here the diode D2 is reverse biased and turns off.

The voltage and current equation are the same as that of the *R* load circuit. In this simulation model, the peak amplitude is set for V_p = 230 V operated at 50 Hz. The phase delay is calculated at α = 30° for the thyristors and the load is considered as R = 10 Ω and L = 10 mH. After simulation, the voltage and current relations are obtained from the waveforms shown in Figure 3.30.

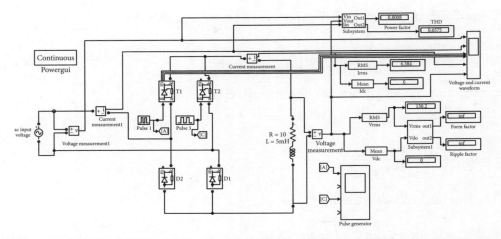

Figure 3.29 Simulink Model of a Single-Phase Semiconverter Half-Controlled Bridge Rectifier with *RL* Load.

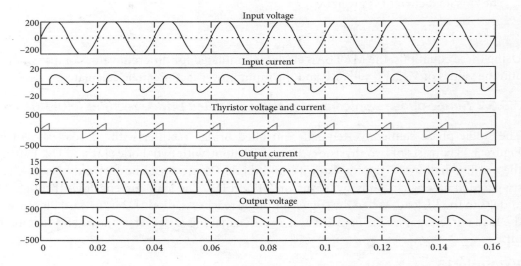

Figure 3.30 Switching Waveforms of a Single-Phase Semiconverter Half-Controlled Bridge Rectifier with *RL* Load.

3.5.3.3 Single-phase semiconverter half-controlled bridge rectifier with RLE load

The single-phase semiconverter half-controlled bridge rectifier with *RLE* load is shown in Figure 3.31. Assuming the inductance to be large, the load current is assumed to be continuous. During the positive half-cycle, T1 is triggered at an angle α, T1 and D1 turns on and the load current flows through the *RLE* load. During the negative half-cycle, T2 will be forward biased only when source voltage is more than *E*. During the $\pi + \alpha$ period, when the source voltage exceeds *E*, T2 is turned on. In this power circuit, the voltage and current values remain positive.

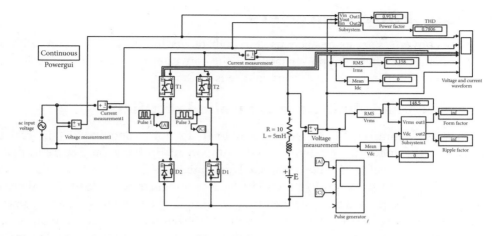

Figure 3.31 Simulink Model of a Single-Phase Semiconverter Half-Controlled Bridge Rectifier with *RLE* Load.

Example 16

The voltage and current equation are the same as that of the *R* load circuit. In this simulation model, the peak amplitude is set for V_p = 200 V operated at 50 Hz. The phase delay is calculated at α = 30° for the thyristors and the load is considered as R = 10 Ω, L = 10 mH, and E = 100 V. After simulation, the voltage and current relations are obtained from the waveforms shown in Figure 3.32.

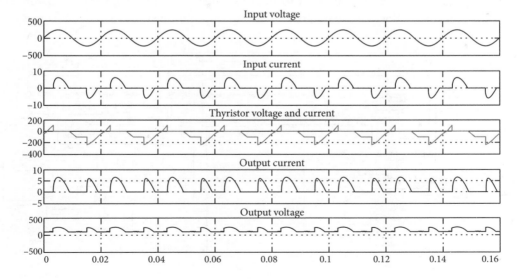

Figure 3.32 Switching Waveforms of a Single-Phase Semiconverter Half-Controlled Bridge Rectifier with *RLE* Load.

3.6 Three Pulse Converters

Three pulse converters are extensively used in high-power industry applications with one quadrant operation, where the power factor of the converter is inversely proportional to the phase angle. The frequency of output voltage is $3F_s$.

3.6.1 Three-phase half-wave-controlled rectifiers

Three-phase half-controlled rectifiers with R load do not generally exist in real life converter systems due to the existence of DC components in the supply current, that is, the input current waveforms have a regular DC value. This converter is a combination of three single-phase half-controlled rectifiers feeding a regular active load. The thyristor T1 is connected in series with one of the phase windings. The thyristors T2 and T3 are in sequence with the thyristor T1. The neutral point of the circuit is connected to one end of the load, while the other end is connected to the cathode point of the circuit. During the period $\omega t = \alpha$, when the thyristor is triggered, the phase voltage V_{an} appears across the load. Thyristor T2 conducts during the time period $\omega t = \alpha + \pi$ and thyristor T3 conducts during the period $\omega t = \alpha + 2\pi$. When the consecutive thyristors become active at various time periods, the previously conducting thyristor is set to the off condition. When thyristor T1 conducts, the next input cycle starts as the thyristor T3 gets turned off due to reverse bias.

3.6.1.1 Three-phase half-wave-controlled rectifier with R load

In the three-phase half-controlled rectifier circuit with R load, the firing angle is set to be greater than 30°, the load appears to be a discontinuous load and the thyristors are naturally commutated when the polarity of the corresponding supply voltage reverses.

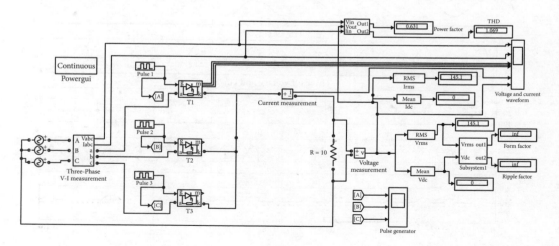

Figure 3.33 Simulink Model of a Three-Phase Half-Wave-Controlled Rectifier with R Load.

Example 17

In this simulation model, the peak amplitude is set for $V_p = 230$ V for each phase with a phase shift of 120° operated at 50 Hz. The phase delay is calculated at $\alpha = 30°$ for the thyristors and the load is considered as $R = 10\ \Omega$. After simulation, the voltage and current relations obtained from the waveforms shown in Figure 3.34 are as follows.

Considering the waveforms, the average load voltage is given by

$$V_{DC} = \frac{3\sqrt{3}}{2\pi} V_m (\cos \alpha) \tag{3.44}$$

The maximum output voltage can be obtained when $\alpha = 0$.

The average load current is given by

$$I_{DC} = \frac{3\sqrt{3}}{2\pi R} V_m (\cos \alpha) \tag{3.45}$$

The rms load voltage is given by

$$V_{rms} = V_m [(1/2) + (3\sqrt{3}/8\pi (\cos 2\alpha))]^{1/2} \tag{3.46}$$

The rms load current is given by

$$I_{rms} = V_{rms}/R \tag{3.47}$$

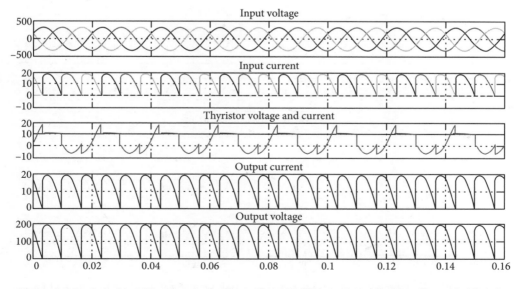

Figure 3.34 Switching Waveforms of a Three-Phase Half-Wave-Controlled Rectifier with R Load.

3.6.1.2 Three-phase half-wave-controlled rectifier with RL load

Refer to Figure 3.35 for a three-phase half-wave-controlled rectifier circuit with *RL* load, where the load current is assumed to be a ripple-free constant with a high inductive value through the thyristor T1.

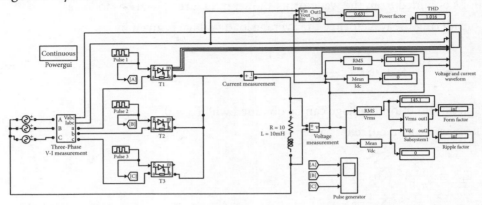

Figure 3.35 Simulink Model of a Three-Phase Half-Wave-Controlled Rectifier with *RL* Load.

Example 18

In this simulation model, the peak amplitude is set for $V_p = 200$ V for each phase with a phase shift of 120° operated at 50 Hz. The phase delay is calculated at $\alpha = 30°$ for the thyristors and the load is considered as $R = 10\ \Omega$ and $L = 10$ mH. After simulation, the voltage and current relations, which are similar to a three-phase half-wave-controlled rectifier with *R* load, obtained from the waveforms shown in Figure 3.36.

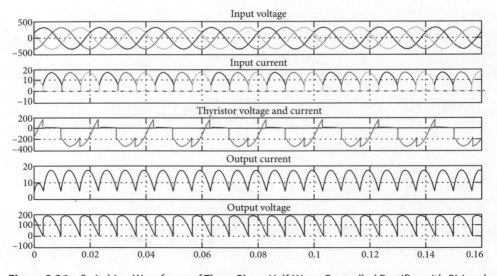

Figure 3.36 Switching Waveforms of Three-Phase Half-Wave-Controlled Rectifier with *RL* Load.

3.6.2 Three-phase half-controlled bridge rectifier with *RL* load

The three-phase half-controlled bridge rectifier is a circuit that consists of three thyristors in the upper arm and three diodes in the lower arm. The Simulink model of the three-phase half-controlled bridge rectifier with *RL* load is shown in Figure 3.37. The diode corresponding to the phase whichever is more negative conducts, that is, the diodes conduct for 120° each. Similarly, the firing pulses for the thyristors are set and said to be operated for two modes, namely, continuous conduction mode and discontinuous conduction mode.

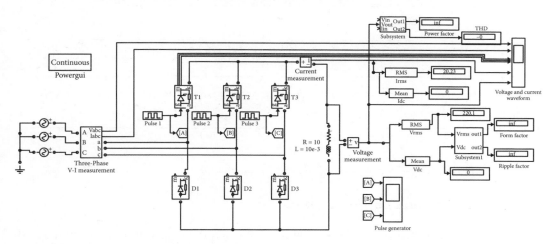

Figure 3.37 Simulink Model of a Three-Phase Half-Controlled Bridge Rectifier with *RL* Load.

Example 19

In this simulation model, the peak amplitude is set for $V_p = 200$ V for each phase with a phase shift of 120° operated at 50 Hz. The phase delay is calculated at $\alpha = 30°$ for the thyristors and the load is considered as $R = 10\ \Omega$ and $L = 10$ mH. After simulation, the voltage and current relations are obtained from the waveforms shown in Figures 3.38 and 3.39. The voltage and current equations can be derived as follows:

The average load voltage,

$$V_{DC} = \frac{3\sqrt{3}}{2\pi} V_m (1 + \cos\alpha) \tag{3.48}$$

Average load current,

$$I_{DC} = \frac{3\sqrt{3}}{2\pi R} V_m (1 + \cos\alpha) \tag{3.49}$$

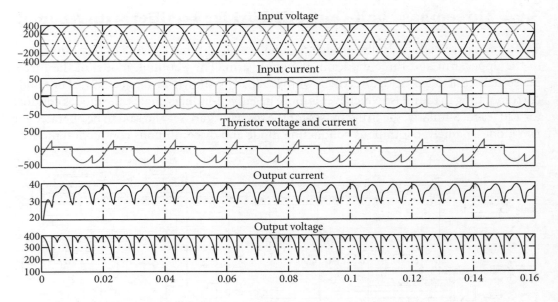

Figure 3.38 Switching Waveforms for Continuous Conduction Mode, $a = 30°$.

3.6.2.1 Continuous conduction mode

The continuous conduction mode is set at $\alpha = 30$, when the thyristor T1 is triggered first. When phases 1–2 have the highest priority, T1 is turned on. As a result, the load current would flow from phase 1, through *RL* load to diode D2, and finally through phase 2.

In a similar way, when phases 1–3 have the highest priority, T1 continues to conduct, but the current shifts its path from D2 to D3. In this mode, the output voltage will never become negative. The output voltages are measured every 120°.

3.6.2.2 Discontinuous conduction mode

In this simulation model, the peak amplitude is set for $V_p = 200$ V for each phase with a phase shift of 120° operated at 50 Hz. The phase delay is calculated at $\alpha = 90°$ for the thyristors and the load is considered as $R = 10\ \Omega$ and $L = 10$ mH. After simulation, the voltage and current relations are obtained from the waveforms shown in Figure 3.39.

The discontinuous conduction mode is set for $\alpha > 60°$. From the switching waveforms shown in Figure 3.39, the output voltage becomes zero during a part and rises to the maximum, whereas when α increases, the duration of the input current decreases. The thyristor current is identical to that of the load current because the thyristor conducts all the times. This effect can be eliminated using an FWD.

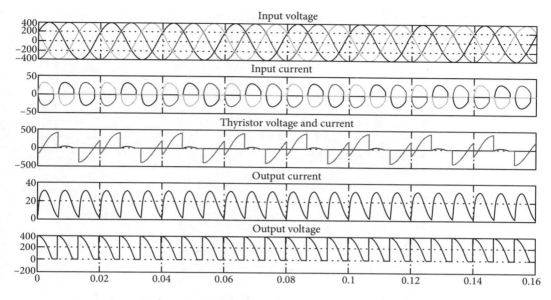

Figure 3.39 Switching Waveforms for Discontinuous Conduction Mode, $a = 90°$.

3.7 Six Pulse Converters

The AC–DC converter that comprises six thyristors works under a three-phase controlled voltage supply known as a six pulse converter. Six pulse converters are essentially utilized as a part of modern robotization. The substitution is extremely direct and contortions in current are drastically reduced. Any thyristor say (T1, T3, T5) from the upper end and any thysistor say (T2, T4, T6) from the lower end will lead for 120° of the input cycle.

In six-pulse converters, the thyristors are terminated in the order T1 → T6 → T2 → T4 → T3 → T5 → T1 with a customary interim period between each termination. Thus, thyristors on the same stage leg are activated after an interim of 180° and hence, cannot direct simultaneously. The thyristors T1T2, T2T3, T3T4, T4T5, T5T6, and T6T1 operate under continuous conduction mode. Every conduction mode is of 60° length of time and shows up in the grouping mentioned earlier. Six pulse converters work with R and RL loads.

3.7.1 Six pulse converter with R load

A six pulse converter operating with R load is shown in Figure 3.40. The thyristors are prompted in the sequence T1 → T_2 → T_3 → T_4 → T_5 → T_6 → T_1 with regular intervals among every triggering. For the six pulse converter to be operated in R load, the different values of

α are set to 0°, 30°, 45°, 60° and 90°. The circuit is said to be operated under inversion mode for α values greater than 90°.

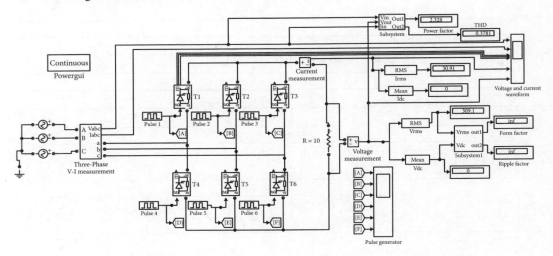

Figure 3.40 Simulink Model of a Six Pulse Rectifier with R Load.

Example 20

For six pulse conduction to take place, each thyristor has to be triggered twice during its conduction cycle. The six pulse full bridge converter with R load circuit is the most commonly used converter for large power drive applications. The circuit consists of two groups, namely, positive (T1, T2, T3) and negative groups (T4, T5, T6). The positive group is turned on during positive supply and the negative group is turned on during negative supply. When the circuit is connected to the supply, triggering gate pulses will be applied to the thyristors in an orderly sequence. Hence, to initiate the functioning of the circuit, two thyristors must be triggered at the same time. When the voltage V_a is the highest positive, the thyristor T_1 is forward biased and it stays ready to conduct on any prompt during this period if it receives a pulse signal arranged on its gate terminal. Similarly, as shown in Figure 3.42, the thyristor T_1 or some supplementary thyristor stays on for 120°. When the voltage V_b is the highest, the thyristor T_3 is forward biased and when V_c is the highest positive voltage, the thyristor T5 is forward biased. On the other hand, when V_a is the highest negative voltage, the thyristor T4 is forward biased; during V_b, thyristor T6 is forward biased and for V_c, T2 conducts.

During the negative half-cycle, the output voltage and current stays in the upper cycle without reaching zero. From this analysis, we can reason that there are two thyristors in conduction during the time of supply voltage. The pulse generated from all the six thyristors is shown in Figure 3.41.

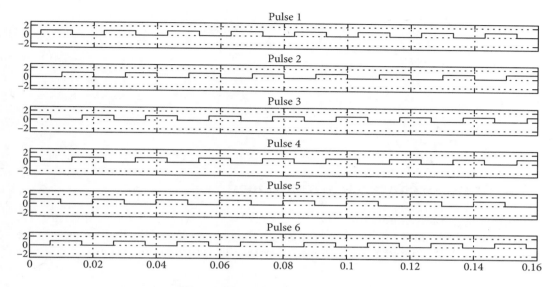

Figure 3.41 Pulse Generator Waveform.

In this simulation model, the peak amplitude is set for $V_p = 200$ V for each phase with a phase shift of 120° operated at 50 Hz. The phase delay is calculated at $\alpha = 30°$ for the thyristors and the load is considered as $R = 10\ \Omega$. After simulation, the voltage and current relations obtained from the circuit are as follows.

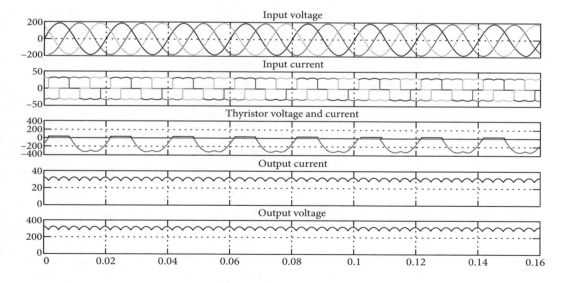

Figure 3.42 Switching Waveforms for a Six Pulse Converter with R Load.

Average load voltage:

$$V_{DC} = \frac{3\sqrt{3}}{\pi} V_m \cos\alpha. \qquad (3.50)$$

Average load current:

$$I_{DC} = \frac{3\sqrt{3}}{\pi R} V_m \cos\alpha. \qquad (3.51)$$

3.7.2 Six pulse converter with *RL* load

When thyristor T1 is conducting, the voltage across A-C will reverse bias thyristor T3, which makes thyristor T3 turn off. At this stage, the voltage across the load is same as the voltage across A-B. This stage continues till thyristor T6 is turned on.

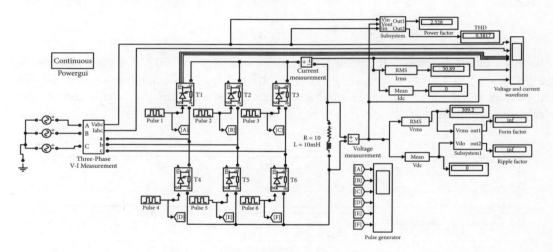

Figure 3.43 Simulink Model of a Six Pulse Rectifier with *RL* Load.

Example 21

While turning on T6, voltage across the line B-C reverse biases thyristor T5. Finally, T5 gets turned off, and here the voltage across the line A-C is the same as the voltage across T1 and T6. The sequences in which the thyristors are triggered are $T_1 \to T_6 \to T_2 \to T_4 \to T_3 \to T_5 \to T_1$. The pulse generated from all the six thyristors are shown in Figure 3.44.

In the simulation model, shown in Figure 3.43, the peak amplitude is set for $V_p = 200$ V for each phase with a phase shift of 120° operated at 50 Hz. The phase delay is calculated for $\alpha = 30°$ for the thyristors and the load is considered as $R = 10\ \Omega$ and $L = 10$ mH. The simulation results of a six pulse rectifier with *RL* load is shown in Figure 3.45.

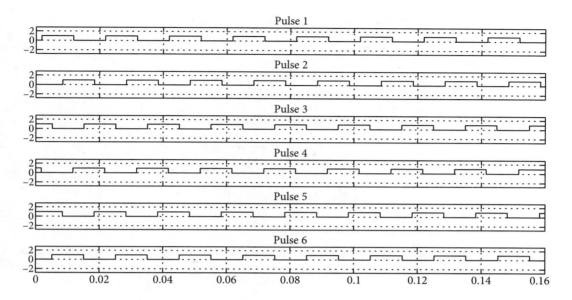

Figure 3.44 Pulse Generator Waveforms.

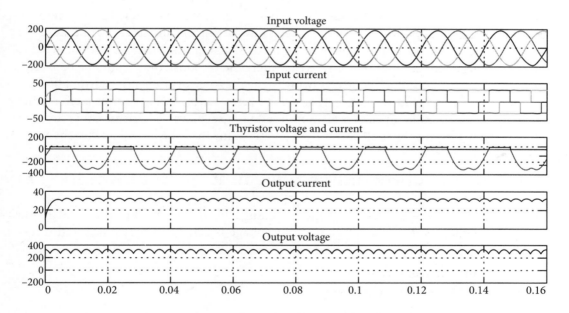

Figure 3.45 Switching Waveforms Model of a Six Pulse Rectifier with *RL* Load.

3.8 Dual Converter

The dual converter shown in Figure 3.46 can be operated in two different modes of operation.

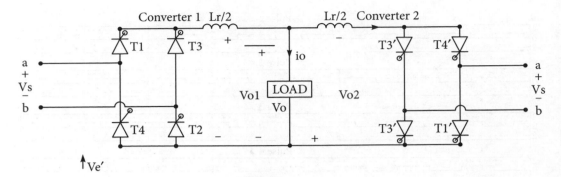

Figure 3.46 Dual Converter.

Noncirculating Current Mode of Operation

In this mode, only one converter is operated at a time. When converter 1 is on, $0 < \alpha < 90°$. V_{DC} is positive and I_{DC} is positive. When converter 2 is on, $0 < \alpha < 90°$. V_{DC} is negative and I_{DC} is negative.

Circulating Current Mode of Operation

Without depending on the load values, the circulating current is maintained over a complete conduction range of both the converters. The flow of current in either way because the converters act as a rectifier and an inverter. A better dynamic response can be obtained due to continuous conduction of the converters, where the response time from one converter to another converter is much shorter. Owing to the presence of circulating current flowing through the converters, the load current falls to zero; hence, there requires a circulating current reactor to build a peak circulation to attain a safe level. Thyristor converters are rated to carry forward the peak current at a greater value than the peak load current.

3.9 Role of Source Inductance in Rectifier Circuits (Ls)

Source inductance causes the outing and incoming silicon controlled rectifiers (SCRs) to conduct together. At this instant, the output voltage is zero for single-phase and three-phase converters.

The role of source inductance is to

- Lower the meant output voltage
- Distort the output voltage and current waveforms
- Modify the performance parameters of converters

Angular period The time period when the outgoing and incoming SCRs conduct together is known as the overlap period.

Commutation angle Both incoming and outgoing SCRs conduct at the overlap angle known as the commutation angle.

3.10 Applications of Controlled Rectifiers

Rectifiers are fundamentally utilized as a part of devices where the control of high power, perhaps coupled with high voltage, are required. Their operation makes them suitable for utilization in medium to high voltage AC power control applications, for example, light darkening, controllers, and engine control. These controllers are likewise utilized for correction of high-power AC in high-voltage–current power transmission and welding innovations.

> **Summary**
>
> After reading this chapter, the reader will have a sound knowledge about the phase-controlled rectifiers used in the power electronics field. The readers will get a brief idea about how to start designing and simulating power electronic circuits. The waveforms and the mathematical expressions provide details about the validation of the circuits.

Solved Examples

1. A 240 V at 50 Hz supply feeds a highly inductive load of 50 Ω resistance through a half-controlled thyristor bridge. When the firing angle $\alpha = 45°$, determine its load power.

   Solution

   $V_{av} = (V_m/\pi)(1 + \cos\alpha) = [(\sqrt{2} \times 240)/(\pi)(1 + \cos 45)] = 184.4$ V

   $I_{av} = V_{av}/R = 184.4/50 = 3.69$ A

   $= 3.69 \times \sqrt{[(180 - 45)/180]}$

   $= 3.2$ A

   $P = 3.2 \times 3.2 \times 50 = 512$ W

2. A full-wave fully-controlled bridge has a highly inductive load with a resistance of 55 Ω, and a supply of 110 V at 50 Hz. Determine the value of load power for a firing angle $\alpha = 75°$.

   Solution

   $V_{av} = [2V_m/(\pi)]\cos\alpha$

   $= [(2 \times \sqrt{(2 \times 110)}/3.14] \times \cos 75$

   $= 99 \cos 75$

   $= 25.6$ V

$$I_{av} = V_{av}/R$$

$$= 25.6/55$$

$$= 0.446 \text{ A} = I_{rms}$$

$$P = I_{rms} \times I_{rms} \times R$$

$$= 0.446 \times 0.446 \times 55$$

$$= 10.9 \text{ W}$$

3. A 240 V at 50 Hz supply feeds a highly inductive load of 50 Ω resistance through a thyristor full control bridge, when the firing angle $\alpha = 45°$. Calculate the load power.

Solution

$$V_{av} = (2V_m/\pi) \times \cos\alpha$$

$$= [(2 \times 339)/3.14] \cos 45$$

$$= 152.6 \text{ V}$$

$$I_{av} = V_{av}/R$$

$$= 152.6/50$$

$$= 3.05 \text{ A} = I_{rms}$$

$$P = \text{Square of } I_{rms} \times R$$

$$= 3.04 \times 3.04 \times 50$$

$$= 466 \text{ W}$$

4. A thyristor half-wave-controlled converter has a supply voltage of 240 V at 50 Hz and a load resistance of 100 Ω. When the firing delay angle is 30, calculate the average value of load current.

Solution

$$V_{av} = (\sqrt{2} \times 240)/(2\pi) \times (1 + \cos 30)$$

$$= 100.8 \text{ V}$$

$$I_{av} = V_{av}/R$$

$$= 100.8/100$$

$$= 126 \text{ mA}$$

5. For a single-phase thyristor converter with R load and delay angle $\alpha = \pi/2$, determine (a) the rectification efficiency, (b) the FF, and (c) the RF.

Solution

Delay angle $\alpha = \dfrac{\pi}{2}$,

$$V_{DC} = \dfrac{V_m}{2\pi}(1 + \cos\alpha) = 0.1592 V_m$$

$$I_{DC} = 0.1592 \frac{V_m}{R}$$

$$V_n = \frac{V_{DC}}{V_{dm}} = 0.5(1+\cos\alpha) = 0.5$$

$$V_{rms} = \frac{V_m}{2}\left[\frac{1}{\pi}\left(\pi - \alpha + \frac{\sin 2\alpha}{2}\right)\right]^{1/2} = 0.3536 V_m$$

$$I_{rms} = 0.3536 \frac{V_m}{R}$$

$$P_{DC} = V_{DC}I_{DC} = \frac{(0.1592 V_m)^2}{R}$$

$$P_{AC} = V_{AC}I_{AC} = \frac{(0.3536 V_m)^2}{R}$$

(a) The rectification efficiency

$$\eta = \frac{P_{DC}}{P_{AC}} = \frac{(0.1592 V_m)^2}{R} = 20.27 \text{ per cent}$$

(b) The FF

$$FF = \frac{V_{rms}}{V_{DC}} = \frac{0.3536 V_m}{0.1592 V_m} = 2.221 \text{ or } 222.1\$$$

(c) $RF = \sqrt{FF^2 - 1} = \sqrt{(222.1^2 - 1)} = 1.983 \text{ or } 198.2 \text{ percent}$

6. A single-phase full converter with an RL load has $L = 6.5$ mH, $R = 0.5$ Ω, and $E = 10$ V. The input voltage is $V_s = 120$ V at (rms) 60 Hz. Determine (a) the load current I_{Lo} at $\omega t = \alpha = 60°$, (b) the average thyristor current I_A, (c) the rms thyristor current I_R, (d) the rms output current I_{rms} and (e) the average output current I_{DC}.

Solution

$$\alpha = 60°, R = 0.5 \text{ Ω}, L = 6.5 \text{ mH}, f = 60 \text{ Hz}, \omega = 2\pi \times 60 = 377 \text{ rad/s}$$

$$V_s = 120 \text{ V}, \theta = \tan^{-1}(\omega L/R) = 78.47°.$$

(a) The steady state load current at $\omega t = \alpha$, $I_{Lo} = 49.34$ A

(b) The numerical integration of i_L in the following equation yields I_A

$$i_L = \frac{\sqrt{2}V_s}{Z}\sin(\omega t - \theta) - \frac{E}{R} + \left[I_{Lo} + \frac{E}{R} - \frac{\sqrt{2}V_s}{Z}\sin(\alpha - \theta)\right]e^{(R/L)(\alpha/\omega - t)}.$$

$I_A = 44.05$ A

(c) By numerical integration of i_L^2 between the limits $\omega t = \alpha$ to $\pi + \alpha$, we get the rms thyristor current as $I_R = 63.71$ A.

(d) The rms output current $I_{rms} = \sqrt{2} I_R = \sqrt{2} \times 63.71 = 90.1$ A.

(e) The average output current $I_{DC} = 2 I_A = 2 \times 44.04 = 88.1$ A.

7. A single-phase dual converter is operated from a 120 V, 60 Hz supply and the load resistance is R = 10 Ω. The circulating inductance is $L_r = 40$ mH, delay angles are $\alpha_1 = 60°$ and $\alpha_2 = 120°$. Calculate the peak circulating current and the peak current of converter 1.

Solution

$\omega = 2\pi \times 60 = 377$ rad/s, $\alpha_1 = 60°$, $V_m = \sqrt{2} \times 120 = 169.7$ V, $f = 60$ Hz, $L_r = 40$ mH

For $\omega t = 2\pi$ and $\alpha_1 = \pi/3$,

$$I_r(\max) = \frac{2V_m}{\omega L_r}(1 - \cos \alpha_1) = \frac{169.7}{377 \times 0.04} = 11.25 \text{ A}$$

The peak load current is $I_p = \frac{169.71}{10} = 16.97$ A. The peak current of converter 1 is (16.97 + 11.25) = 28.22 A.

8. SCRs with peak forward voltage rating of 1000 V and average on-state current rating of 40 A are used in single-phase midpoint converters and single-phase bridge converters. Find the power that these two converters can handle. Use a factor of safety of 2.5.

Solution

Maximum voltage across SCR in a single-phase midpoint converter is $2V_m$. Therefore, this converter can be designed for a maximum voltage of $\frac{1000}{2 \times 2.5} = 200$ V. Hence,

Maximum average power that the midpoint converter can handle

$$= \left(\frac{2V_m}{\pi} \cos \alpha\right) I_{TAV} = \frac{2 \times 200}{\pi} \times 40 \times \frac{1}{1000} = 5.093 \text{ kW}$$

SCR in a single-phase bridge converter is subjected to a maximum voltage of V_m. Therefore, maximum voltage for which this converter can be designed is $\frac{1000}{2.5} = 400$ V. Hence,

Maximum average power rating of bridge converter $= \frac{2 \times 400}{1000 \times \pi} \times 40 = 10.186$ kW.

9. A three-phase half-wave-controlled converter is fed from a three-phase, 400 V at 50 Hz source and is connected to a load taking a constant current of 36 A. Thyristors have a voltage drop of 1.4 V. (a) Calculate value of load

voltage for a firing angle of 30° and 60°. (b) Determine average and rms current ratings as well as peak inverse voltage (PIV) of thyristors. (c) Find the average power dissipated in each thyristor.

Solution

(a) Here, average output voltage

$$V_0 = \frac{3V_{ml}}{2\pi}\cos\alpha - v_T; \; V_{ml} = \sqrt{2}\times 400 \text{ V}; \text{ and } v_T = 1.4 \text{ V}$$

For a firing angle of 30°, $V_0 = \frac{3\sqrt{2}\times 400}{2\pi}\cos 30° - 1.4 = 232.474$ V

For a firing angle of 60°, $V_0 = \frac{3\sqrt{2}\times 400}{2\pi}\cos 60° - 1.4 = 133.63$ V

(b) Average current rating of SCR $I_{TA} = \frac{I_0}{3} = \frac{36}{3} = 12$ A

rms current rating of SCR $I_{Tr} = \frac{I_0}{\sqrt{3}} = \frac{36}{\sqrt{3}} = 20.785$ A

PIV of SCR $= \sqrt{3}V_{mp} = V_{ml} = \sqrt{2}\times 400 = 565.6$ V

(c) Average power dissipated in each SCR $= I_{TA}v_T = 12\times 1.4 = 16.8$ W

10. A three-phase full converter bridge is connected to a supply voltage of 230 V per phase and a frequency of 50 Hz. The source inductance is 4 mH. The load current on the DC side is constant at 20 A. If the load consists of a DC source of internal emf 400 V with internal resistance of 1 Ω, then calculate, (a) firing angle delay and (b) overlap angle in degrees.

Solution

(a) Converter output voltage $= E + I_0R = 400 + 20\times 1 = 420$ V

$$V_0 = \frac{3V_{ml}}{\pi}\cos\alpha - \frac{3\omega L_s I_0}{\pi}$$

$$\therefore 420 = \frac{3\sqrt{6}\times 230}{\pi}\cos\alpha - \frac{3(2\pi\times 50)4}{1000\times\pi}\times 20$$

$$\alpha = 34.382°$$

Therefore, firing angle delay is 34.382°.

$$420 = \frac{3\sqrt{6}\times 230}{\pi}\cos(\alpha+\mu) + \frac{3(2\pi\times 50)4}{1000\times\pi}\times 20 \quad \alpha+\mu = \cos^{-1}\frac{396\times\pi}{3\sqrt{6}\times 230} = 42.602°$$

$$\mu = 42.602 - 34.382 = 8.22°$$

Therefore, overlap angle in degrees = 8.22°.

11. The load on the DC side of an uncontrolled single-phase half-wave rectifier consists of a resistance of 20 Ω in series with an inductance of 100 mH. The input is from a 240 V (sinusoidal 60 Hz) AC. Determine the input DC component of the output voltage.

Solution

We have $X = \omega L = 2\pi \times 60 \times 0.1 = 37.7 \, \Omega$

$$\phi = \tan^{-1}\frac{X}{R} = \tan^{-1}\frac{37.7}{20} = 62.05°$$

$$V_m = \sqrt{2} \times 240 = 339.4 \text{ V}.$$

Being an uncontrolled rectifier, the switching element is a diode, and so the firing angle is zero. To obtain the instant θ_1 at which the diode commutates, we put $i = 0$ to equate the expression on the right side of the following equation to zero.

$$i = \frac{V_m}{Z}\left[\sin(\theta + \alpha - \phi) - e^{-(R/X)\theta}\sin(\alpha - \phi)\right]$$

$$0 = \frac{V_m}{Z}\left[\sin(\theta_1 - \phi) + e^{-(R/X)\theta_1}\sin\phi\right]$$

Numerical substitution for the quantities with angles in radians in the previous equation leads to

$$0 = 7.95\left[\sin(\theta_1 - 1.083) + 0.883e^{-0.5305\theta_1}\right]$$

Solving this gives

$$\theta_1 = 4.314 \text{ rad} = 247.2°$$

The conduction angle in the negative half-cycle for this instant of commutation will be

$$\sigma = \theta_1 - 180° = 67.2.$$

The DC component of the output voltage for this conduction angle in the negative half-cycle can be given as

$$V_{DC} = \frac{V_m}{2\pi}(1 + \cos\sigma)$$

Substituting the values in this equation gives,

$$V_{DC} = 75 \text{ V}$$

12. A three-phase diode rectifier has a no load DC voltage of 250 V. The leakage reactance per phase as seen from the secondary of the transformer is 0.4 Ω. Ignore all other causes of DC voltage drop.
    (a) Determine the DC terminal voltage of the rectifier when it supplies 50 A.
    (b) If the switching elements were thyristors and we have brought the no-load DC voltage to 100 V by phase control, what will be the DC terminal voltage when supplying the same DC load current of 50 A.

Solution

(a) The fictitious equivalent resistance in the DC equivalent circuit is $3 \times 0.4/2\pi = 0.19 \, \Omega$. The resulting voltage drop for 50 A is $50 \times 0.19 = 9.5$ V. The converter terminal voltage on load is $250 - 9.5 = 240.5$ V.

(b) The equivalent resistance value is not dependant on the firing angle, and therefore will be the same as in (a). The voltage drop for the same DC current will also be the same. Therefore, for this load current, the terminal voltage will be $V_d = 100 - 9.5 = 90.5$ V.

13. A double three-phase rectifier with an interphase transformer has diodes as the switching elements. The DC voltage at the transition load is 240 V. The full-load DC current is 80 A. Determine the maximum instantaneous voltage across the interphase transformer and its rms current rating.

Solution

The AC phase voltage (rms) corresponding to three-pulse midpoint operation is $\dfrac{240}{1.17} = 205$ V.

Peak value of phase voltage $= \sqrt{2} \times 205 = 289.9$ V.

The peak value of the interphase transformer voltage is $\dfrac{1}{2} \times 289.9 = 145$ V.

The rms current rating is $\dfrac{1}{2} I_d = 40$ A.

14. A three-phase diode bridge rectifier is fed from a 440 V three-phase 60 Hz bus. Determine the frequency and amplitude of the fundamental component of the ripple voltage in the output.

Solution

The output voltage waveforms will consist of 60° segments of the AC line voltage waveform, which are centred at the peak. Therefore, we can express one repetitive period as follows:

$$v = V_m \cos\theta \text{ in the interval } \theta = -\frac{1}{6}\pi \text{ to } \theta = \frac{1}{6}\pi$$

We shall write this in terms of β such that $\beta = 6\theta$, where β will be our phase angle measure in terms of the fundamental ripple frequency.

$$v = V_m \cos\frac{1}{6}\beta \text{ in the interval } \beta = -\pi \text{ to } \beta = \pi$$

The amplitude of the fundamental Fourier component will be given by

$$V_{r1} = \frac{1}{\pi}\int_{-\pi}^{\pi} V_m \cos\frac{1}{6}\beta \cos\beta \, d\beta$$

After integration and evaluation, this becomes

$$V_{r1} = 0.0546 V_m$$

where V_m, the peak magnitude of the AC line voltage, is $\sqrt{2} \times 440 = 622.3$ V. This gives the amplitude of the fundamental component of the ripple voltage as 34.0 V.

Its frequency is $6 \times 60 = 360$ Hz.

15. The specification sheet for an SCR gives maximum rms on state current as 35 A. If this SCR is used in a resistance circuit, compute the average on state current rating for a half-sine-wave current for conduction angles of (a) 180°, (b) 90°, and (c) 30°.

Solution

$$I_{av} = \frac{1}{2\pi}\int_{\theta_1}^{\pi} I_m \sin\theta\, d\theta = \frac{I_m}{2\pi}(1+\cos\theta_1)$$

$$I_{rms} = \left[\frac{1}{2\pi}\int_{\theta_1}^{\pi} I_m^2 \sin^2\theta\, d\theta\right]^{\frac{1}{2}}$$

$$= \left[\frac{I_m^2}{2\pi}\int_{\theta_1}^{\pi}\left\{\frac{1}{2}-\frac{\cos^2\theta}{2}\right\}\right]^{\frac{1}{2}}$$

$$= \left[\frac{I_m^2}{2\pi}\left\{\frac{\theta}{2}-\frac{\sin^2\theta}{4}\right\}_{\theta_1}^{\pi}\right]^{\frac{1}{2}}$$

$$= \left[\frac{I_m^2}{2\pi}\left\{\frac{\pi-\theta_1}{2}+\frac{\sin^2\theta_1}{4}\right\}\right]^{\frac{1}{2}}$$

(a) For 180° conduction angle, $\theta_1 = 0°$

$$I_{av} = \frac{I_m}{2\pi}\left[1+\cos 0°\right] = \frac{I_m}{\pi}$$

$$I_{rms} = \left[\frac{I_m^2}{2\pi}\left\{\frac{\pi}{2}-\frac{1}{4}(0)\right\}\right]^{\frac{1}{2}} = \frac{I_m}{2}$$

$$FF = \frac{I_{rms}}{I_{av}} = \frac{I_m}{2}\cdot\frac{\pi}{I_m} = \frac{\pi}{2}$$

$$I_{TAV} = \frac{I_{rms}}{FF} = \frac{35\times 2}{\pi} = 22.282\text{ A}$$

(b) For 90° conduction angle, $\theta_1 = 90°$

$$I_{av} = \frac{I_m}{2\pi}\left[1+\cos\theta_0\right] = \frac{I_m}{2\pi}$$

$$I_{rms} = \left[\frac{I_m^2}{2\pi}\left\{\frac{\pi}{4}+\frac{1}{4}(0)\right\}\right]^{\frac{1}{2}} = \frac{I_m}{2\sqrt{2}}$$

$$FF = \frac{I_{rms}}{I_{av}} = \frac{I_m}{2\sqrt{2}}\cdot\frac{2\pi}{I_m} = \frac{\pi}{\sqrt{2}}$$

$$I_{TAV} = \frac{I_{rms}}{FF} = \frac{35\times\sqrt{2}}{\pi} = 15.755\text{ A}$$

(c) For 30° conduction angle, $\theta_1 = 150°$.

$$I_{av} = \frac{I_m}{2\pi}\left[1+(-0.866)\right] = 0.021 I_m$$

$$I_{rms} = \left[\frac{I_m^2}{2\pi}\left\{\frac{\pi}{12}+\frac{1}{4}(-0.866)\right\}\right]^{\frac{1}{2}} = 0.085 I_m$$

Form factor(FF) = $\dfrac{0.085 I_m}{0.021 I_m} = 3.98$

$$I_{TAV} = \frac{35}{3.98} = 8.79 \text{ A}$$

16. The specification sheet for an SCR gives maximum rms on state current as 35 A. If this SCR is used in a resistance circuit, compute the average on state current rating for a half-rectangular-wave current for conduction angles of (a) 180°, (b) 90° and (c) 30°.

Solution

Conduction angle = $\dfrac{T}{\eta T} \times 360$

$\eta = \dfrac{\text{Conduction angle}}{360}$

$I_{av} = \dfrac{I \times T}{\eta T} = \dfrac{I}{\eta}$

$I_{rms} = \left[\dfrac{I^2 \times T}{\eta T}\right]^{\frac{1}{2}} = \dfrac{I}{\sqrt{\eta}}$

(a) For 180° conduction angle, $\eta = \dfrac{360}{180} = 2$

$I_{av} = \dfrac{I}{2}$ and $I_{rms} = \dfrac{I}{\sqrt{2}}$

$FF = \dfrac{I}{\sqrt{2}} \times \dfrac{2}{I} = \sqrt{2}$

$I_{TAV} = \dfrac{35}{\sqrt{2}} = 24.75 \text{ A}$

(b) For 90° conduction angle, $\eta = \dfrac{360}{90} = 4$

$I_{av} = \dfrac{I}{2}$ and $I_{rms} = \dfrac{I}{\sqrt{4}} = \dfrac{I}{2}$

$FF = \dfrac{I}{2} \times \dfrac{4}{I} = 2$

$$I_{TAV} = \frac{35}{2} = 17.5 \text{ A}$$

(c) For 30° conduction angle, $\eta = \frac{360}{30} = 12$

$$I_{av} = \frac{I}{12} \text{ and } I_{rms} = \frac{I}{\sqrt{12}}$$

$$FF = \frac{I}{\sqrt{12}} * \frac{12}{I} = \sqrt{12}$$

$$I_{TAV} = \frac{35}{\sqrt{12}} = 10.10 \text{ A}$$

17. Determine the commutating components for an auxiliary commutation method if I_L = 10 A, E_{DC} = 100 V, and t_q = 50 ms.

Solution

$$C = \frac{I_{Lmax} \cdot t_q}{E_{DC}}$$

$$C = \frac{10 * 50 * 10^{-6} *}{E_{DC}} = 5\,\mu F$$

Also,

$$L_{min} = \frac{E_{DC}^2}{I_L^2} \times C$$

$$= \left[\frac{100}{10}\right]^2 \times 5 \times 10^{-6} = 0.5 \text{mH}$$

$$L_{max} = \frac{0.01T^2}{\pi^2 C}$$

Assume operating frequency of the circuit to be 400 Hz.

$T = 2.5\,\mu s$

$L_{max} = 1.26$ mH

18. In a class D commutation circuit, E_{DC} = 200 V, L = 10 mH, and C = 50 mF. Determine the minimum on time of SCR 1 and peak value of capacitor current.

Solution

Minimum on time of SCR 1 is given by

$$t_{on(min)} = \pi\sqrt{LC} = 69\,\mu s$$

Peak SCR current is given by

$$I_p = E_{DC}\sqrt{\frac{C}{L}}$$

$$= 200\sqrt{50*10^{-6}/10*10^{-6}}$$

$$= 447.21 \text{ A}$$

14. The thyristor in the given figure has a latching current level of 50 mA and is fired by a pulse of width 50 ms. Show that without R, SCR will fail to remain on when the firing pulse ends. Find the maximum value of R to ensure firing. Neglect the SCR volt drop and assume that the initial value of rate of rise of current remains constant over the entire pulse width

Solution

For the given load,

$$\tau = \frac{L}{R} = \frac{0.5}{20} = 0.025 \text{ s}$$

Maximum value of steady state current:

$$= \frac{100}{20} = 5 \text{ A}$$

Without R, SCR current i will grow exponentially as

$$i = I_0 * \left(1 - e^{\frac{-t}{\tau}}\right)$$

$$= 5\left[1 - e^{\left(\frac{-t}{0.025}\right)}\right]$$

$t_{on} = t = 5 \ \mu sec$

$$i = 5\left[1 - e^{\left(\frac{-50*10^{-6}}{0.025}\right)}\right]$$

$$= 9.99 * 10^{-3}$$

$$= 10 \text{ mA}$$

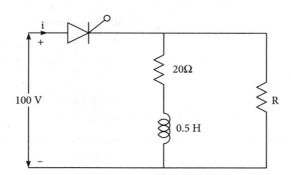

Objective Type Questions

1. Determine the average output voltage of a single-phase half-wave-controlled rectifier with 400 sin 314t input supply and firing angle 60°
   (a) 400/K
   (b) 300/K
   (c) 240/K
   (d) 360/K

2. What could be the range of firing angle of a single-phase one-pulse-controlled converter with input source 400 sin 314t, with load counter emf of 200 V?
   (a) 30° to 150°
   (b) 60° to 180°
   (c) 60° to 120°
   (d) 30° to 180°

3. Calculate the peak inverse value of the thyristor of a single-phase midpoint converter with a 230/200 V transformer on the secondary side.
   (a) 100 V
   (b) 141.4 V
   (c) 200 V
   (d) 282.8 V

4. The output voltage equation of a single-phase full converter is
   (a) $\dfrac{1}{\pi}\int_{\alpha}^{\pi+\alpha} V_m \cos\theta\, d\theta$
   (b) $\dfrac{1}{\pi}\int_{0}^{\pi+\alpha} V_m \cos\theta\, d\theta$
   (c) $\dfrac{1}{\pi}\int_{\alpha-(\pi/2)}^{\alpha+(\pi/2)} V_m \cos\theta\, d\theta$
   (d) $\dfrac{1}{2\pi}\int_{\alpha-(\pi/2)}^{\alpha+(\pi/2)} V_m \cos\theta\, d\theta$

5. The equation of average value of a single-phase semiconverter is
   (a) $\dfrac{1}{\pi}\int_{\alpha}^{\pi} V_m \cos\theta\, d\theta$
   (b) $\dfrac{1}{\pi}\int_{(\pi/2)-\alpha}^{(\pi/2)+\alpha} V_m \cos\theta\, d\theta$
   (c) $\dfrac{1}{\pi}\int_{\alpha-(\pi/2)}^{\alpha+(\pi/2)} V_m \cos\theta\, d\theta$
   (d) $\dfrac{1}{\pi}\int_{\alpha-(\pi/2)}^{\pi} V_m \cos\theta\, d\theta$

6. The SCR conducts for how many radians in case of continuous conduction mode?
   (a) $(\pi - \alpha)$ radians
   (b) π radians
   (c) α radians
   (d) $(\pi - \alpha)$ radians

7. In a single-phase converter under discontinuous load current and extinction angle, the SCR conducts for
   (a) α radians
   (b) $(\beta - \alpha)$ radians
   (c) β radians
   (d) $(\alpha + \beta)$ radians

8. Calculate the load current under discontinuous mode in case of a single-phase full converter for α and $\pi/3$ as firing and extinction angles.
   (a) $(\beta - \alpha) < \pi$
   (b) $(\beta - \alpha) > \pi$
   (c) $(\beta - \alpha) = \pi$
   (d) $(\beta - \alpha) = 3\pi/2$

9. At which value does the FWD conduct in a single-phase converter with discontinuous mode of $\beta > \pi$?
   (a) α
   (b) $\beta - \pi$
   (c) $\pi + \alpha$
   (d) β

10. At which value does the FWD conduct in a single-phase converter with discontinuous mode of $\beta < \pi$?
    (a) α
    (b) $\pi - \beta$
    (c) $\beta - \pi$
    (d) Zero degree

11. In a single-phase semiconverter, for discontinuous conduction and extinction angle $\beta < \pi$, each SCR conducts for the period
    (a) α
    (b) β
    (c) $\pi - \alpha$
    (d) $\beta - \alpha$

12. In a single-phase semiconverter, for discontinuous conduction and extinction angle $\beta > \pi$, each SCR conducts for the period
    (a) $\pi - \alpha$
    (b) $\beta - \pi$
    (c) α
    (d) β

13. When DC load is surpassed using an FWD
    (a) Reversal of load voltage prevention takes place
    (b) Load current transfers away from load
    (c) (a) and (b)
    (d) None of the above

14. In case of a single-phase full converter, the average voltage is 133 V and 325 V. Calculate the firing angle.
    (a) 35°
    (b) 45°
    (c) 130°
    (d) 70°

15. The converter operating in both three phase and six phase is called
    (a) Three-phase semiconverter
    (b) Six-phase full converter
    (c) Three-phase semiconverter
    (d) Three-phase full converter

16. In a three-phase semiconverter, the FWD conducts for less than or equal to 60° for
    (a) 30°
    (b) 60°
    (c) 90°
    (d) 0°

17. In a three-phase semiconverter, the FWD conducts for less than or equal to 120° and extinction angle of 110° for
    (a) 10°
    (b) 30°
    (c) 50°
    (d) 70°

18. The SCRs of a three-phase semiconverter are fired at intervals of
    (a) 60°
    (b) 90°
    (c) 120°
    (d) 150°

19. The SCRs of a three-phase full converter are fired at intervals of
    (a) 30°
    (b) 60°
    (c) 90°
    (d) 120°

20. In case of a three-phase semiconverter, the frequency of the ripple output voltage depends on
    (a) Firing angle and load resistance
    (b) Firing angle and supply frequency
    (c) Firing angle and load inductance
    (d) Only on load circuit parameters

21. What is the average value of the thyristor when the load current is ripple free in case of three-phase full converter?
    (a) 1/2
    (b) 1/3
    (c) 1/4
    (d) 1/5

22. What is the average value of the thyristor when the load current is ripple free in case of one-phase full converter?
    (a) 1/2
    (b) 1/3
    (c) 1/4
    (d) 1/5

23. The number of SCRs conducting in case of single-phase rectifier operation at overlap condition is
    (a) 1
    (b) 2
    (c) 3
    (d) 4

24. The output voltage frequency is equal to ------ in case of three-phase full converter.
    (a) Supply frequency f
    (b) $2f$
    (c) $3f$
    (d) $6f$

25. Which converter requires neutral point connection?
    (a) Three-phase semiconductor
    (b) Three-phase full converter
    (c) Three-phase half-wave converter
    (d) Three-phase converter with diodes

26. The ripple voltage frequency of three-phase half-controlled bridge converter depends on
    (a) Firing angle
    (b) Load inductance
    (c) Load resistance
    (d) Supply frequency

27. Determine the power dissipated by the load for a half-wave-controlled circuit with RL load of 90° conduction with an applied voltage of 800 V.
    (a) 1800 W
    (b) 81 W
    (c) 52.36 W
    (d) 0 W

28. In a single-phase full-wave SCR circuit with RL load
    (a) Power is delivered to the source for delay angles of less than 90°
    (b) The SCR changes from inverter to converter at $\alpha = 90°$
    (c) The negative DC voltage is maximum at $\alpha = 180°$
    (d) To turn off the thyristor, the maximum delay angle must be less than 180°.

29. The ripple voltage frequency of three-phase half-controlled semiconverter depends on (I) firing angle, (II) load resistance, (III) supply frequency, (IV) load inductance
    (a) I, II, and IV
    (b) II, III, and IV
    (c) I and II
    (d) I and III

30. In a three-phase semiconverter
    I There is zero degree conduction of FWD for firing angle less than or equal to 60°
    II FWD conducts for 50° under 120° firing angle and 110° extinction angle
    III SCRs are triggered at an interval of 60°
    (a) I and II
    (b) II and III
    (c) I and III
    (d) I, II, and III

31. The output voltage value of a three-phase-controlled bridge rectifier with an increase in overlap angle
    (a) Decreases
    (b) Increases
    (c) Does not change
    (d) Depends on load inductance

32. The average output of a three-phase half-wave rectifier for input voltage of 200 V is
    (a) 233.91 V
    (b) 116.95 V
    (c) 202.56 V
    (d) 101.28 V

33. The response of commutated mode operation under inverter mode is
    (a) It draws both real and reactive power from the AC supply
    (b) It draws real power from the AC supply
    (c) It delivers both real and reactive power to AC supply
    (d) It draws reactive power from AC supply

34. What is the displacement power factor of the rectifier when the single-phase fully controlled converter feeds constant current into the load of 30°?
    (a) 1
    (b) 0.5
    (c) $\dfrac{1}{\sqrt{3}}$
    (d) $\dfrac{\sqrt{3}}{2}$

35. Determine the rms current value through the thyristor of a three-phase fully controlled converter with a constant DC load of 150 A.
    (a) 50 A
    (b) 100 A
    (c) $\dfrac{150\sqrt{2}}{\sqrt{3}}$
    (d) $\dfrac{150}{\sqrt{3}}$

36. Assume that the DC output current is constant for a six pulse rectifier bridge connected to 50 Hz supply, the lowest harmonic component in the AC source line current is
    (a) 100 H
    (b) 150 Hz
    (c) 250 Hz
    (d) 300 Hz

37. The type of converter that can operate in both three pulse and six pulse operating modes are
    (a) One-phase full converter
    (b) Three-phase half-wave converter
    (c) Three-phase semiconverter
    (d) Three-phase full converter

38. The output voltage equation of a three-phase full converter during overlap condition is
    (a) Zero
    (b) Source voltage
    (c) Source voltage minus inductance drop
    (d) Average value of conducting phase voltages.

Review Questions

1. Explain the role of FWD in rectifier circuits.
2. Mention some of the merits of FWDs in power electronics circuits.
3. Define delay angle.
4. In which way are bridge converters better than midpoint converters?

5. Define commutation angle.
6. Classify the methods of firing used in line commutated converters.
7. Derive the expression for average and rms value of sinlge phase semiconverters.
8. Explain the role of input power factor in power circuits?
9. Mention some of the merits and demerits of six pulse converters.
10. Classify the types of rectifier circuits in power electronics.
11. Explain the different types of controlled rectifier in detail.
12. Differentiate between half-controlled bridge and fully-controlled bridge rectifiers.
13. Explain the working principle of a one pulse converter with different loads and assume the firing angle to be 30°.
14. Explain the working principle of a two pulse converter with R and RL load with bridge configurations. Assume $\alpha = 60$.
15. Explain three-phase half-bridge converter with RL load under continuous and discontinuous modes of operation in detail. Assume $\alpha = 30$.
16. Explain the concept of a six pulse converter using R and RL load in detail.

Practice Questions

1. A resistive load of 7 Ω is connected through a single-pulse converter to a single-phase voltage source of 230 V at 50 Hz. Calculate the following parameters using MATLAB/Simulink: Average DC voltage and DC current, rms voltage and rms current, FF, RF, input power factor, and THD for α = 30, 60, 90.

2. A single-phase full-wave converter is operated from a 230 V at 50 Hz, for RL load of 5 Ω and 10 mH. Consider the delay angle = 0, 30, 45, 60, 90. Determine, using MATLAB/Simulink, the following factors:
   Average DC voltage, average DC current, rms voltage, rms current, FF, RF, input power factor, and THD.

3. A single-phase half-controlled converter is connected to an RLE load of R = 4 Ω, L = 0.1 H, and E = 20 V. Consider the supply voltage to be 220 V at 50 Hz and delay angle = 30° and 60°. At the point when the load current is said to be constant, calculate the average load voltage and current and THD using MATLAB/Simulink.

4. A three-phase half-wave controller has a supply voltage of 120 V/phase. Determine the average and rms voltages MATLAB/Simulink for delay angle of 0°, 30°, 45°, 60°, 90°, assuming continuous load current for RL load of R = 5 Ω and L = 1 mH.

5. A three f fully controlled converter is connected to three f AC source of 400 V, frequency of 50 Hz, which could operate at a firing angle of α = 60. Using MATLAB/Simulink, calculate the average DC voltage and DC current, rms voltage and rms current, FF, RF, input power factor, and THD for α = 30°, 60°, 90°. Assume R = 5 Ω and L = 0.2 mH.

6. A six pulse converter is connected to a 220 V at 50 Hz AC supply. If the inductance at the AC side is .001 H and the commutation angle is 45°, calculate the load current at the DC side using MATLAB/Simulink.

7. A single-phase full-wave rectifier is connected to a supply power with an impedance load value of 3 Ω with 230 V AC supply at a firing angle of about 30°. Calculate the average load, voltage, and the load current using MATLAB/Simulink.

8. Assuming the load to be continuous, calculate the average load voltage and load current of a three-phase bridge rectifier with an RL load of $R = 5$ and $L = 0.55$ mH at a source voltage of 230 V and 50 H using MATLAB/Simulink.

9. A six pulse converter is connected to a 220 V at 50 Hz AC supply. If the load is taken as $R = 10\ \Omega$ and the commutation angle is 60°, calculate the load current at the DC side using MATLAB/Simulink.

10. Determine the rms value of the load current and its voltage for a single-phase fully-controlled rectifier with R load of 10 Ω, for the input voltage to be 100 V and 50 Hz with the firing angle as 30° using MATLAB/Simulink.

11. With relevant waveform, explain the concept of three-phase half-wave-controlled converter with RL load.

12. With necessary circuit and waveforms, explain the principle of operation of a three-phase-controlled bridge rectifier feeding RL load and derive the expression for the average output DC voltage.

13. With relevant waveforms, explain the operation of a three-phase semiconverter with RLE load.

14. What is the role of source inductance in single-phase rectifier circuits? Explain with relevant examples.

15. Explain the concept of a dual converter with its two operating modes in detail.

16. Explain the effect of source inductance in the operation of a three-phase fully-controlled converter, indicating clearly the conduction of various thyristors during one cycle with relevant waveforms.

17. Derive an expression for harmonic factor, displacement factor and power factor of a single-phase semiconverter from the fundamental principle.

18. A three-phase fully-controlled rectifier is connected to a three-phase AC supply of 230 V, 60 Hz. Load current is continuous and has a negligible ripple. If the average load current I_{DC} = 150 A and the commutating inductance L_c = 0.1 mH. Determine the overlap angle when α = 10°.

19. A three-phase half-wave rectifier is operated from a three-phase star connected 208 V, 60 Hz supply. Load resistance = 10 Ω. If it is required to obtain an average output voltage 50 per cent of maximum possible output voltage. Calculate (i) delay angle, (ii) rms value of output current, (iii) average value of output current, (iv) thyristor average and rms current, (v) efficiency, (vi) total utilization factor and (vii) supply power factor.

4

DC Choppers Using MATLAB (DC–DC Converters)

Learning Objectives

- ✓ To describe the need, control strategies and functionality of choppers
- ✓ To list out the various blocks required for simulation
- ✓ To explain the different types of choppers using MATLAB/Simulink.
- ✓ To examine the operations of step-up and step-down chopper.
- ✓ To explain the different commutation techniques using MATLAB/Simulink.
- ✓ To enumerate the various applications of choppers

4.1 Introduction

DC–DC converters are much required these days as numerous modern applications use DC voltage. The execution of these applications will be enhanced if we could utilize a variable DC supply. It will help enhance controllability of, for example, gears. Metro autos, trolley transports, battery worked vehicles are other appliances which may work better with variable DC supply. With a chopper, it is possible to control and fluctuate a consistent DC voltage. The chopper, which is a static power device, works by changing from settled power levels to variable levels. It is basically a high-speed switch, which connects and disconnects the load from source at a high rate to get the desired output.

In hardware, a chopper circuit is utilized to describe various sorts of electronic exchanging devices that are part of power control and signal applications. Basically, a chopper is an electronic switch that is utilized to intrude on one sign under the control of an alternate sign. Choppers can expand or diminish the DC voltage level at its inverse side. In this way, choppers fill the same need in DC circuit that rectifiers do if there should arise an occurrence of AC current. So, they are otherwise called DC transformers. Exchanging devices using these choppers are either on or off; so choppers have very few losses and circuits using choppers are very efficient. However, the load current is very irregular and a smoothing or high-exchanging frequency may be needed to avoid undesirable impacts. In signal processing circuits, utilization of choppers has become very important to prevent electronic drifts. The standard chopper can be enhanced using synchronous demodulators that basically undo the 'chopping' process.

4.2 Choppers and their Classification

As mentioned earlier, a static device that converts settled or fixed DC input voltage or power into variable voltage or power is referred to as a chopper. The chopper meets the standards of an AC transformer, and thus its characteristics are identical. Since choppers involve one-phase conversion, they are more efficient. Just like the transformer, the chopper can step up or even step down the fixed DC voltage. Chopper circuits are highly efficient and responsive with smooth control on all four quadrant operations. Choppers may be classified based on a few parameters.

1. On the basis of input and output voltage levels:
   - Step-down chopper
   - Step-up chopper
2. On the basis of circuit operation:
   - First quadrant
   - Two quadrant
   - Four quadrant
3. On the basis of commutated technique:
   - Voltage commutated
   - Current commutated
   - Load commutated

Chopper circuits can be used in the following devices:

- Switched mode power supplies (SMPS) and speed controllers
- Electronic amplifiers of class D
- Channel filters of switched reluctance

For all chopper designs working from a settled DC input voltage, the output voltage is controlled by occasional opening and shutting of the switches that are part of the chopper circuit. The following distinctive are used systems specifically to control output voltage:

- Pulse width modulation system
- Frequency modulation

4.3 Control Strategies of Chopper

The control strategy of a chopper is classified into two types, namely:

- Time ratio control (TRC)

  In the TRC method, the output DC voltage can be controlled by the following methods:
  - Pulse width modulation or constant frequency system
  - Variable frequency control or frequency modulation

- Current limit control

4.3.1 Pulse width modulation or constant frequency system

Figure 4.1 shows the principle of a pulse width modulation system.

In a pulse width modulation system, the t_{on}, that is, the turn-on time is varied by keeping the chopping frequency f and the chopping period T constant. The output voltage is varied by varying the turn-on time t_{on}.

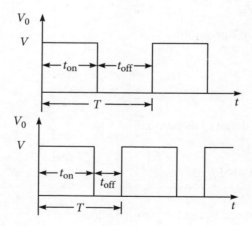

Figure 4.1 Principle of Pulse Width Modulation System.

4.3.2 Variable frequency control or frequency modulation

Figure 4.2 shows the principle of variable frequency control or frequency modulation.

In the frequency modulation technique, the chopping frequency f is varied by keeping the turn-on and turn-off time constant. To obtain the range of the output voltage, the frequency is varied over a wide range. The major demerit of this system is the harmonics that occurs in the load side; also for T_{off}, the load current may become discontinuous.

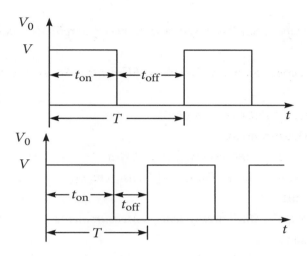

Figure 4.2 Principle of Variable Frequency Control.

4.3.3 Current limit control

In this method, the turn on and the turn off of the circuit is determined by the previously set value of the load current. The moment the load current reaches the upper limit, the switch is turned off. At this instant then, the load current freewheels and begins to decay exponentially. When it falls to the lower limit, the switch is set to the on condition and the load current begins to increase. This method involves feedback looping and triggering, and hence, this concept is destined to be complex.

4.4 Selection of Components from the Simulink Library Browser

Before we begin MATLAB simulation, we must choose the correct components. Listed are the components and their location:

New File

>>*Type Simulink on Matlab Command Window>>File>>New>>Model*

AC Voltage Source

>>*Libraries>>Simscape>>SimpowerSystems>>Electrical Sources>>AC Voltage Source*

DC Source

>>*Libraries>>Simscape>>SimPowerSystems>>Electrical Sources>>DC Voltage Source*

Thyristor/MOSFET
>>Libraries>>Simscape>>SimPowerSystems>>PowerElectronics>>Thyristor/MOSFET
Diode
>>Libraries>>Simscape>>SimPowerSystems>>PowerElectronics>>Diode
Series *RLC* Branch
>>Libraries>>Simscape>>SimpowerSystems>>Elements>>Series RLC Branch
Pulse Generator
>>Libraries>>Sources>>Pulse Generator
Three-Phase V–I Measurement
>>Libraries>>Simscape>>SimPowerSystems>>Extra Library>>Measurements>>Three-Phase V–I Measurement
Voltage Measurement
>>Libraries>>Simscape>>SimPowerSystems>>Measurements
Current Measurement
>>Libraries>>Simscape>>SimPowerSystems>>Measurements
Mean
>>Libraries>>Simscape>>SimPowerSystems>>Extra Library>>Measurement>>Mean Value
RMS
>>Libraries>>Simscape>>SimPowerSystems>>Extra Library>>Measurement>>RMS Value
Subsystem
>>Simulink>>Commonly Used Blocks>>Subsystem
THD
>>Libraries>>Simscape>>SimPowerSystems>>Extra Library>>Measurement>>Total Harmonic Distortion
From
>>Simulink>>Signal Routing>>From
Goto
>>Simulink>>Signal Routing>>Goto
Display
>>Simulink>>Sinks>>Display
Scope
>>Simulink>>Sinks>>Scope

Ground
>>Libraries>>Simscape>>SimPowerSystems>>Application libraries>>Elements>>Ground

4.5 Principle of Operation of a Step-down Chopper

Choppers are classified into two types based on their input and output voltage levels.

Step-down chopper: The output voltage is less than the input voltage.

Step-up chopper: The output voltage is greater than the input voltage.

The step-down chopper consists of a DC voltage source supplied to an R load. In the circuit shown in Figure 4.3, the thyristor acts as the switch. When the thyristor is on, the supply voltage appears across the load and when the thyristor is off, the voltage across the load will be zero.

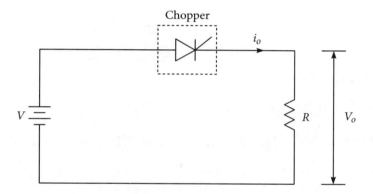

Figure 4.3 Operation of Step-down Chopper with R Load.

In case of a step-down chopper with an RLE load (Figure 4.4), a chopped load voltage is obtained from a constant DC supply of magnitude V_s. During the on condition, the load voltage is equal to the source voltage and during the off condition, the load current flows through the freewheeling diode (FWD), where the load current is continuous.

The output voltage of a step-down chopper is given by

$$V_o = dV_s \tag{4.1}$$

In terms of frequency, the load voltage is given by

$$V_o = f T_{on} V_s \tag{4.2}$$

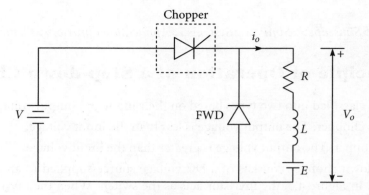

Figure 4.4 Step-down Chopper Circuit with *RLE* Load.

Output power

$$P_o = V_o I_o \tag{4.3}$$

Effective input resistance

$$R = V/I \text{ and } R_i = R/d \tag{4.4}$$

The output voltage can be varied by varying the duty cycle.

4.6 Principle of Operation of a Step-up Chopper

In the step-up chopper, the average output voltage is greater than the input supply voltage. The inductor *L* is placed in series with the source voltage, which is considered very essential. Figure 4.5 shows the working principle of a step-up chopper.

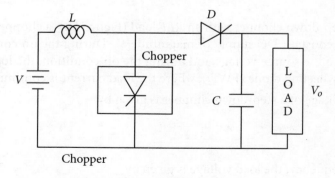

Figure 4.5 Step-up Chopper Circuit.

The step-up chopper circuit is mainly used to obtain a load voltage greater than the input supply voltage. The inductance and the capacitance values are selected in such a way that the desired output voltage and current are obtained. During its operation, when the chopper is turned on, the inductor is connected across the supply. In this state, the current through the inductor stores a certain amount of energy. When the chopper turns to the off condition, the current through the inductor is forced to flow through the diode D and the load. Finally, the current tends to decrease resulting in reverse polarity of the induced EMF (electromagnetic field) in L.

Therefore, the voltage across the load is given by

$$V_o = V + L(di/dt) \tag{4.5}$$

The capacitor used in the circuit provides a continuous output voltage, while the diode used in the circuit controls the flow from the capacitor to the source; step up choppers are used for braking operation.

When the chopper is on:

Voltage across inductor $L = V$.

Therefore, energy stored in the inductor $= VIT_{on}$

When chopper is off:

Voltage across inductor $L = V_o - V$

Therefore, energy stored in the inductor $= V_o - VIT_{on}$

Neglecting losses in the inductor,

$$V_o = V(T/T - T_{on}) \tag{4.6}$$

where $T = T_{on} + T_{off}$

Therefore,

$V_o = V(1/1 - d)$, where d represents the duty cycle $= T_{on}/T$

4.7 Performance Parameters of Step-up and Step-down Choppers

- The thyristor requires a certain minimum time to turn on and turn off.
- Duty cycle d can be varied only between a minimum and maximum value, limiting the minimum and maximum value of the output voltage.

$$\text{Duty cycle} = \left(\frac{T_{on}}{T_{on} + T_{off}}\right) = \left(\frac{T_{on}}{T}\right) \tag{4.7}$$

where T_{on} is turn on time, T_{off} is turn off time and T is total time period.

For a step-up chopper:

$$\text{Average output voltage across load } V_o = \frac{V_s}{1-\alpha} \tag{4.8}$$

$$\text{Energy supplied by inductor} = W_{out} = (V_o - V_s)\frac{(I_{min} + I_{max})}{2} \times T_{off} \tag{4.9}$$

For a step-down chopper:

$$\text{Average output voltage across load } V_o = V_s\left(\frac{T_{on}}{T}\right) = fT_{on}V_s = \alpha V_s \tag{4.10}$$

where V_s = supply voltage; $\frac{1}{T} = f$

$$\text{Average output current through load } I_o = \alpha\frac{V_s}{R} \tag{4.11}$$

$$\text{Root mean square (rms) value of output voltage} = \sqrt{\alpha}V_s \tag{4.12}$$

$$\text{RMS value of thyristor current} = \frac{\sqrt{\alpha}V_s}{R} \tag{4.13}$$

$$\text{Effective input resistance of chopper} = \frac{R}{\alpha}$$

The minimum and maximum values of load current is given by

$$I_{max} = \frac{\frac{V_s}{R}\left[1 - e^{-\frac{T_{on}}{T_a}}\right]}{\left[1 - e^{-\frac{T}{T_a}}\right]} - \left(\frac{E}{R}\right)$$

$$I_{min} = \frac{\frac{V_s}{R}\left[e^{\frac{T_{on}}{T_a}} - 1\right]}{\left[e^{\frac{T}{T_a}} - 1\right]} - \left(\frac{E}{R}\right)$$

$$\text{Ripple } \Delta I = \frac{V_s}{R} \left[\frac{\left(1-e^{-\frac{T_{on}}{T_a}}\right)\left(1-e^{-\frac{T_{off}}{T_a}}\right)}{\left[1-e^{-\frac{T}{T_a}}\right]} \right]$$

$$\text{Per unit ripple} = \frac{\left(1-e^{-\frac{\alpha T}{T_a}}\right)\left(1-e^{-\frac{(1-\alpha)T}{T_a}}\right)}{\left[1-e^{-\frac{T}{T_a}}\right]}$$

Maximum value of ripple current is given by $(\Delta I)_{max} = \dfrac{V_s}{4fL}$ \hfill (4.14)

$$\text{Ripple factor } = \frac{\text{AC ripple voltage}}{\text{DC voltage}} = \sqrt{\left(\frac{1}{\alpha}\right) - 1} \qquad (4.15)$$

The load current ripples are inversely proportional to the chopping frequency f. To reduce the ripple current, the frequency should be as high as possible. The features of step-down and step-up choppers are compared in Table 4.1.

Table 4.1 Features of Step-down and Step-up Choppers

S. No.	Parameters	Step-down Chopper	Step-up Chopper
1	Voltage output	0 to V volts	V to +∞ volts
2	Chopper position switch	Series	Parallel
3	Output voltage equation	$V_{LDC} = D \times V$ volts	$V_o = V/(1-D)$ volts
4	Inductance value	Not applicable	Required
5	Usage	Motoring	Regenerative braking
6	Quadrant	Single	Single
7	Operation	Quadrant 1	Quadrant 1
8	Application	Motor speed control	Battery charging

4.8 Chopper Configuration

In chopper circuits, unidirectional power semiconductors are utilized. If the power semiconductor devices are arranged in a sequence, choppers are constructed to work in one or more of the four quadrants. Based on the quadrants they work in, there are five types of choppers–A, B, C, D, and E. These categories can be analyzed using MATLAB/Simulink.

4.8.1 Type A chopper

It is a solitary quadrant chopper whose operation is limited in the first quadrant. The circuit and its performance are demonstrated in Figures 4.6 and 4.7.

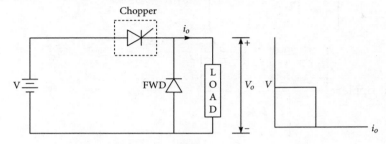

Figure 4.6 Type A Chopper.

At the point when chopper is on, supply voltage V is applied over the load. At the point when chopper is off, $V_o = 0$ and the load current keeps on streaming in the same bearing through the FWD. The normal estimations of output voltage and current are always positive. It is utilized to control the rate of the DC chopper. The output current obtained in a step-down RL load can be utilized to examine the execution of type A chopper.

In a first quadrant or type A chopper, when chopper Switch1 is on, output voltage is equal to the supply voltage, that is, $V_o = V_s$ and current I_o flows in the direction shown by the arrow in Figure 4.7, that is, positive direction. The effect of turn on of the switch results in the measurement of V_o and I_o, as shown in Figure 4.7.

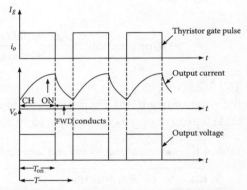

Figure 4.7 Performance of Type A Chopper.

Thus, both current and voltage are taken as positive and load power is certain, which means voltage is conveyed from source to load. When the switch is in the off position, the current starts freewheeling through the diode; the output voltage is zero but output current is certain. Thus, in type A chopper, the normal estimation of the output response is said to be positive always. Type A choppers are step-down choppers, where the normal estimation of the output response is not the input source voltage. They are generally used in motor control operation.

Example 1

The MATLAB/Simulink model of a type A chopper is shown in Figure 4.8. The Simulink model is generated by extracting the various blocks available in the library browser. After compilation, the input and the output responses of the prescribed circuit can be validated when the MATLAB model is run.

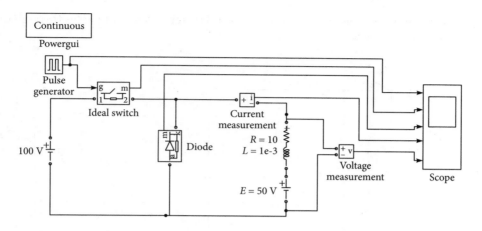

Figure 4.8 Simulation Model of a Type A Chopper.

The time taken for simulation of the results is set at 0.16 s; the load value is set for $R = 10\ \Omega$; and the supply voltage is $V_{supply} = 100$ V. The set of results for different firing angles while operating with $E = 50$ V and RL load, where $R = 10$ W and $L = 1$ mH are provided. Refer to Figure 4.9 for the simulation waveform of a type A chopper.

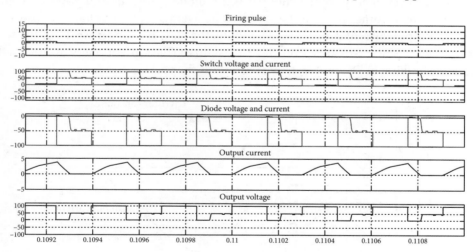

Figure 4.9 Simulation Waveform of a Type A Chopper.

4.8.2 Type B chopper

The type B chopper is a solitary quadrant chopper working in the second quadrant. The MATLAB/Simulink model is demonstrated in Figure 4.10.

The unique feature of type B chopper includes the usage of DC voltage source E at the load terminal. When the circuit is on, the initial response of the output voltage is zero. The current starts streaming in the inverse heading as shown in Figure 4.10 at the point when chopper is off, which surpasses the source voltage V_s. Hence, current passes through diode D and is treated as negative.

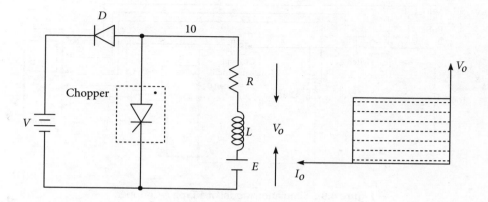

Figure 4.10 Type B Chopper.

At the point when the chopper is in the on position, E drives a current through L and R in a heading inverse to that shown in Figure 4.11. During the on time of the chopper, the inductance L stores energy. When chopper is off, diode D directs, and part of the energy level in inductor L comes back to the supply. The average output voltage is positive and the average output current is negative. Therefore, the class B chopper works in the second quadrant where power streams from load to source. Class B choppers are utilized for regenerative braking of DC motor; they are step-up choppers.

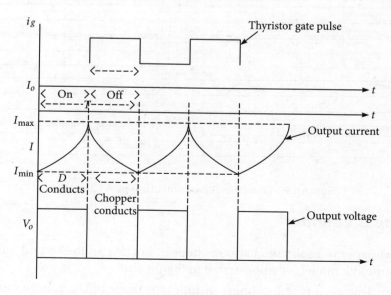

Figure 4.11 Performance of a Type B Chopper.

In a type B chopper, as mentioned earlier, current I_o is constantly negative; V_o is certain (occasionally zero).

Example 2

The MATLAB/Simulink waveform of a type B chopper is shown in Figure 4.12. The Simulink model is generated by extracting the various blocks available in the library browser. After compilation, the input and the output responses of the prescribed circuit can be validated when the MATLAB model is run. The time taken for simulation of the results is set at 0.16 s, the load value is set for $R = 10$ W and the supply voltage is $V_{supply} = 100$ V. Refer to Figure 4.13 for the simulation waveform of a type B chopper operating at $E = 50$ V and RL load, where $R = 10$ ohm and $L = 1$ mH.

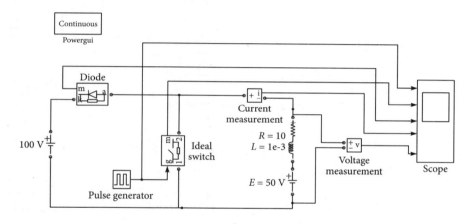

Figure 4.12 Simulation Model of a Type B Chopper.

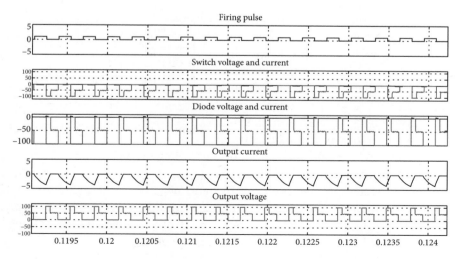

Figure 4.13 Simulation Waveform of a Type B Chopper.

4.8.3 Type C chopper (regenerative chopper)

Type C chopper is a combination of a type A chopper and a type B chopper connected in parallel as shown in Figures 4.14 and 4.15.

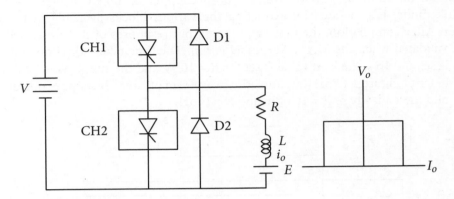

Figure 4.14 Type C Chopper.

This is a two quadrant chopper whose operation is bounded between the first and second quadrant. In the first quadrant, CH1 is on and D2 conducts, whereas in the second quadrant, CH2 is on and D1 conducts. When CH1 is on, the load current is positive. The output voltage is equivalent to V and the load gets power from the source. When CH1 is off, the energy in inductance L strengthens current to course through diode D2 and the output voltage is zero. Current keeps on streaming in the positive direction. When CH2 is set off, the voltage E powers current to stream in the inverse heading through L and CH2. The output voltage is zero. On closing CH2, the energy in the inductance drives current through diode D1 and the supply output voltage is V, the input current gets to be negative and control streams from load to source. Average output voltage is positive. Average output current can take both positive and negative values. Choppers CH1 and CH2 should not to be turned on all the while as it would short circuit the supply. Class C chopper can be utilized both for DC engine control and regenerative braking of DC motors. It can be used as a step-up or step-down chopper.

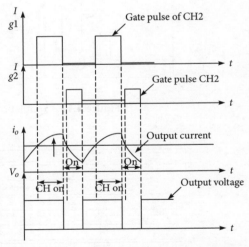

Figure 4.15 Performance of a Type C Chopper.

When the metal–oxide–semiconductor field-effect transistor (MOSFET) switch is on, current moves through R and inductor L will be charged. Thus, there will be output voltage V_o and current I_o. When MOSFET switch 1 is in the off position, D2 will be activated and current I_o will move through the same bearing with zero output voltage. In this way, MOSFET switch 1 behaves like a type A chopper of the first quadrant. When the second MOSFET switch is turned on, the output voltage becomes zero and the output current streams towards the reverse bearing as indicated in Figure 4.16, where the inductor will be energized to a level which surpasses the estimation of the supply voltage V_s and diode D1. Successively, the output voltage and output current are converse with each other in opposite directions. The second MOSFET switch acts like a type B chopper, which operates in the second quadrant. Type C choppers are mainly applied in forward motoring and forward braking concepts.

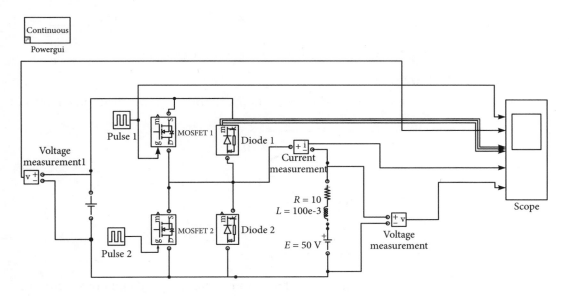

Figure 4.16 MATLAB/Simulink Model of a Type C Chopper.

Example 3

The MATLAB/Simulink waveform of a type C chopper is shown in Figure 4.16. The Simulink model is generated by extracting the various blocks available in the library browser. After compilation, the input and the output responses of the prescribed circuit can be validated when the MATLAB model is run. Refer to Figure 4.17 for the simulation waveform. The time taken for simulation of the results is set at 0.16 s, the load value is set for $R = 10\,\Omega$ and the supply voltage is $V_{supply} = 100$ V.

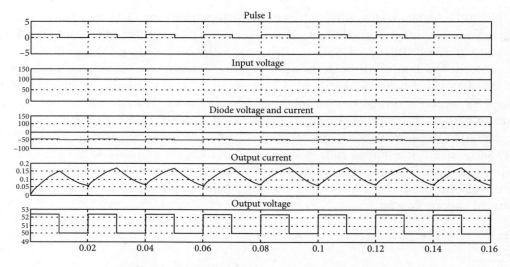

Figure 4.17 Simulation Waveform of a Type C Chopper.

4.8.4 Type D chopper

Figure 4.18 represents the schematic of a type D chopper, which is a two quadrant chopper.

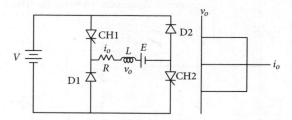

Figure 4.18 Type D Chopper.

When both CH1 and CH2 are activated, the output voltage $V_o = V$ and output current courses through the load. When CH1 and CH2 are killed, the load current keeps on streaming in the same heading through load, D1 and D2, because of the energy stored in the inductor L. Output voltage $V_o = -V$. There will be average load voltage if chopper's on time is more than the off time. Normal output voltage will be negative if $T_{ON} < T_{OFF}$. Hence, the heading of load current is always positive, but the stack voltage can be positive or negative.

When the two choppers are on, the output voltage V_o will be equivalent to V_s. Even at the instant when $V_o = -V_s$ and the choppers are switched off, the diodes D1 and D2 will conduct. The turn on time T_{on} will be more in comparison with the turn off time and the output voltage will be normally positive, which is demonstrated in the waveform shown in Figure 4.19. As the diodes and choppers conduct current just in one bearing, the leading of load current will be always positive.

DC Choppers Using MATLAB (DC–DC Converters)

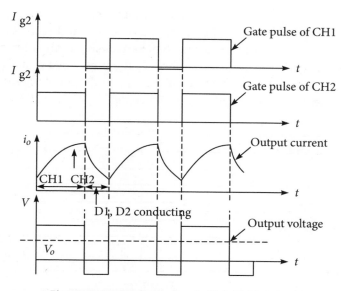

Figure 4.19 Performance of a Type D Chopper.

Example 4

Output responses are certain as the power flows from the source to load and in this manner, the forward quadrant operation of the type D chopper is obtained.

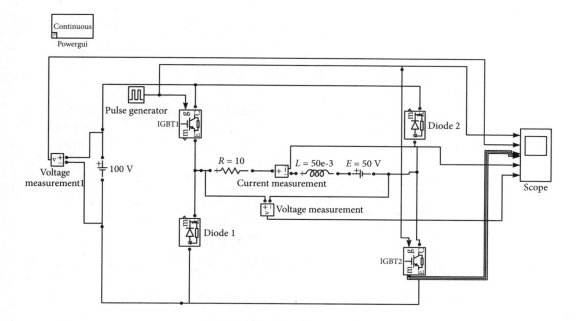

Figure 4.20 MATLAB/Simulink Model of a Type D Chopper.

The MATLAB/Simulink model of the type D chopper is shown in Figure 4.20. The MATLAB/Simulink waveform of a type D chopper is shown in Figure 4.21. The Simulink model is generated by extracting the various blocks available in the library browser. After compilation, the input and the output responses of the prescribed circuit can be validated when the MATLAB model is run. The time taken for simulation of the results is set at 0.16 s, the load value is set at $R = 10\ \Omega$ and the supply voltage is $V_{supply} = 100$ V.

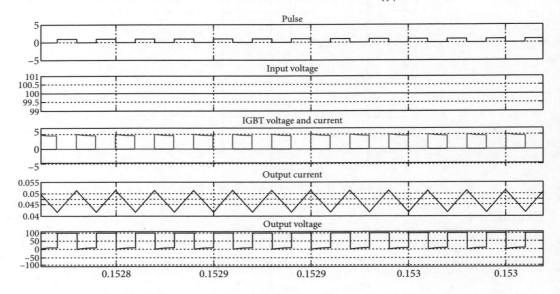

Figure 4.21 Simulation Waveform of a Type D Chopper.

4.8.5 Type E chopper

The forth quadrant operating chopper or the type E chopper comprises four semiconductor switches and four diodes coordinated in antiparallel direction. The switches take part in all the quadrants and are said to be dynamic. Figure 4.22 represents the type E chopper.

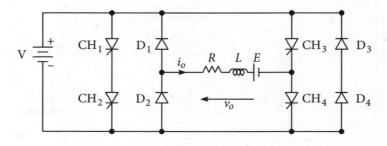

Figure 4.22 Type E Chopper.

The Simulink model of a type E chopper is shown in Figure 4.23. During the first quadrant operation of the circuit, chopper 4 will be on. Chopper 3 will be turned off and chopper 1 will be activated. As choppers 1 and 4 are on, the load voltage V_o will be equivalent to the source voltage V_s and the load current I_o will start to stream through the circuit. V_o and I_o will be positive as the first quadrant operation occurs. When chopper 1 is turned off, the positive current freewheels through chopper 4 and the diode D2. The type E chopper works as a step-down chopper in the first quadrant. For this situation, chopper 2 will be operational and the other three are kept off. As the chopper 2 is on, negative current will begin moving through the inductor L–CH2–E and D4. Energy is put away in the inductor L as chopper 2 is on. When chopper 2 is off, the current will be bolstered once again to the source through diodes D1 and D4. Here, $(E + L \cdot di / dt)$ will be more than the source voltage V_s. In the second quadrant, the chopper will act as a step-up chopper as the power is returned back from load to source. In the third quadrant operation, chopper 1 will be kept off, chopper 2 will be on, and chopper 3 is operated.

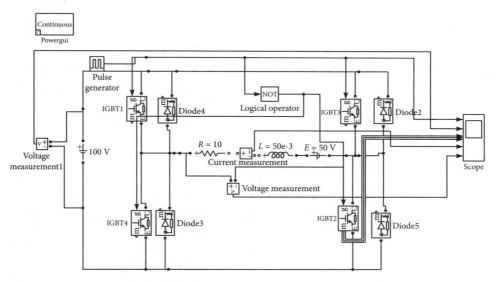

Figure 4.23 MATLAB/Simulink Model of a Type E Chopper.

For third quadrant operation, most of the load is switched. As chopper 3 is on, the load gets joined with the source V_s and V_o; I_o will be negative in this third quadrant operation. The chopper works as a step-down chopper. When chopper 4 is turned on, positive current begins to move through chopper 4–D2–E and the inductor L will store energy. As chopper 4 is turned off, the current flows to the source through diodes D2 and D3. The operation will be in the fourth quadrant as the load voltage is negative; however, the load current is certain. In this quadrant, the chopper works as a step-up chopper as the power is sustained back from load to source.

Example 5

The MATLAB/Simulink waveform of a type E chopper is shown in Figure 4.24. The Simulink model is generated by extracting the various blocks available in the library browser. After compilation, the input and the output responses of the prescribed circuit can be validated when the MATLAB model is run. The time taken for simulation of the results is set at 0.16 s, the load value is set for $R = 10\ \Omega$ and the supply voltage is $V_{supply} = 100$ V.

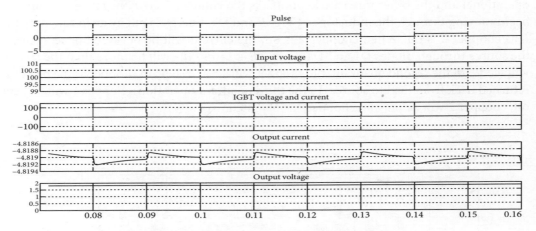

Figure 4.24 Simulation Waveform of a Type E Chopper.

4.9 Switching Mode Regulators

Switching mode controllers are categorized into three essential DC–DC converter circuits, called buck, boost and buck–boost. In these circuits, a power semiconductor device is utilized as an ideal switch that is turned on by a gate pulse at its terminal. The ideal switch is associated in arrangement with a load to a DC supply; or, a positive (forward) voltage is connected in the middle of the anode and cathode terminals. The switch turns off when the current drops underneath the holding current, or a converse (negative) voltage is connected in the middle of the anode and cathode terminals. In this way, the switch is constraint commutated, for which extra circuit has to be used – usually an alternate thyristor is utilized.

4.9.1 Buck converter

A buck converter, shown in Figure 4.25, comprises a switch, which is a power semiconductor device. Likewise, a freewheeling diode is utilized to permit the load current to course through it when the switch is turned off. The load is a resistive load. At times, a battery or back emf is associated in arrangement with the load, which is inductive. Because of the load inductance, the load current must be permitted a way, which is given by the diode.

Without this diode, the high emf of the inductance caused by the diminishing load current may damage the exchanging device. If the exchanging device is utilized is a thyristor, this circuit is a step-down chopper, because the output voltage is typically lower than the data voltage. This DC–DC converter is termed as a buck converter.

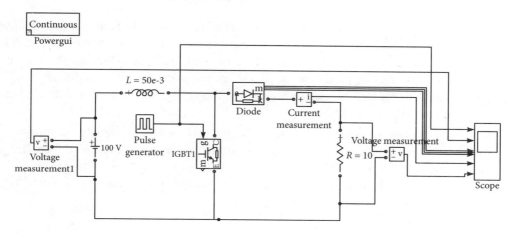

Figure 4.25 MATLAB/Simulink Model of a Buck Converter.

Example 6

Starting with the switch open in the off position, the current in the circuit is 0. When the switch is shut, the value of the current will start to increase with the inductor delivering a contradicting voltage because of the evolving current at the terminal side. The net voltage over the load gets diminished, because the voltage drop balances the source.

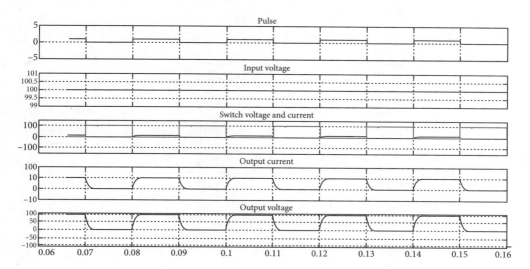

Figure 4.26 Simulation Waveform of a Buck Converter.

Over the long run, the load voltage expands when the rate of change of current and the inductor voltage drop. At this instance, the inductor energizes slowly in the attractive field. When the switch opens, while the current value is changing, a sudden voltage drop occurs across the inductor; hence, the net voltage at the load terminal will be dependent on the supply source.

The source voltage increases thus the current value will reduce. The current that changes with respect to time delivers an inductor voltage, thus helping the source voltage. The energized inductor favours the current to move through the load. During this time, the inductor is releasing its energy into whatever is left of the circuit. If switch is shut again before the inductor completely releases its energy, the voltage at the load will dependably be more noteworthy than zero.

The Simulink model is generated by extracting the various blocks available in the library browser. After compilation, the input and the output responses of the prescribed circuit can be validated when the MATLAB model is run. The time taken for simulation of the results is set at 0.16 s, the load value is set for $R = 10\ \Omega$, $L = 50$ mH and the supply voltage is $V_{supply} = 100$ V.

4.9.2 Boost converter

A DC–DC power converter whose output response is more prominent than its input voltage is referred to as a boost or step-up converter. It is a switched mode controller which has in its circuit no less than two power semiconductors, say a diode and a switch, and a storage component, a capacitor, inductor, or the two in blend. Channels are regularly added to the output of the converter to decrease the output voltage ripple. The MATLAB/Simulink model of a boost converter is shown in Figure 4.27.

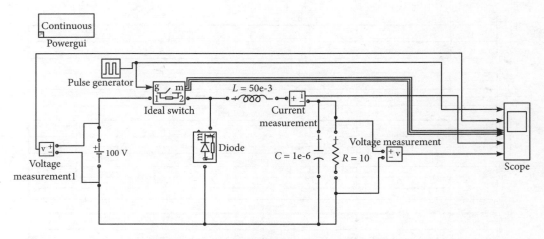

Figure 4.27 MATLAB/Simulink Model of a Boost Converter.

Example 7

The principle that drives the boost converter is the tendency of an inductor to oppose changes in current by making and decimating an effective output. In a boost converter, the output voltage is constantly higher than the input voltage. The simulation waveform of a boost converter is shown in Figure 4.28. Consider the switch is closed at any instant. The current streams through the inductor L, which stores energy for effective operation of the circuit.

At an alternate instant, when the circuit switch is opened, the current through the circuit will decrease as the impedance value is high. The existing electric field maintains the current streaming towards the load and the terminal will be reached accordingly. Thus, using the two energy sources in arrangement, the capacitor will be charged through the diode D. The inductor will not release its energy in the intermediate charging stages, and the load terminal will experience more prominent voltage value than the source when the switch is said to be opened. Laterally, the capacitor is also charged in parallel to this voltage level. When the switch is said to be closed, the capacitor is ready enough to supply the energy to the load terminal. During this stage, the diode in the blocking state keeps the capacitor from releasing energy through the switch module. Thus, the switch should be sufficiently opened to keep the capacitor from excessive release of energy.

The Simulink model is generated by extracting the various blocks available in the library browser. After compilation, the input and the output responses of the prescribed circuit can be validated when the MATLAB model is run. The time taken for simulation of the results is set at 0.16 s, the load value is set for $R = 10\ \Omega$, $L = 50$ mH with $C = 1\ \mu F$ and the supply voltage is $V_{supply} = 100$ V.

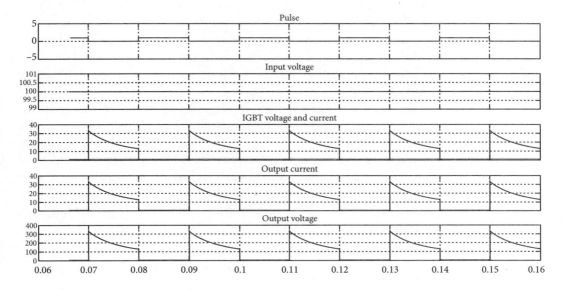

Figure 4.28 Simulation Waveform of a Boost Converter.

4.9.3 Buck–boost converter

A converter of this kind results in an output voltage magnitude that is more prominent than or not exactly the data voltage size. The output responses of the circuits are much bigger than the input voltage, which goes down to just about zero. The MATLAB/Simulink model of a buck–boost converter is shown in Figure 4.29.

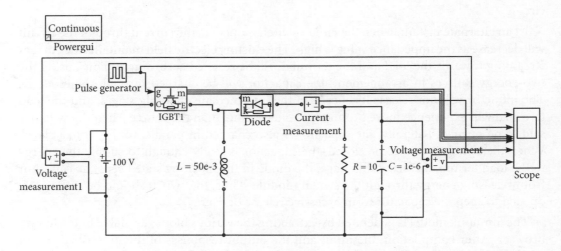

Figure 4.29 MATLAB/Simulink Model of a Buck–Boost Converter.

Example 8

The simulation waveform is shown in Figure 4.30. The essential principle of the buck–boost converter is basic. During the on state of the converter circuit, the input voltage source and the inductor L are associated with each other, which results in aggregating energy in L; the capacitor supplies energy to the load terminal. Similarly, during the off state of the converter circuit, the inductor interlinks with the load terminal and energy is interchanged from L to C and to R. Resembling the buck and boost converter, the circuit operation of the buck part is best seen regarding the inductor's 'reluctance' to permit immediate change in the current. Even at the initial state, the current through the inductor is zero, since the switch is kept open. When the switch is closed, the blocking diode initiates the current streaming into the circuit coursing through the inductor. However, the inductor does not take part in immediate current change, as it will initially keep the current at its lower state by dropping the majority of the voltage supplied by the source. The inductor will then allow the current to increase by reducing the voltage drop at its terminal over the long run. During this time, the inductor will store energy as an effective ground.

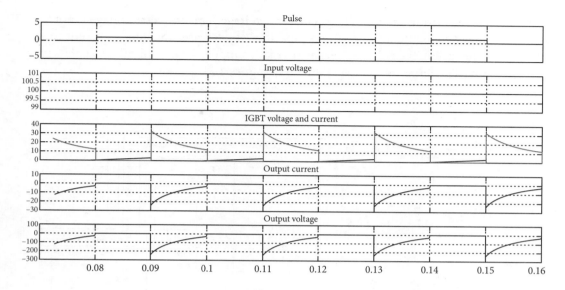

Figure 4.30 Simulation Waveform of a Buck–Boost Converter.

The Simulink model is generated by extracting the various blocks available in the library browser. After compilation, the input and the output responses of the prescribed circuit can be validated when the MATLAB model is run. The time taken for simulation of the results is set at 0.16 s, and the load value is set for $R = 10\ \Omega$, $L = 50$ mH with $C = 1\ \mu F$ and the supply voltage is $V_{supply} = 100$ V.

4.9.4 Cuk converter

The cuk converter is a buck converter followed by a boost converter whose output voltage is either greater or lower than the input voltage. The major drawback of the cuk converter is that the circuit is difficult to stabilize, and thus, there are more possibilities to cause damages. The cuk converters are mostly neglected.

4.10 Chopper Commutation

Chopper commutation techniques are classified based on three methods, namely, voltage commutation, current commutation, and load commutation.

4.10.1 Voltage-commutated chopper

The voltage-commutated chopper is utilized as part of high-power circuits where load uncertainties are not extensive. It is otherwise called a parallel-capacitor turn-off chopper or drive-commutated chopper or traditional chopper. In this chopper, the leading thyristor is commutated by using a pulse of huge converse voltage. This converse voltage is normally connected by exchanging a pre-charged capacitor. It decreases the anode current to zero quickly. At that point, the voltage over the thyristor supports in the culmination of its turn-off procedure. The Simulink model is shown in Figure 4.31.

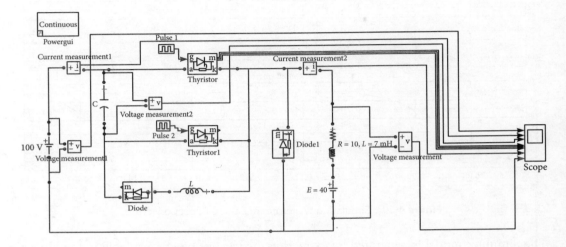

Figure 4.31 MATLAB/Simulink Model of a Voltage-Commutated Chopper.

- Minimum turn on time of chopper $= \pi\sqrt{LC}$ s

  Minimum duty cycle of voltage-commutated chopper $\alpha_{min} = \pi f \sqrt{LC}$

  The output current $I_o = \dfrac{CV_s - (-V_s)}{2t_c}$

where $c = \dfrac{I_o t_c}{V_s}, L \geq \left(\dfrac{V_s}{I_o}\right)^2 c$

The current through the circuit is $i_c = \dfrac{V_s}{\omega_o L}\sin\omega_o t$

Peak capacitor current $I_{cp} = \dfrac{V_s}{\omega_o L} = V_s\sqrt{\dfrac{C}{L}}$

The minimum time required for change in the polarity of capacitor from V_s to $-V_s$

$t_1 = \dfrac{\pi}{\omega}; t_1 = \pi\sqrt{LC}$

In a voltage-commutated chopper, average value of output voltage is given by

$$V_o = \frac{V_s}{T}(T_{on} + 2t_c)$$

$$= \frac{V_s}{T}\left(2\frac{cV_s}{I_o}\right)$$

The working of this chopper begins by charging the capacitor c. The capacitor is assumed to have $+V_c$ voltage. At $t = 0$, fundamental thyristor T1 is shut. So current moves through V_s–T1–load–$(-V_s)$. To begin with the capacitor, the current climbs from 0 to its maximum. As I_c reduces to zero, capacitor is charged to voltage $-V_s$ at $t = t_1$. When T1 is to be turned off and assistant thyristor TA is activated, the capacitor voltage $-V_c$ shows up crosswise over and is turned off because of opposite tendency condition. Presently, the load current goes through V_s, C, TA and load. The voltage V_c changes from $-V_c$ to $+V_c$. At t_2, thyristor TA is turned off. At $t = t_3$, FWD is turned on and the cycle repeats.

Example 9

The chopper can start only if the capacitor is charged to $+V_s$. So in the simulation, an initial voltage of about 100 V is set. Based on the design considerations, L and C values are designed. The value of C can be calculated based on the turn off time of the main thyristor T1. The value of L can be calculated based on the turn off time of the auxiliary thyristor. Here, the auxiliary thyristor is turned on when the main thyristor has to be turned off.

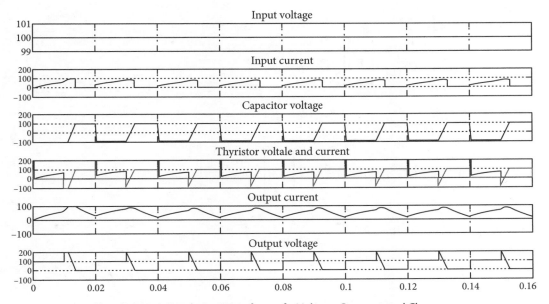

Figure 4.32 Simulation Waveform of a Voltage-Commutated Chopper.

The simulation waveform of the voltage-commutated chopper is shown in Figure 4.32. The Simulink model is generated by extracting the various blocks available in the library browser. After compilation, the input and the output responses of the prescribed circuit can be validated when the MATLAB model is run. The time taken for simulation of the results is set at 0.16 s, the load value is set for $R = 10\ \Omega$, $L = 7$ mH with $E = 40$ V and the supply voltage is $V_{supply} = 220$ V.

4.10.2 Current-commutated chopper

In case of a current-commutated chopper, T2 is turned on, i.e., T2 is switched on at the gate and turned on capacitor c is charged to $-V_{DC}$ as T2 is ON, the inductor L and capacitor C arrangement circuit is shorted; thus, the capacitor is charged to negative voltage $-V_{DC}$. The MATLAB/Simulink model of a current-commutated chopper is shown in Figure 4.33.

The values of L and C are given by

$$L = \frac{V_s t_c}{X I_o \left[\pi - 2\sin^{-1}\left(\frac{1}{x}\right)\right]}$$

$$C = \frac{X I_o t_c}{V_s \left[\pi - 2\sin^{-1}\left(\frac{1}{x}\right)\right]}$$

where $X = \dfrac{I_{Cp}}{I_o}$

$$t_c = \left[\pi - 2\sin^{-1}\left(\frac{1}{x}\right)\right]\sqrt{LC}$$

$$t_c = I_{Cp}\sin\omega t$$

At $\omega t = \theta_1, i_c = I_o = I_{Cp}\sin\theta_1, \theta_1 = \sin^{-1}\left[\dfrac{I_o}{I_{Cp}}\right] = \sin^{-1}\left[\dfrac{1}{x}\right]$

Peak capacitor voltage $= V_s + I_o\sqrt{\dfrac{L}{C}}$

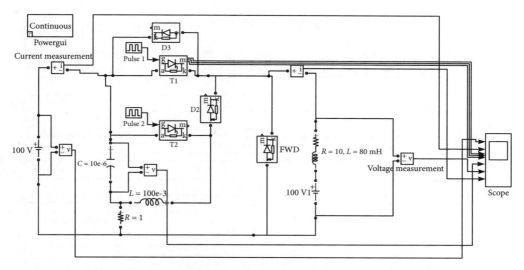

Figure 4.33 MATLAB/Simulink Model of a Current-Commutated Chopper.

Example 10

The capacitor c is completely charged, and current switches, thus T2 is turned off, when the capacitor is charged. It gives $-V_{DC}$ voltage opposite inclination to thyristor T2, so T2 is turned off. Hence, we can say that T2 was turned on to charge the thyristor to negative voltage. Presently, the current streams reverse through thyristor T1 and the capacitor, charged to negative V_{DC}, pushes current I_c inverse to T1. The capacitor current does not course through diode, as the forward voltage of the diode is more than that of the drop crosswise over thyristor T1. Significantly, the current can switch the thyristor because the thyristor gives lesser impedance compared to the diode. T1 is turned off by I_C. So now the impedance offered by T1 is more than a diode, and the current I_c moves through diode. The current I_c gradually diminishes and when $I_c = I_o$, the diode D1 is no more forward one-sided. As the diode is turned off, the capacitor C is charged to $+V_{DC}$.

The simulation waveform of a current-commutated chopper is shown in Figure 4.34. The Simulink model is generated by extracting the various blocks available in the library browser. After compilation, the input and the output responses of the prescribed circuit can be validated when the MATLAB model is run. The time taken for simulation of the results is set at 0.16 s, the load value is set for $R = 10\ \Omega$, $L = 80$ mH with $E = 100$ V, and the supply voltage is $V_{supply} = 100$ V.

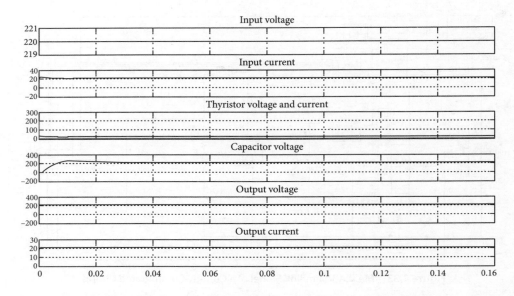

Figure 4.34 Simulation Waveform of a Current-Commutated Chopper.

4.10.3 Load-commutated chopper

In load commutation, the circuit consists of four thyristors and a commutating capacitor C. Thyristors T1 and T2 form one pair and T3 and T4 form another pair, which conduct the load alternately. When one pair acts as the main thyristor, the other pair acts as the auxiliary thyristor. Initially, the capacitor is charged to reverse voltage due to the conduction of thyristors T3 and T4.

Example 11

The capacitor is charged with $+V_{DC}$ voltage by triggering chopper T2. Hence, current flows through $+V_s$–C–T2–load–$(-V_s)$. After C is charged, T2 is turned off as current from C decreases. Now T1 is triggered. So C discharges through T1. After C is completely discharged, C is charged to $-V_{DC}$. A conducting thyristor is turned off when the load current flowing through the thyristor either becomes zero due to the nature of load circuit parameters or is transferred to another device from the conducting thyristor. Now when T2 is triggered again, voltage across T1 becomes $-V_{DC}$. So T1 is turned off. Now C starts discharging through T2. So total voltage across the load is $2V_{DC}$. But C slowly discharges to 0. Now V_s charges C to $+V_{DC}$. After C is charged, the charging current becomes zero and so T2 is turned off and the cycle repeats. The simulation model is shown in Figure 4.35.

The Simulink model is generated by extracting the various blocks available in the library browser. After compilation, the input and the output responses of the prescribed circuit can be validated when the MATLAB model is run. The time taken for simulation of the results is set at 0.16 s, the load value is set for $R = 1\ \Omega$, $L = 1$ mH with $E = 8$ V, and the supply voltage is $V_{supply} = 220$ V. Refer to Figure 4.36 for simulation waveform.

DC Choppers Using MATLAB (DC–DC Converters) 237

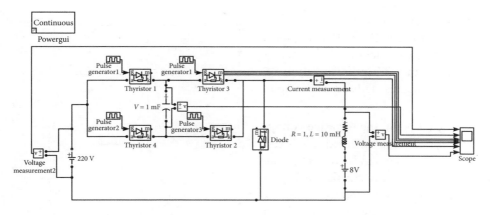

Figure 4.35 Simulation Model of a Load-Commutated Chopper.

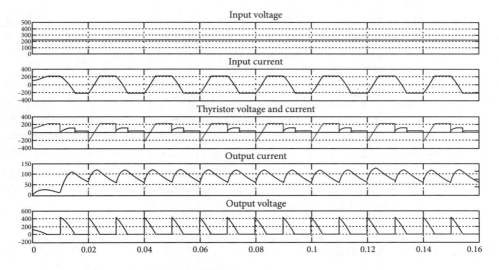

Figure 4.36 Simulation Waveform of a Load-Commutated Chopper.

4.11 Jones Chopper

Jones chopper is an example of a class D chopper in which a charged capacitor is exchanged by an auxiliary silicon controlled rectifier (SCR) to commutate the principle SCR. In this circuit, SCR1 is the principle switch and SCR2 is the auxiliary switch that is of lower limit than SCR1 and is utilized to commutate SCR1 by a converse voltage created over the capacitor C. The additional component of the circuit is the tapped autotransformer T through a bit of which the load current streams. Refer to Figure 4.37 for a Jones circuit.

When SCR1 is in the on position, capacitor C releases through SCR1, L1 and D1. This release current does not move through L2 and back to the battery due to the transformer activity of T. The load current is grabbed by SCR1 and the FWD i,e D1 is opposite one-sided. As the capacitor voltage swings negative, the converse predisposition on diode D2 diminishes. This proceeds up to a period.

When SCR2 is on, the negative voltage on capacitor C is connected crosswise over SCR1 and it gets to be off. The load current that is typically consistent begins to stream in SCR2 and capacitor C. The capacitor C is charged emphatically at first up to a voltage equivalent to supply voltage V_{DC}. The FWD gets to be forward conducting and starts to pickup load current. What is more, capacitor current begins to diminish. After this, inductance L2 is constrained into the capacitor C.

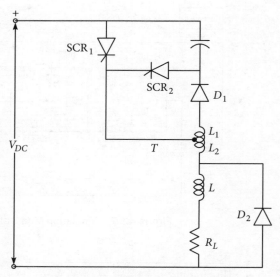

Figure 4.37 Jones Chopper Circuit.

The capacitor starts discharging; the capacitor current keeps diminishing as the current through SCR2 diminishes and the thyristor is switched off. The cycle repeats when SCR1 is again turned on. The principle point of interest of a Jones chopper over the other circuits is that it permits the utilization of higher voltage and lower microfarad commutating capacitor. This is because the energy obtained from inductor L2 can be constrained into the commutating capacitor instead of just charging the capacitor by supply voltage. In this circuit, there is no starting problems and any part of the SCR can be turned on initially as there is incredible adaptability in control.

4.12 Morgan Chopper

Figure 4.38 shows the circuit outline of a Morgan's chopper. In this circuit, thyristor is said to be the principle thyristor, where capacitor C, saturable reactor SR and diode D together structure the compensation circuit. SR when energized is thought to be negligible, henceforth can be dismissed. During operation the inductance is thought to be low. When the thyristor is set to the off condition, the capacitor C will be charged to the supply voltage E_{DC}, and the reactor is set in the positive saturable condition. At time $t = \alpha$, when thyristor is turned on, the capacitor voltage shows up over the reactor and the flux is driven by the positive immersion towards the negative side. The capacitor voltage remains basically consistent with the same.

When the centre flux achieves negative immersion, the capacitor releases through the thyristor and the post part of the thyristor, consequently shaping a reverberation circuit. Here, the releasing time of the capacitor is similarly short and the inversion of the maximum of the limit takes place rapidly. After this, the capacitor voltage is negative, which urges the reactor in the opposite course and the centre is driven from negative immersion towards positive immersion.

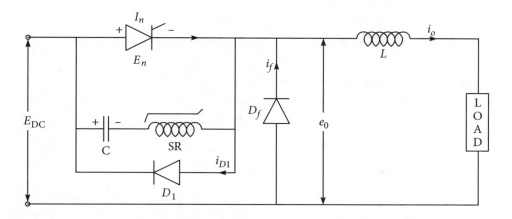

Figure 4.38 Morgan Chopper.

After a fixed interim of time, the flux achieves the positive immersion after which the capacitor releases rapidly through the thyristor in the converse direction. The discharging current first goes through the thyristor when it is turned off and then through diode D. Thus, when the thyristor is turned off, the load current courses through the Df. During the ON period the normal output voltage can be modified by shifting the working frequency. The on period can also be controlled by fluctuating the voltage–time characteristics of the reactor by method for the current through it. It can be controlled when we need to move the reactor from the positive side to the negative side. Consequently, the set up of a straight reactor is favourable when the turn off time is low and high during on time. The circuit expense is low, which is an extra benefit.

4.13 AC Choppers

The AC voltage magnitude can be changed by two methods, namely, step-up and step-down using transformers. We can also change the magnitude of AC voltage by means of a solid-state switch. In this method, the AC input voltage is switched on and off periodically by means of a suitable switch, which is known as an AC chopper.

4.14 Source Filter

The major drawback of DC choppers is that the supply current pulsates due to chopper operation, which results in harmonics leading to undesirable effects such as distortion, heating, interference and fluctuation. To rectify such problems, a filter can be used in chopper circuits. To obtain a perfect DC current, the capacitor should be large. To reduce harmonics, an LC filter can be used instead of a large capacitor.

The inductive reactance is given by $X_L = 2\pi FL$

The inductive capacitance is given by $X_C = 1/2\pi FC$

The nth harmonic current is given by $I_n = X_C I_{ch}/(n^2 X_L - X_C)$

where

F is the chopper frequency

I_n is the rms nth harmonic of the supply current

I_{ch} is the rms nth harmonic of the chopper

4.15 Multiphase Chopper

Multiphase DC–DC converters are used for high-voltage high-control applications. A summed up converter is arranged such that the cells and voltage doublers are associated in parallel or in arrangement to increase the output voltage and/or the output power. Besides, non-resisting diminished device voltage and current evaluations by association, the multiphase converter has the accompanying advantages: high stride-up voltage pick up with altogether decreased transformer turn proportion, low-level current swell because of interleaving impact, zero-voltage exchanging turn on of switches and zero-current exchanging turn off of diodes, no extra clipping and start-up circuits required, high-segment accessibility and simple heat circulation because of the utilization of numerous little segments, and adaptability in device determination bringing about improved configuration.

4.16 Applications of Choppers

Choppers has become widely and some of the customized applications are listed here:

- SMPS (that include DC–DC converters)
- DC motor speed controllers and variable frequency drives
- Battery-operated charger, electric cars and appliances
- DC motor speed controllers and voltage boosters
- Electronic amplifiers of class D
- Switched capacitor filters

DC Choppers Using MATLAB (DC–DC Converters)

> ## Summary
> At the end of this chapter, the users will be able to identify the need and the various functionalities of chopper circuits. The different modes and categories of chopper are enumerated. Various real-time applications that occur in today's world are listed. The commutation circuits and the regulation modes add an additional value to real-time applications.

Solved Problems

1. In a chopper circuit which is operating at TRC at a frequency of 2 kHz with 460 V supply. Calculate the conduction period of the thyristor in each cycle when the load voltage is 460 V.

    Solution

    $$V = 460 \text{ V}, \quad V_{DC} = 350 \text{ V}, \quad f = 2 \text{ kHz}$$

    Chopping period $T = \dfrac{1}{f}$

    $$T = \dfrac{1}{2 \times 10^{-3}} = 0.5 \text{ ms}$$

    Output voltage $V_{DC} = \left(\dfrac{T_{ON}}{T}\right) V$

    Conduction period of thyristor

    $$T_{ON} = \dfrac{T \times V_{DC}}{V}$$

    $$T_{ON} = \dfrac{0.5 \times 10^{-3} \times 350}{460}$$

    $$T_{ON} = 0.38 \text{ ms}$$

2. A step-up chopper operates at an input supply of 200 V and output voltage of 600 V. Considering the conducting time to be 200 µs, calculate the chopping frequency, new output voltage when the pulse width is halved for constant frequency.

    Solution

    $$V = 200 \text{ V}, \quad t_{ON} = 200 \text{ µs}, \quad V_{dc} = 600 \text{ v}$$

    $$V_{DC} = V\left(\dfrac{T}{T - t_{ON}}\right)$$

    $$600 = 200\left(\dfrac{T}{T - 200 \times 10^{-6}}\right)$$

Solving for T,

$$T = 300 \ \mu s$$

Chopping frequency

$$f = \frac{1}{T}$$

$$f = \frac{1}{300 \times 10^{-6}} = 3.33 \ \text{kHz}$$

Pulse width is halved

$$\therefore \quad t_{ON} = \frac{200 \times 10^{-6}}{2} = 100 \ \mu s$$

Frequency is constant

$$\therefore \quad f = 3.33 \ \text{kHz}$$

$$T = \frac{1}{f} = 300 \ \mu s$$

Output voltage $= V\left(\dfrac{T}{T - t_{ON}}\right)$

$$= 200 \left(\frac{300 \times 10^{-6}}{300 - 100 \times 10^{-6}}\right) = 300 \ V$$

3. A chopper is operated with R load of 20 Ω and input voltage of 220 V. During the on state of the chopper, the voltage drop is 1.5 V and the frequency is 10 kHz. Determine the average output voltage and the chopper on time, when the duty cycle is 80 per cent.

Solution

$$V_s = 220 \ V, \quad R = 20 \ \Omega, \quad f = 10 \ \text{kHz}$$

$$d = \frac{t_{ON}}{T} = 0.80$$

V_{ch} = Voltage drop across chopper = 1.5 V

Average output voltage

$$V_{DC} = \left(\frac{t_{ON}}{T}\right) V_s - V_{ch}$$

$$V_{DC} = 0.80(220 - 1.5) = 174.8 \ V$$

Chopper on time $t_{ON} = dT$

Chopping period $T = \dfrac{1}{f}$

$$T = \dfrac{1}{10 \times 10^3} = 0.1 \times 10^{-3}\,s = 100\,\mu s$$

Chopper on time

$$t_{ON} = dT$$

$$t_{ON} = 0.80 \times 0.1 \times 10^{-3}$$

$$t_{ON} = 0.08 \times 10^{-3} = 80\,\mu s$$

4. With a chopping frequency of 250 Hz and average load current 30 A, calculate the on and off periods of the chopper when the load resistance is 2 Ω and supply voltage = 110 V.

   Solution

   $I_{dc} = 30\,A,\ f = 250\,Hz,\ V = 110\,V,\ R = 2\,\Omega$

   Chopping period $T = \dfrac{1}{f} = \dfrac{1}{250} = 4 \times 10^{-3} = 4\,ms$

   $$I_{DC} = \dfrac{V_{DC}}{R} \quad \text{and} \quad V_{DC} = dV$$

   $$\therefore\ I_{DC} = \dfrac{dV}{R}$$

   $$d = \dfrac{I_{DC} R}{V} = \dfrac{30 \times 2}{110} = 0.545$$

   Chopper on period

   $$t_{ON} = dT = 0.545 \times 4 \times 10^{-3} = 2.18\,ms$$

   Chopper off period

   $$t_{OFF} = T - t_{ON}$$

   $$t_{OFF} = (4 \times 10^{-3}) - (2.18 \times 10^{-3})$$

   $$t_{OFF} = 1.82 \times 10^{-3} = 1.82\,ms$$

5. For a chopper with resistive load of 10 Ω, the input voltage is 200 V. The voltage drop is 2 V during the on time of the chopper and the frequency of the chopper is 1 kHz. Calculate for duty cycle 60 per cent, the average output voltage, rms value of output voltage, effective input resistance of chopper and chopper frequency.

Solution

$V = 200$ V, $R = 10\ \Omega$, Chopper voltage drop $V_{ch} = 2$ V

$d = 0.60, f = 1$ kHz

Average output voltage

$$V_{DC} = d(V - V_{ch})$$

$$V_{DC} = 0.60(200 - 2) = 118.8\ \text{V}$$

RMS value of output voltage

$$V_0 = \sqrt{d}\,(V - V_{ch})$$

$$V_0 = \sqrt{0.6}\,(200 - 2) = 153.37\ \text{V}$$

Effective input resistance of chopper is

$$R_i = \frac{V}{I_s} = \frac{V}{I_{DC}}$$

$$I_{DC} = \frac{V_{DC}}{R} = \frac{118.8}{10} = 11.88\ \text{A}$$

$$R_i = \frac{V}{I_s} = \frac{V}{I_{DC}} = \frac{200}{11.88} = 16.83\ \Omega$$

Output power is

$$P_0 = \frac{1}{T}\int_0^{dT} \frac{v_0^2}{R}\,dt = \frac{1}{T}\int_0^{dT} \frac{(V - V_{ch})^2}{R}\,dt$$

$$P_0 = \frac{d(V - V_{ch})^2}{R}$$

$$P_0 = \frac{0.6(200 - 2)^2}{10} = 2352.24\ \text{W}$$

Input power

$$P_i = \frac{1}{T}\int_0^{dT} V i_0\,dt$$

$$P_i = \frac{1}{T}\int_0^{dT} \frac{V(V - V_{ch})}{R}\,dt$$

$$P_i = \frac{dV(V - V_{ch})}{R}$$

$$P_i = \frac{0.6 \times 200(200 - 2)}{10} = 2376 \text{ W}$$

Chopper efficiency

$$\eta = \frac{P_o}{P_i} \times 100$$

$$\eta = \frac{2352.24}{2376} \times 100 = 99 \text{ per cent}$$

6. A chopper controlled electric train is powered from a 1500 V DC supply. The power semiconductor switching element has a minimum effective on time of 40 μs. During starting and slow speed running, the output of the chopper has to go as low as 15 V. What is the highest chopper frequency to satisfy this requirement?

Solution

The minimum duty cycle needed is

$D = 15/1500 = 0.01$

The minimum possible on time $T_{on} = 40$ μs

$D T_s = T_{on}$

$T_s = T_{on}/D = 40/0.01 = 4000$ μs

Therefore the maximum possible chopper frequency is $f = (1/T_s) = 1/(4000 \text{ μs}) = 250$ Hz.

7. A chopper has a DC source voltage $V_s = 230$ V, load resistance = 10 Ω. Take a voltage drop of 2 V across the chopper when it is on. For a duty cycle of 0.4, calculate (a) average and rms values of output voltage and (b) chopper efficiency.

Solution

$V_s = 230$ V, $R = 10$ Ω, $V_{ch} = 2$ V, $D = 0.4$

(a) When chopper is on, the output voltage is $(V_s - 2)$ volts and during the time chopper is off, output voltage is zero.

Average value of output voltage,

$V_o = D (V_s - 2) = 0.4 (230 - 2) = 91.2$ V

RMS value of output voltage,

$$V_{rms} = \sqrt{D}(V_s - V_{ch}) = \sqrt{0.4}(230 - 2) = 144.2 \text{ V}$$

(b) Power delivered to the load,

$$P_o = \frac{V_{rms}^2}{R} = \frac{144.2^2}{10} = 2079.36 \text{ W}$$

Power input to chopper

$$P_i = V_s I_o = 230 \times \frac{91.2}{10} = 2097.6 \text{ W}$$

Chopper efficiency,

$$\text{Efficiency} = \frac{P_o}{P_i} = \frac{2079.36}{2097.6} = 99.13 \text{ per cent}$$

8. The DC–DC converter shown in Figure S1 has the following parameters:

$V_s = 50$ V

$D = 0.4$

$L = 400$ μH

$C = 100$ μF

$f = 20$ kHz

$R = 20$ Ω

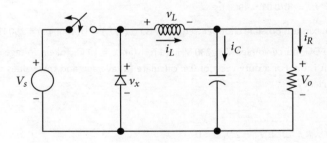

Figure S1 DC-DC converter

Solution

The inductor current is assumed to be continuous, and hence, the output voltage is computed as

$V_o = V_s D = (50)(0.4) = 20$ V

The maximum inductor current

$$I_{max} = V_o \left(\frac{1}{R} + \frac{1-D}{2Lf} \right)$$

$$= 20\left[\frac{1}{20} + \frac{1-0.4}{2(400)(10^{-6})(20)(10^3)}\right]$$

$$= 1.5 \text{ A}$$

The minimum inductor current

$$I_{min} = V_o\left(\frac{1}{R} - \frac{1-D}{2Lf}\right)$$

$$= 20\left[\frac{1}{20} - \frac{1-0.4}{2(400)(10^{-6})(20)(10^3)}\right]$$

$$= 0.25 \text{ A}$$

Note that the minimum inductor current is positive. Hence, the assumption of continuous conduction is valid.

The output voltage ripple is given by

$$\frac{\Delta V_o}{V_o} = (1-D)/8LCf^2$$

$$= 0.469 \text{ per cent}$$

9. Design a buck converter to produce an output voltage of 18 V across a 10 Ω load resistor. The output voltage ripple must not exceed 0.5 per cent. The DC supply is 48 V. Design for continuous inductor current. Specify the duty ratio, the switching frequency, the values of the inductor, and capacitor. Assume ideal components.

Solution

The duty ratio for continuous conduction current is given by

$$D = V_o / V_s$$

$$= 18/48 = 0.375$$

Fix the switching frequency arbitrarily to 40 kHz,

$$L_{min} = (1-D)(R)/2f = (1-0.375)(10)/2(40000)$$

$$= 78 \text{ μH}$$

Let the inductor be 25 per cent larger than L_{min} to ensure continuous current operation.

$$L = 1.25\, L_{min} = 1.25\ (78\ \mu H) = 97.5\ \mu H$$

The capacitor is selected as

$$C = \frac{1-D}{8L\left(\dfrac{\Delta V_o}{V_o}\right)f^2}$$

$$= \frac{1 - 0.375}{8(97.5)(10^{-6})(0.005)(40000)^2}$$

$$= 100 \ \mu F$$

10. Power supplies for telecommunication applications may require high currents at low voltages. Design a buck converter that has an input voltage of 3.3 V and an output voltage of 1.2 V. The output current varies between 4 and 6 A. The output voltage ripple must not exceed 2 per cent. Specify the inductor value such that the peak-to-peak variation in inductor current does not exceed 40 per cent of the average value.

Solution

The duty ratio is determined by

$$D = \frac{V_o}{V_s} = \frac{1.2}{3.3} = 0.364$$

The average inductor current is the same as the output current. Analyzing the circuit for an output current of 4 A,

$$I_L = I_o = 4 \ \text{A}$$

$$L = \frac{V_s - V_o}{\Delta i_L \ f} = 0.955 \ \mu H$$

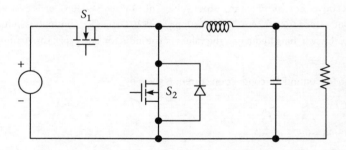

Figure S2 A Synchronous Buck Converter.

Analyzing the circuit for an output current of 6 A,

$$I_L = I_o = 6 \ \text{A}$$

$$\Delta i_L = (40\%)6 = 2.4 \ \text{A}$$

$$L = \frac{V_s - V_o}{\Delta I_L \ f} = 0.636 \ \mu H$$

Since 0.636 µH would be too small for the 4 A output, use L = 0.955 µH, which would be rounded to 1 µH. Using L = 1 µH, the minimum capacitance is determined as

$$C = \frac{1-D}{8L\left(\frac{\Delta V_o}{V_o}\right)f^2} = 0.16\,\mu F$$

11. The chopper shown in Figure S3 is switching at a frequency of $f = 1$ kHz, with a duty cycle of 50 per cent. Determine

    (a) The DC load current I_d
    (b) The peak-to-peak ripple current

    Solution

    The DC component of the output voltage $V_o = D V_s = 50$ V

    So the DC component of the output current is $\quad I_d = V_o/R = 10$ A

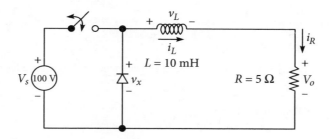

Figure S3 DC-DC converter

The chopper period is $T = 1/f = 1$ ms

$T_{on} = 0.5$ ms, $T_{off} = 0.5$ ms.

The time constant $\tau = L/R = 2$ ms.

$$I_{max} = \frac{V_s}{R}\left(\frac{1-e^{-\frac{T_{on}}{\tau}}}{1-e^{-\frac{T}{\tau}}}\right) = 11.24\text{ A}$$

$$I_{min} = \frac{V_s}{R}\left(\frac{e^{\frac{T_{off}}{\tau}}-e^{-\frac{T}{\tau}}}{1-e^{-\frac{T}{\tau}}}\right) = 8.75\text{ A}$$

So the peak-to-peak ripple current $I_{pp} = I_{max} - I_{min} = 2.49$ A

and the percentage peak-to-peak ripple current is given by $100 \times I_{pp}/I_d = 100 \times (2.49/10) = 24.9$ per cent.

12. In a battery powered car, the battery voltage is 120 V. It is driven by a DC motor and employs chopper control. The resistance of the motor circuit is 0.2 Ω. During braking, the configuration of the chopper is changed to a voltage step-up mode. While going down the hill at certain speed, the induced emf of the machine is 110 V and the braking current is 10 A. Determine the chopper duty cycle. Assume continuous current operation.

Solution

$$V_o = V_s - I_d R = 110 - 10 \times 2 = 108 \text{ V}$$

For step-up chopper

$$V_o = (1 - D) V_s$$

$$108 = (1 - D) 120$$

$$D = 0.1 \text{ or } 10 \text{ per cent.}$$

13. Design a boost converter that will have an output of 30 V from a 12-V source. Design for continuous inductor current and an output ripple voltage of less than one per cent. The load is a resistance of 50 Ω. Assume ideal components for this design.

Solution

To determine the duty cycle,

$$D = 1 - V_s/V_o = 1 - (12/30) = 0.6$$

If the switching frequency is selected at 25 kHz, then the minimum inductance for continuous current is determined from

$$L_{min} = \frac{D(1-D)^2}{2f} = \frac{0.6(1-0.6)^2 \, 50}{2(25000)} = 96 \, \mu H$$

To provide a margin to ensure continuous current, let $L = 120 \, \mu H$. Note that L and f are selected somewhat arbitrarily and that other combinations will also give continuous current.

$$I_L = \frac{V_s}{(1-D^2)R} = \frac{12}{(1-0.6^2)50} = 1.5 \text{ A}$$

$$\frac{\Delta i_L}{2} = V_s D \frac{T}{2L} = 1.2 \text{ A}$$

Hence, $I_{max} = 1.5 + 1.2 = 1.7$ A

$$I_{min} = 1.5 - 1.2 = 0.3 \text{ A}$$

The minimum capacitance required to limit the output ripple voltage to 1 per cent is determined from

$$C \geq \frac{D}{R\left(\frac{\Delta V_o}{V_o}\right)f} = 48 \, \mu F$$

14. A step-up chopper is used to deliver load voltage of 500 V from a 200 V DC source. If the blocking period of the thyristor is 80 μs, compute the required pulse width.

Solution

$$V_o = 500 \text{ V}$$

$$V_s = 220 \text{ V}$$

$T_{off} = 80\ \mu s$

$T_{on} = ?$

$V_o = V_s(T_{on} + T_{off})/T_{off}$

$500 = 220(T_{on} + 80\ \mu s)/80\ \mu s$

$T_{on} = 101.6\ \mu s$

15. What is the effective input resistance of a chopper, whose load resistance is 25 Ω, duty cycle is 30 per cent and chopping frequency is 1 kHz?

Solution

$R = 25\ \Omega$

$D = 0.3$

$f = 1\ kHz$

Effective input resistance R_{eff}

$R_{eff} = R/D = 25/0.3 = 83.33\ \Omega$

Objective Type Questions

1. What is the value of per unit ripple to be maximum for a DC chopper, when the duty cycle is
   - (a) 0.2
   - (b) 0.5
   - (c) 0.7
   - (d) 0.8

2. For a step-up chopper, the output value is given by
   - (a) $V_s(1 + a)$
   - (b) $V_s/(1 - a)$
   - (c) $V_s(1 - a)$
   - (d) $V_s/(1 + a)$

3. The average output value of type A chopper is given by
   - (a) $V_s a$
   - (b) $(1 - a)V_s$
   - (c) V_s/a
   - (d) $V_s/(1 - a)$

4. What is the duty cycle value when the chopper frequency is 200 Hz and the on time is 2 ms?
   - (a) 0.4
   - (b) 0.8
   - (c) 0.6
   - (d) None of these

5. In case of DC motors, the chopper control provides variation in
   - (a) Input voltage
   - (b) Frequency
   - (c) Both (a) and (b)
   - (d) None of these

6. Which is the commutation technique that gives the best performance in chopper?
   - (a) Voltage commutation
   - (b) Current commutation
   - (c) Load commutation
   - (d) Supply commutation

7. What could be the efficient percentage of operation of choppers?
   (a) 50–55 per cent
   (b) 65–72 per cent
   (c) 82–87 per cent
   (d) 92–99 per cent

8. DC chopper waveforms are generally
   (a) Continuous input and discontinuous output
   (b) Discontinuous input and discontinuous output
   (c) Continuous input and continuous output
   (d) Discontinuous input and discontinuous output

9. While controlling the average output voltage of a chopper, the chopping frequency in the PWM (pulse width modulation) method becomes
   (a) Varied
   (b) constant
   (c) Either of these
   (d) None of these

10. What will be the value of the commutating capacitor and the turn off time of one thyristor with chopping frequency of 2 kHz and duty cycle of 0.4 with load current of 50 A fed from a 200 V DC source?
    (a) 25 mF, 50 ms
    (b) 50 mF, 50 ms
    (c) 25 mF, 25 ms
    (d) 50 mF, 25 ms

11. Determine the range of the duty cycle where a DC battery is charged from 200 V source with an internal emf of 90–120 V. The internal resistance is about 1 W.
    (a) 15–65
    (b) 65–8
    (c) 8–95
    (d) None of these

12. For a chopper with 100 V input, calculate the average output voltage and the ripple factor when the load voltage consists of rectangular pulses of 1 ms and overall cycle of 3 ms.
    (a) 25 V
    (b) 50 V
    (c) 33.33 V
    (d) None of these

13. For a step-down chopper, the ratio of V_o/V_s is given by
    (a) D
    (b) $1 - D$
    (c) $1/1 - D$
    (d) $D/1 - D$

14. The output ripple in case of polyphase choppers
    (a) Decreases
    (b) Increases
    (c) Remains same
    (d) Has low frequency

15. The basic characteristics of chopper drives are
    (a) Slow response with smooth control
    (b) Fast response with smooth control
    (c) Less efficient with fast response
    (d) None of these

16. In automobiles, the chopper can be used for
    (a) Speed control
    (b) Braking
    (c) Both (a) and (b)
    (d) None of these

17. The load voltage of DC choppers are directed by
    (a) Thyristors of the circuit
    (b) Load driven by the circuit
    (c) Voltage of the circuit
    (d) None of these
18. By keeping the average load current constant, if it is desired to reduce the ripple content of load current, the action needed to be taken is to
    (a) Keep duty cycle constant and increase the chopping frequency
    (b) Keep duty ratio and chopping frequency equal
    (c) Decrease chopping frequency
    (d) Decrease the duty cycle
19. A voltage-commutated chopper has the following parameters: V_s = 200 V, load circuit parameter 1 W, 2 mH, 5 V commutation circuit parameter: L = 25 mH, C = 50 mF. For constant load current at 100 A, the effective on period and peak current through the main thyristor are respectively
    (a) 1000 ms, 200 A
    (b) 700 ms, 382.8 A
    (c) 700 s, 282.8 A
    (d) 1000 ms, 382.8 A
20. What will be the average output voltage and output current for type A chopper with source voltage 100 V, an on period = 100 ms, an off period = 150 ms and R = 2 W, L = 5 mH and E = 10 V?
    (a) 40 V, 15 A
    (b) 66 V, 28.3 A
    (c) 60 V, 25 A
    (d) 40 V, 20 A
21. Referring to the circuit shown in Figure 4.35, the maximum current value in the SCR will be
    (a) 200 A
    (b) 170.7 A
    (c) 141.4 A
    (d) 707.7

Review Questions

1. What are DC choppers?
2. Mention some of the applications of a DC chopper.
3. Mention some of the merits and demerits of a DC chopper.
4. Differentiate step-up and step-down chopper.
5. Write down the expressions for average output voltage of a step-up and step-down chopper.
6. Define the term duty cycle.
7. Classify the types of control strategies used in chopper circuits.
8. Define the term TRC.
9. Classify the types of TRC.
10. What is meant by frequency modulation control?
11. Explain the concept of PWM control used in DC chopper.
12. What is meant by FM control in a DC chopper?
13. What is meant by PWM control in a DC chopper?
14. What are the different types of choppers with respect to commutation process?

15. What is meant by voltage commutation?
16. What is meant by current commutation?
17. What is meant by load commutation?
18. What are the advantages of current-commutated chopper?
19. What are the advantages of load-commutated chopper?
20. What are the disadvantages of load-commutated chopper?
21. Discuss the main classification of DC–DC converters.
22. Explain with appropriate waveforms, the different control strategies used for obtaining variable output voltage from a DC chopper.
23. State some applications of DC chopper.
24. Brief about the operation of a step-down DC–DC converter with constant output voltages.
25. Explain the power circuit diagram of a buck converter in detail.
26. With necessary diagrams and waveforms, explain the operation of a buck–boost converter.
27. Mention the merits and demerits of a boost converter.

Practice Questions

1. A step-up chopper with a pulse width of 150 ms is operating on 220 V DC supply. Compute the load voltage if the blocking period of the device is 20 ms.
2. A boost regulator has an input voltage of $V = 5$ V. The switching frequency is 50 Hz. If $L = 150$ mH and $C = 220$ μF, calculate the output voltage and current for a run time of 20 ms.
3. A buck–boost regulator has an input voltage of $V = 12$ V. The duty cycle is 0.25 and the switching frequency is 50 Hz. The inductance $L = 150$ μH and filter capacitance is $C = 220$ μF. Determine the average output voltage, peak-to-peak ripple voltage and current.
4. A buck converter has an input voltage of $V = 5$ V. The switching frequency is 50 Hz. If $L = 10$ mH and $C = 20$ μF, calculate the output voltage and current for a run time of 0.16 ms.
5. Consider the following parameters:
   $R = 10$ Ω, $L = 10$ mH, $C = 2$ μF, $V_s = 100$ V, time = 20 ms. Discuss type A, type B, type C, type D and type E chopper using MATLAB/Simulink.
6. For a boost regulator with input voltage of 6 V, $C = 430$ μF, $L = 250$ mH and switching frequency of 20 kHz, calculate the average output voltage and average current for $\alpha = 0.6$.
7. A DC chopper is connected to an inductive load with a resistance of 10 Ω. For a 20 ms time period, the supply voltage is 300 V. Estimate the average load voltage and current.
8. A step-up chopper is used to deliver the load voltage of 500 V from a 220 V DC source. If the blocking period of the switch is 80 ms, calculate the pulse width required. Compare the results using MATLAB/Simulink.
9. A step-down chopper with a pulse width of 150 ms is operated on 230 V DC supply. Calculate the load voltage and load current when the blocking period of the device is 40 ms. Compare the results using MATLAB/Simulink.

10. A boost regulator has an input voltage of 5 V. The switching frequency is 25 kHz. If $L = 15$ mH and $C = 22\,\mu F$, determine the duty cycle, ripple current and ripple voltage. Compare the same with the results using MATLAB/Simulink.

11. Explain the concept of the buck converter with neat diagrams and waveforms. Also derive the expression for peak-to-peak ripple voltage across the capacitor load voltage.

12. Describe the working principle of a boost converter with necessary circuit and waveforms.

    (i) Explain the various control strategies of a chopper.

13. For a buck converter with input voltage of 12 V and output voltage of 5 V, design a filter component with switching frequency of 25 kHz and peak-to-peak ripple voltage of 20 mV and ripple current of 0.8 A.

14. A DC–DC chopper is operated at 200 V and load of 15 W with a voltage drop of 1.5 V with chopping frequency of 10 kHz. For a duty cycle of 80 per cent, determine the rms and average output voltage, and the on time of the chopper.

15. Describe the working principle of a buck–boost converter with necessary circuit and waveforms.

16. Explain zero voltage switching resonant converters.

5

Inverters Using MATLAB (DC–DC Converters)

> **Learning Objectives**
>
> - ✓ To understand inverters and their types
> - ✓ To explain the basic simulation model of voltage source inverters
> - ✓ To determine the performance parameters of inverters
> - ✓ To understand the principle of pulse width modulation
> - ✓ To explain three-phase inverters
> - ✓ To learn harmonic reduction techniques
> - ✓ To explain the basic simulation models of current source inverters
> - ✓ To understand resonant inverters

5.1 Introduction

In general, converter devices that transform DC power into AC power are called inverters at a desired output frequency and voltage, where the output voltage could be established at an alternative or variable frequency. They are low contorted sine-wave devices, where the output voltage of the inverters contain harmonics at whatever point it is non-sinusoidal. These harmonics can be lessened by using relevant control plans.

The output voltage waveform of the inverter can be in the form of a semisquare or a square. The voltage may be adequate for low- as well as medium-power applications. For high-power applications, low twisted sinusoidal wave structures are essential. The frequency response of the inverter circuit can be controlled by the proportion at which the power semiconductor switches are turned on and off by the inverter control setup.

The harmonic nature of the output voltage can be minimized or lessened altogether by using power semiconductor devices. The separating of harmonics is not achievable when the output frequency changes over a wide range; AC wave structures with low sound is vital. When AC output voltage of the inverter is given to the transformer or AC engine, this output voltage must be shifted in conjunction with frequency to maintain favourable conditions. The DC power input to the inverter may be a battery, a power device, solar-based cells or other DC sources. But in most modern applications, it is sustained by a rectifier.

5.2 Inverters and their Classification

Inverters can be classified based on the following parameters.

- input source
- output voltage
- technique for substitution
- association with other devices

5.2.1 Classification based on input source

Input-based source inverters are of two types.

- Voltage source inverters (VSIs)
- Current source inverters (CSIs)

In VSI, the input to the supplier are given by a free DC voltage source, whereas in CSI, the voltage source is initially changed into a current source and afterwards used to supply voltage to the inverter.

5.2.2 Classification based on output voltage

Inverters can be classified based on output voltage:

- Square-wave inverter
- Quasi-square-wave inverter
- Pulse width modulation (PWM) inverters

A square-wave inverter creates a square-wave AC voltage of consistent magnitude. The output voltage of this sort of inverter must be changed by controlling the input DC voltage. The square-wave AC output voltage of an inverter is sufficient for low- and medium-power applications. The PWM utilizes an exchanging plan inside the inverter to alter the state of the output voltage structure.

5.2.3 Classification based on technique for substitution

Based on the strategy for replacement, silicon controlled rectifier (SCR) inverters are chiefly arranged into two types:

- Line-commutated inverters
- Force-commutated inverters

In line-commutated inverters, the AC circuits' line voltage is accessible over the device; when the current in the SCR experiences a zero characteristic zero, the device is turned off. This procedure is known as the commutation process and inverters based on this principle are known as line-commutated inverters.

In force-commutated inverters, since the supply does not experience zero point, some outside source is needed to commutate the devices. This process is known as the forced replacement process and the inverters included in this procedure are called force-commutated inverters. These inverters are further characterized into the following types.

- Auxiliary-commutated inverters
- Complementary-commutated inverters

5.2.4 Classification based on associations with other devices

Based on the association of the thyristor and the commutating parts, inverters can be characterized into three categories.

- Arrangement inverters
- Parallel inverters
- Span inverters
  - Half-bridge inverter
  - Full-bridge inverter

5.3 Selection of Components from Simulink Library Browser

Before we discuss MATLAB simulation, the choice of components is more important. Listed are the components and their location:

New File

>>Type Simulink on Matlab Command Window>>File>>New>>Model

AC Voltage Source
>>Libraries>>Simscape>>SimPowerSystems>>Electrical Sources>>AC Voltage Source
DC Source
>>Libraries>>Simscape>>SimPowerSystems>>Electrical Sources>>DC Voltage Source
Thyristor/MOSFET
>>Libraries>>Simscape>>SimPowerSystems>>PowerElectronics>>Thyristor/MOSFET
Diode
>>Libraries>>Simscape>>SimPowerSystems>>PowerElectronics>>Diode
Series *RLC* Branch
>>Libraries>>Simscape>>SimPowerSystems>>Elements>>Series RLC Branch
Pulse Generator
>>Libraries>>Sources>>Pulse Generator
Three-Phase *V–I* Measurement
>>Libraries>>Simscape>>SimPowerSystems>>Extra Library>>Measurements>>Three-Phase V–I Measurement
Voltage Measurement
>>Libraries>>Simscape>>SimPowerSystems>>Measurements
Current Measurement
>>Libraries>>Simscape>>SimPowerSystems>>Measurements
Mean
>>Libraries>>Simscape>>SimPowerSystems>>Extra Library>>Measurement>>Mean Value
RMS
>>Libraries>>Simscape>>SimPowerSystems>>Extra Library>>Measurement>>RMS Value
Subsystem
>>Simulink>>Commonly Used Blocks>>Subsystem
THD
>>Libraries>>Simscape>>SimPowerSystems>>Extra Library>>Measurement>>Total Harmonic Distortion
From
>>Simulink>>Signal Routing>>From
Goto
>>Simulink>>Signal Routing>>Goto

Display
>>*Simulink*>>*Sinks*>>*Display*

Scope
>>*Simulink*>>*Sinks*>>*Scope*

Ground
>>*Libraries*>>*Simscape*>>*SimPowerSystems*>>*Application Libraries*>>*Elements*>>*Ground*

5.4 Voltage Source Inverters

As mentioned earlier an inverter is known as a VSI if the input DC voltage stays consistent. The input DC voltage is made accessible over the load by controlling the metal–oxide–semiconductor field-effect transistor (MOSFET). The input DC source has little or negligible impedance. This input DC voltage is changed into a square-wave AC input voltage. VSIs using MOSFET employs self-commutated base or gate-driven signals. VSIs are used to control the rate of electric engines by changing the frequency and the voltage; they comprise a rectifier, DC connection and an output converter. They are accessible in the low-voltage and medium-voltage range.

5.4.1 Single-phase voltage source inverters

The single-phase VSI holds two voltage sources of equal voltage, which are connected in series. Two switches are connected in series. The connecting point of the two voltage sources and two switches will be the terminals of the load. The load can be either an *R* load or an *RL* load. The diodes are connected antiparallel to each switch. This will help during freewheeling if *RL* load is connected. In the following sections, single-phase half-bridge and full-bridge VSIs are analyzed.

5.4.1.1 Single-phase half-bridge inverter with R load

In the single-phase half-bridge inverter, the input DC voltage is divided into two. The half-bridge inverter comprises one shaft of switches, whereas the full-bridge inverter has two such legs. Each leg of the inverter comprises two arrangements associated with electronic switches. Each of these switches comprises a MOSFET-controlled switch crosswise over which an uncontrolled diode is placed in a parallel way. These switches are equipped for leading bidirectional current yet they have to withstand extremity of voltage. The intersection point of the switches in each one leg of the inverter serves as an output point for the load. In the middle of the circuit, the single-phase load is associated between the mid-point of the input DC supply and the intersection point of the two switches. Refer to Figure 5.1 for the simulation diagram of a single-phase half-bridge inverter with *R* load.

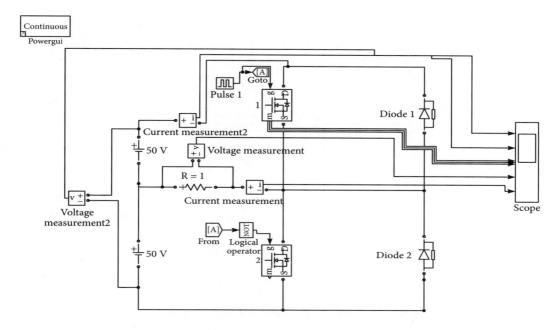

Figure 5.1 Simulation Diagram of a Single-Phase Half-Bridge Inverter with *R* Load.

MATLAB Example 1

In this simulation model, the amplitude is set for $V_{DC} = 100$ V operated at 50 Hz. The phase delay is calculated for $\alpha = 30°$ for MOSFETS and the load is considered as $R = 1\ \Omega$. After simulation, the voltage and current relations obtained from the circuit are shown here.

In a single-phase half-bridge inverter, diodes D1 and D2 are associated with changing the direction of the current for R load. The switch S1 is connected with the drive from 0 to $\pi/2$. In any case, diode D1 leads from 0 to $\pi/2$. The output current I_o diminishes from maximum negative towards zero. Thus, S1 is converse one-sided and it does not lead till D1 stops directing S1. Presently, the switch S1 conducts from $\pi/2$ to π. The output current increases from zero to I_{max}, where the current is supplied by DC supply. At $\pi/2$, S1 is turned off and base drive to S2 is connected. At the same time, S2 does not direct. To maintain the load current in the same bearing at S2, diode current turns into zero, and S2 conducts from $\pi/2$ to π. The output current is negative and it increases from zero to $-I_{max}$. The output voltage is also negative. Refer to Figure 5.2 for the simulation waveform for a single-phase half-bridge inverter with *R* load.

The root mean square (rms) output voltage is given by

$$V_{rms} = V_s/2 \tag{5.1}$$

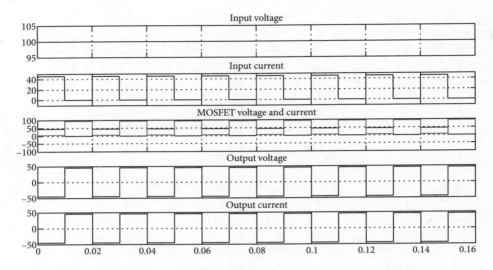

Figure 5.2 Simulation Waveform for a Single-Phase Half-Bridge Inverter with *R* Load.

5.4.1.2 Single-phase full-bridge inverter with R load

The primary disadvantage of the half-bridge inverter is that it requires a three-wire DC supply. This disadvantage can be overcome by the single-phase full-bridge inverter. This circuit requires four MOSFETs; the four diodes are gate signs connected individually to the MOSFETs S1, S2, S3 and S4. The simulation model of a single-phase full-bridge inverter with *R* load is shown in Figure 5.3.

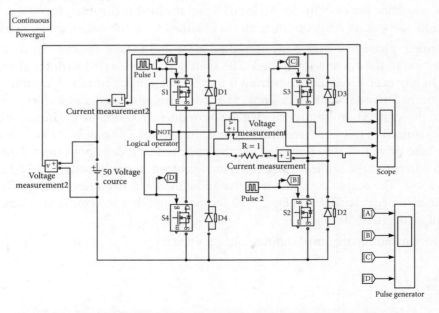

Figure 5.3 Simulation Model of a Single-Phase Full-Bridge Inverter with *R* Load.

In this simulation model, the amplitude is set for $V_{DC} = 100$ V operated at 50 Hz. The phase delay is calculated for $\alpha = 30°$ for the MOSFETs; the load is considered as $R = 1\ \Omega$. After simulation, the voltage and current relations obtained from the circuit are shown here.

MATLAB Example 2

For this device, switches S1 and S2 have to be turned on at the same time at a frequency $f = 1/T$. Switches S3 and S4 have to be turned on out of phase with S1 and S2. Frequency of the output voltage can be controlled by shifting the time period π. When S1 and S2 turns on, there is load voltage and when S3 and S4 turns on, load voltage is negative. The load voltage wave structure is rectangular and is not influenced by the load.

The rms output voltage is given by

$$V_{rms} = V_s \tag{5.2}$$

The output voltage and output current of a single-phase full-bridge inverter is multiplied when compared with a single-phase half-bridge inverter. During inverter operation, two thyristors in the same line, that is, S1 and S4, likewise S2 and S3 should not direct all the while; this will cause short out of the source. The load voltage V_0 and the load current I_0 will be in phase with one another for R load. Refer to Figure 5.4 for the simulation waveform of a single-phase full-bridge inverter with R load.

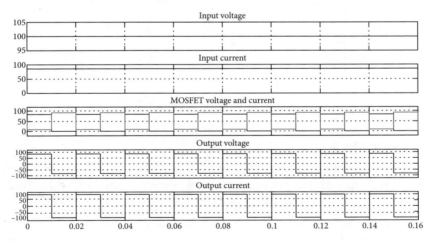

Figure 5.4 Simulation Waveform of a Single-Phase Full-Bridge Inverter with R Load.

5.4.1.3 Single-phase half-bridge inverter with RL load

Let us analyze the operation of a half-bridge inverter having resistive load and inductive load (Figure 5.5). Diodes D1 and D2 are connected across the thyristors to conduct for inductive load. The switch T1 is applied to the drive from 0 to $\pi/2$. But diode D1 conducts from 0 to π. The output current I_o decreases from maximum negative towards zero. Hence, T1 is reverse biased and it does not conduct till D1 stops conducting at T1. The switch T1 conducts from 0 to $\pi/2$. The output current increases from zero to I_{max}. The current is supplied by DC supply. In this simulation model, the amplitude is set for $V_{DC} = 100$ V operated at 50 Hz. The phase delay is calculated for $\alpha = 30°$ for the MOSFETs and the load is considered as $R = 1\,\Omega$, $L = 1$ H. After simulation, the voltage and current relations obtained from the circuit are shown here.

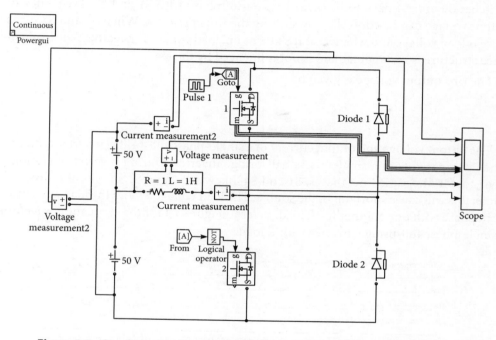

Figure 5.5 Simulation Model of a Single-Phase Half-Bridge Inverter with *RL* Load.

MATLAB Example 3

When $t = \pi/2$, S1 is turned off and base drive is applied to S2. But S2 does not conduct and the load inductance tries to maintain the load current in the same direction. Hence, it generates a large voltage and this voltage polarity forward biases diode D2.

The input current flows against the supply, which flows through D2. This current goes on decreasing and becomes zero at π. During this period, $\pi/2$ to π, energy is supplied by the load inductance to the DC supply. Hence, it is also called feedback operation. At π, diode current becomes zero and starts conducting at π. The output current is negative and it increases from zero to $-I_{max}$. The output voltage is also negative. At π, S2 is turned off and S1 is applied to the base drive. The load inductance generates large voltage with polarity. Hence, diode D1 conducts. The load current is negative and decreases towards zero and thus, the cycle repeats. When S1 and S2 conduct, the energy is supplied by the DC supplies. When diodes D1 and D2 conduct, the energy is supplied by the load inductance to the DC supplies. Hence, diodes are also called feedback diodes. The input current is the combined current due to both the DC supplies. The simulation waveform of a single-phase half-bridge inverter with RL load is shown in Figure 5.6.

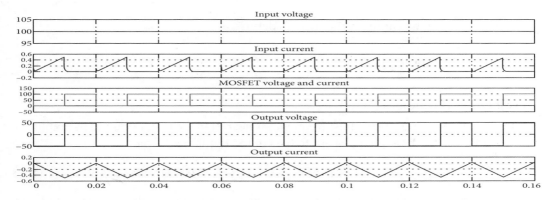

Figure 5.6 Simulation Waveform of a Single-Phase Half-Bridge Inverter with *RL* Load.

5.5 Performance Parameters of Inverters

The output of practical inverters contains harmonics and the harmonic nature of an inverter is assessed by the following parameters:

Harmonic Factor of *n*th Harmonic (HF_n)

HF is the measure of individual harmonic contribution

$$\text{HF} = \frac{V_{on}}{V_{01}} \text{ for } n > 1 \tag{5.3}$$

where,

V_{01} is the rms value of the fundamental component; and
V_{on} is the rms value of the *n*th harmonic component.

Total Harmonic Distortion (THD)

The measure of closeness fit between a waveform and its fundamental component is defined as THD.

$$\text{THD} = \frac{1}{V_{01}} \sqrt{\left(\sum_{n=2,3}^{\infty} V_{on}^2 \right)} \tag{5.4}$$

Distortion Factor (DF)

The THD gives the total harmonic content, rather than the measure of each harmonic component. Distortion factor (DF) is the measure of effective reduction of distortions or harmonics to specify the values in the second-order load filter as

$$\text{DF} = \frac{1}{V_{01}} \sqrt{\left(\sum_{n=2,3\ldots}^{\infty} \left(\frac{V_{on}}{n^2} \right)^2 \right)} \tag{5.5}$$

DF of an individual harmonic component is defined as

$$\text{DF}_n = \frac{V_{on}}{V_{01}.n^2} \text{ for } n > 1 \tag{5.6}$$

Lowest Order Harmonics (LOH)

It is the harmonic component whose frequency is nearest to the major one and its amplitude is more noteworthy than or equivalent to 0.03 of the fundamental part.

5.6 McMurray Inverter (Auxiliary-Commutated Inverter)

The McMurray inverter is a drive-commutated inverter, which depends on an *LC* circuit and a helper switch for replacement in the load circuit. The circuit is a reverberating *LC* circuit and is connected such that it can turn on or off a switch conveying the load current. A single-phase full-bridge inverter utilizing auxiliary commutation is demonstrated in Figure 5.7.

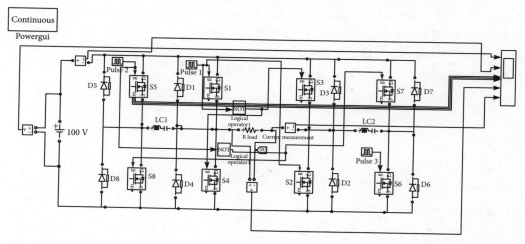

Figure 5.7 Simulation Model of a McMurray Auxiliary Commutated Inverter.

MATLAB Example 4

This compensation plan is known as an auxiliary-commutated inverter as it requires extra switches for turning off every fundamental switch. The circuit comprises the principle switches S1, S2, S3 and S4, the freewheeling diodes (FWDs) D1, D2, D3 and D4, the assistant switches S5, S6, S7 and S8, and the commutating segments *L* and *C*. When the exchanging pair S1 and S2 conducts, a negative voltage is delivered over the load. Making the pair of switches to direct, a substituting voltage is created over the load.

In any circuit, understanding the operation of the commutation procedure is vital; the commutating circuit must be composed from appropriate devices. The operation of McMurray inverters is subdivided into different working modes: the starting mode is when the exchanging pairs S1 and S2 are activated. When thyristors S1 and S2 are turned on, there is positive load voltage. The commutating capacitors C1 and C2 now have voltage of the polarities shown in Figure 5.8, as a result of the substitution of switches S3 and S4. The current waveforms of McMurray inverters are also shown in Figure 5.8.

When switches S5 and S6 are activated, they turn off the primary switches S1, S2 which were directing. At this point, capacitors C1 and C2 begin releasing and consequently, current I_c climbs, joining in the load current from S1 and S2 and not through D1 and D2. As the load current is constant, an increase in I_c causes a related reduction in the current through the switches. At π, the capacitor current I_c climbs and therefore, current through the switches S1 and S2 turns to zero.

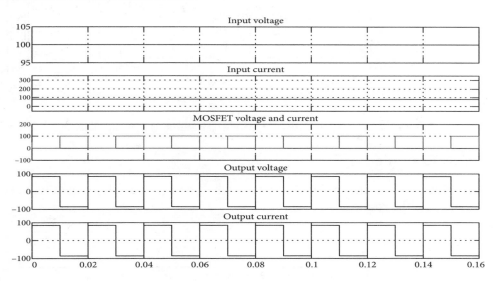

Figure 5.8 Simulation Waveform of McMurray Auxiliary Commutated Inverter.

5.7 Modified McMurray Half-Bridge and Full-Bridge Inverter

The McMurray half-bridge inverter circuit and full-bridge inverter circuit are shown in Figures 5.9 and 5.10. A resistor R, and helper diodes D1, D2, D3 and D4 are added to the essential McMurray circuit. These parts allow the evacuation of capacitor overvoltage towards the end of the compensation process. After the replacement of the switch S1, the extra energy in the capacitor C, less the energy scattered in R, is fed back to the DC supply by a current coursing through L, R, D1, the DC source and D4. Thus, after substitution of switch S2, the extra energy in capacitor C, less the energy disseminated in R, goes to the DC supply by a current coursing through L, D3, V_s, D2 and R. The voltage drop over R and D2 commutates switch S2. Resistor *R* restrains the undershoot of the capacitor voltage during this release interim.

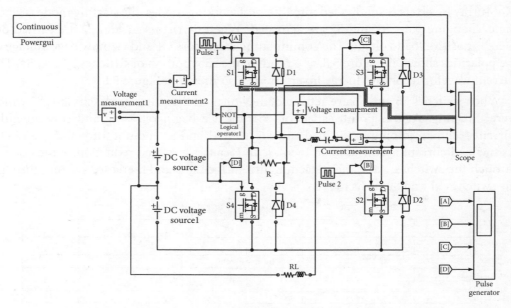

Figure 5.9 Simulation Model of a Modified McMurray Half-Bridge Inverter.

MATLAB Example 5

The principles behind the modified McMurray half-bridge and full-bridge inverters are similar and so is the commutation.

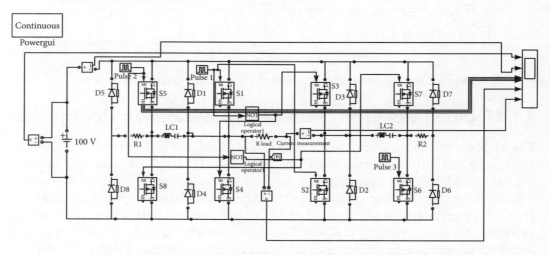

Figure 5.10 Simulation Waveform of Modified McMurray Full-Bridge Inverter.

MATLAB Example 6

If the resistor is sufficiently expansive in the arrangement circuit, the capacitor is released to the DC supply voltage V_S for all values of load current. In this simulation model, the amplitude is set for $V_{DC} = 100$ V each operated at 50 Hz. The phase delay is calculated for $\alpha = 30°$ for the switches and the load is considered as $R = 10\ \Omega$. After simulation, the voltage and current relations obtained from the circuit are shown here (refer Figures 5.11 and 5.12).

5.7.1 Modified McMurray half-bridge inverter

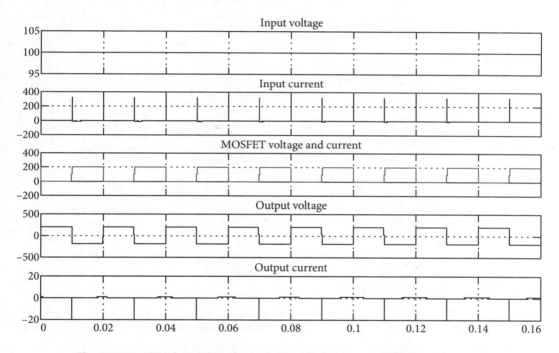

Figure 5.11 Simulation Waveform of a Modified McMurray Half-Bridge Inverter.

5.7.2 Modified McMurray full-bridge inverter

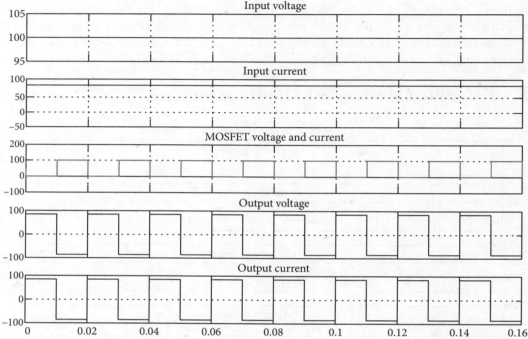

Figure 5.12 Simulation Waveform of a Modified McMurray Full-Bridge Inverter.

5.8 PWM Inverters

PWM is a technique of controlling the output voltage by controlling the pulse width balance inside the inverters. The basic strategies utilized are sinusoidal pulse width adjustments, which will be explained further later. In PWM strategies, the commonly utilized power semiconductor devices are power bipolar junction transistors (BJTs), power MOSFETs and insulated gate bipolar transistors (IGBTs). These devices can be turned on and off by utilizing control signals. It comprises a power circuit and gate-triggering circuit. The power circuit may be a single-phase or three-phase inverter. The comparator checks reference signs and bearer signals. This output is boosted to the activating unit. It gives pulses for the inverter. This pulse passes to the inverter through the enhancer and the isolator. The inverter output voltage can be controlled by fluctuating the on and off time of the devices.

On account of advances in strong state power devices and chip, exchanging power converters are used as part of mechanical application to change over and convey energy to the engine or load. PWM signs are pulses with settled frequency and variable pulse width. There is one pulse of altered size in each PWM period. Besides this, the width of the pulse changes from pulse to pulse by regulating the signal. When a PWM sign is connected to the gate of a power transistor, it causes the turn on and turns off of the transistor to change from one PWM period to another PWM period according to the same adjusting signal. The frequency of a PWM signal must be much higher than that of the tuning signal, with the major frequency such that the energy conveyed to the engine and its load depends, for the most part, on the adjusting signal. The loads of a symmetric PWM sign are constantly symmetric at the central point of each PWM period. The PWM ensures that the same side appears at the end of each PWM period. It has been demonstrated that symmetric PWM signals produce less signals in the output streams and voltages. This is the most well-known strategy for controlling output voltage. The advantages offered by PWM systems are as follows: lower power dissemination; easy to actualize and control; no temperature variation; compatible with today's computerized smaller scale processors; the output voltage can be controlled with no extra parts; and lower order harmonics can be reduced or minimized alongside its output voltage control. As higher order harmonics can be separated effectively, the sifting prerequisites are minimized. The primary disadvantage of this strategy is that SCRs are costly as they should have low turn on and turn off times. PWMs are categorized into two types: (1) single PWM and (2) multiple PWM.

5.8.1 Single pulse width modulation

In single PWM control, there is one and only pulse for each half-cycle and the width of the pulse fluctuates to control the output voltage. Gate signals are produced by single pulse width modulation over the reference sign to the square-wave signal. This procedure passes the reference sign to the zero-intersection circuit, which considers that the positive part of the input sign is the active part of the output signal, that is, square wave and the negative part of the input sign is the negative part of the output signal.

5.8.2 Multiple pulse width modulation

The harmonic nature of the output signal can be decreased by utilizing a few pulses as part of every half-cycle of the output voltage. The gating signs are created by contrasting the reference signal and the triangular carrier wave. The frequency of the reference signal sets the output frequency (f_o) and carrier frequency (F_c), which decide the quantity of pulses per half-cycle, where $p = f_o/f_c$. The variation in modulation index reflects the changes in pulse and output voltage.

MATLAB Example 7

In this system for balance, a few pulses for every half-cycle are utilized as part of the different pulse width modulation (Figure 5.13). As opposed to keeping the width of all pulses the same, the width of each pulse is differed corresponding to the completeness of a sine-wave assessed at the core of the same pulse.

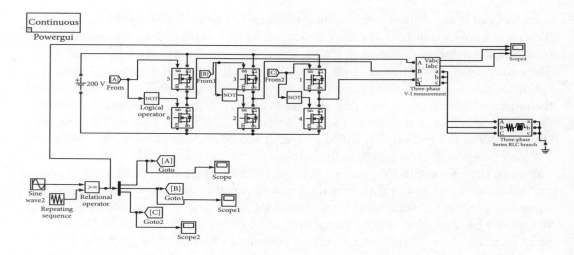

Figure 5.13 Pulse Width Modulation.

As mentioned earlier, gating signs are created by contrasting a sinusoidal reference sign (Figure 5.14) and a unidirectional triangular carrier wave of frequency. The frequency of the reference sign, the focus of the inverter output frequency and its peak value control the adjustment slope and thus, the rms output voltage. The quantity of pulse every half-cycle relies on the bearer frequency. Besides this, the two MOSFETs 1 and 4 cannot lead the immediate output voltage. The zone of each pulse relates roughly to the region of the sinusoidal wave in between the adjoining central nodes of the off periods of the triggering pulse.

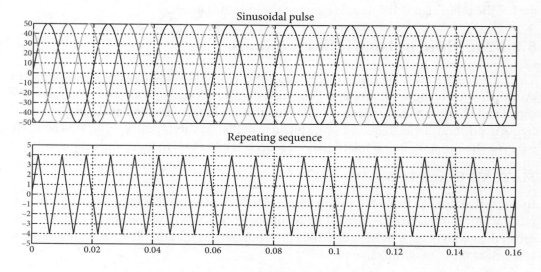

Figure 5.14 Pulse Width Modulation with Reference Signal.

5.9 Three-Phase Bridge Inverter

Three-phase inverters favour three-phase loads (Figure 5.15). The input supply being fed to the three-bridge inverter is the DC supply that can be taken from a rectifier or from a battery. The inverter concept can be simply explained as the change in the firing sequence from one SCR to an other. In general, a three-phase SCR holds six power semiconductor devices, and the mode of conduction can be 120° and 180°.

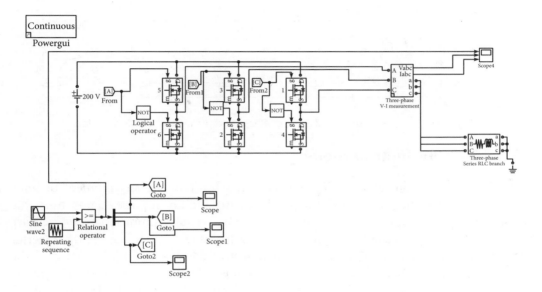

Figure 5.15 Simulation Model of a Three-Phase Bridge Inverter.

5.9.1 180° Conduction mode

In this conduction mode, every MOSFET conducts for 180° (refer to Figure 5.16). At any moment of time, only two MOSFETs stay on. Six substitutions in every cycle are needed. The gating signs and different voltage waveforms of the three-phase bridge inverter with 120° conduction for every MOSFET is demonstrated in Figure 5.17. The simulation parameters includes a voltage source of 100 V. The thyristors are conducted for 30° with phase difference and the three-phase RL load values are set for $R = 10 \ \Omega$ and $L = 13$ mH for both 180° and 120° conduction.

MATLAB Example 8

In Figure 5.15, the inverter operation has been separated into six interims. The 180° mode inverter additionally requires six steps, each of 60°, for finishing one cycle of the output AC voltage.

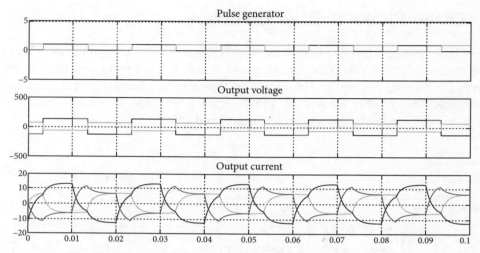

Figure 5.16 Simulation Model of Three-Phase Bridge Inverter 180° Conduction.

5.9.2 120° Conduction mode

In the 120° conduction mode, every SCR conducts for 120°. At any moment of time, only two MOSFETs stay on. Like in the previous case, additional pulse demonstrates the conduction time of every MOSFET. For this situation, six substitutions in every cycle are needed. The gating signs and different voltage waveforms of a three-phase bridge inverter with 120° conduction for every MOSFET is demonstrated in Figure 5.17. In the figure, the inverter operation has been separated into six interims. Like the 180° mode, the 120° mode inverter additionally requires six steps, each of 60°, for finishing one cycle of the output AC voltage.

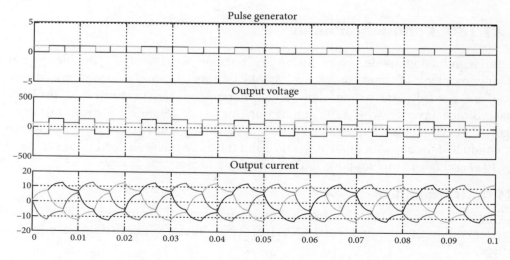

Figure 5.17 Simulation Model of Three-Phase Bridge Inverter 120° Conduction.

It can be observed that two MOSFETs conduct at once, one from the upper circuit and the other from the lower circuit. There are three modes of operation in one half-cycle and comparable circuits for a star-associated load. The line voltage is $V_{AB} = \sqrt{3}v_{an}$ with a phase development of 30°. There is a postponement of $\pi/6$ between the turning of T1 and turning of T4. Subsequently, there will be no circuit of the DC supply through one upper and one lower MOSFET. Whenever two low terminals are associated with DC supply, the third one remains partly open. The potential will rely on load qualities and would be unpredictable.

5.10 Current Source Inverters

Till now, we have analyzed VSIs where the output voltage of the inverter is kept consistent autonomous of the load. However, there are instances where the load current is compelled to fluctuate from positive to negative, and vice versa; to adapt to inductive loads, thyristors or switches with FWDs are needed. Normally, to keep the inverter output voltage consistent, a vast capacitor is added at the DC input side of the inverter.

In a CSI, the current from the DC source is kept at a successfully steady level, independent of load inverter conditions. This is attained by embedding a huge inductance in arrangement with DC supply to allow changes in the inverter voltage even when di/dt is low. The DC input to CSI is received from a settled voltage AC source through a controlled rectifier span, or through a diode span, or through a diode extension and a chopper. The current input to CSI is very nearly surge free; the L-channel is utilized before CSI. As it is a steady current system, the CSI is commonly used to drop at harmonic frequencies with a specific end goal to avoid issues either on exchanging or with harmonic overvoltages. CSIs can be used in the following applications:

(i) Speed control of AC engines
(ii) Induction heating
(iii) Lagging VAR compensation

5.10.1 Single-phase capacitor-commutated current source inverter with R load

The single-phase capacitor-commutated CSI with R load is indicated in Figure 5.18. The recreation model comprises a capacitor in parallel with the load, which is utilized for putting away the charge for commutating switches. Thyristors T1, T2, T3 and T4 structure the power circuit.

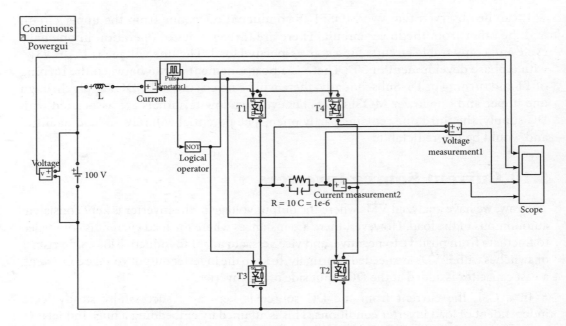

Figure 5.18 Simulation Model of a Single-Phase Capacitor-Commutated Current Source Inverter with *R* Load.

MATLAB Example 9

Thyristors T1 and T2 are activated together and T3 and T4 are activated later. Before $t = 0$, the capacitor voltage is situated at $V_c = -V_1$. At this stage, thyristors T1 and T2 are activated and when T1 and T2 get turned on, the capacitor applies reverse voltage crosswise over T3 and T4 and makes them off. The source current courses through T1 in parallel with *R* and *C*, through T2.

From 0 to $\pi/2$, the thyristor flows are equivalent to the source current and the capacitor voltage changes from $-V_1$ to V_1 through the charging current. When T3 and T4 get turned on, thyristors T1 and T2 get turned off. Presently, the source current moves through T3, *R*, *C* and T4.

In this simulation model, the input DC voltage is set for 100 V and the thyristors T1 and T2 are triggered at $\alpha = 30$; the parallel RC load is set for $R = 1\ \Omega$ and $C = 1\ \mu F$. The simulation waveform for a single-phase capacitor-commutated CSI with *R* load is shown in Figure 5.19.

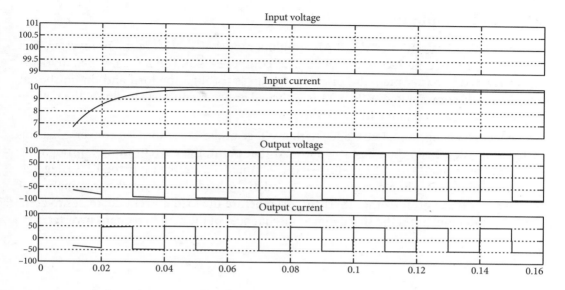

Figure 5.19 Simulation Waveform for a Single-Phase Capacitor-Commutated Current Source Inverter with R Load.

5.11 Resonant Converters

Resonance power converters contain LC whose voltage and current waveforms shift sinusoidally during one or more subintervals of every exchanging period. These sinusoidal waveforms are vast in size, and the values do not make a difference. The resonant converters are classified as follows

- High-frequency AC inverters
- DC–DC resonant converters
- Inverters or rectifiers influencing line frequency AC

5.11.1 Series resonant converters

Series resonant inverters are focused around the swing of the full current. The inverter comprises resounding parts and exchanging power hardware devices. The LC segments are the resonant devices. The reverberating segments and exchanging devices are set in arrangement with the load to structure an underdamped circuit. This circuit is called the full inverter. The current through the exchanging devices falls to zero because of the full qualities of the circuit. This sort of inverter creates a rough sinusoidal waveform. The frequency extends from 20 kHz to 100 kHz. The extent of the resounding parts is not vast because of the exchanging frequency. Some of the basic applications include ultrasonic, heating and fluorescent lighting.

MATLAB Example 10

Figure 5.20 demonstrates the MATLAB Simulink model of a series resonant inverter. The inverter consists of unidirectional switches, *LC* components and an *R* load. When thyristor T1 is triggered, a harmonic pulse of current courses through the load and diminishes at zero; T1 is self-commutated. When T1 is activated and current starts to climb, the voltage crosswise over L1 is certain with extremity.

The incited voltage on L2 now adds to the voltage of the capacitor *C* in converse biasing of thyristor T2; T2 can then be turned off. At that point, the current moves through the *RLC* segments while T1 and T2 are turned off. At time $t = \pi/2$, the thyristor T2 is triggered and a converse resounding current moves through the load. Therefore, the termination of one thyristor turns off the other thyristor, even before the load current decreases to zero. Figure 5.21 demonstrates the simulation consequences of the series resonant inverter.

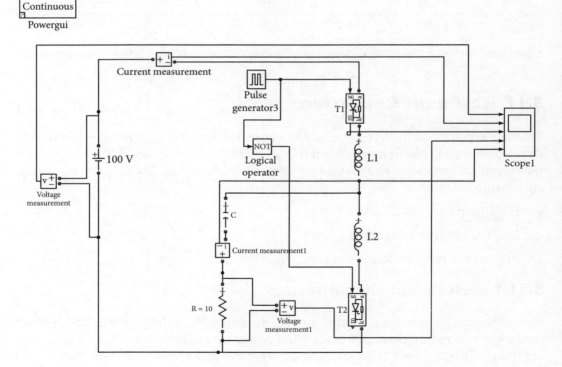

Figure 5.20 Simulation Model of a Series Resonant Inverter.

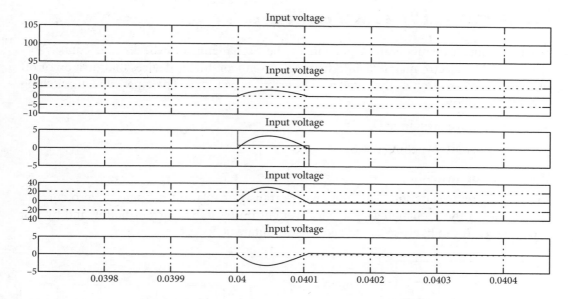

Figure 5.21 Simulation Waveform of a Series Resonant Inverter.

5.11.2 Parallel resonant converters

For a parallel resonant converter (refer to Figure 5.22), the complete tank is still in arrangement. It is called a parallel resonant converter since for this situation, the load is in parallel with the full capacitor. More precisely, this converter should be called full converter with parallel load. Since the transformer side is a capacitor, an inductor is included in the optional side to match the impedance. The simulation for the parallel resonant inverter is left as an exercise.

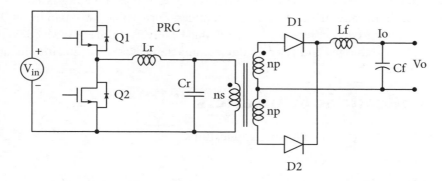

Figure 5.22 Parallel Resonant Inverter.

5.11.3 ZVS and ZCS PWM converters

Soft switching can moderate a portion of the exchanging drawbacks and conceivably decrease EMI (electro magnetic interference) in semiconductor devices that turn on or off at the zero intersection of their voltage or current waveforms. The types of soft switching include the following

- Zero Voltage Switching (ZVS)
- Zero Current Switching (ZCS)

Zero Current Switching

Figure 5.23 represents soft switching, where transistor turn-off happens at zero current. ZCS removes the exchanging problems brought on by IGBT current tailing and by lost inductances. It can likewise be utilized to commutate SCRs.

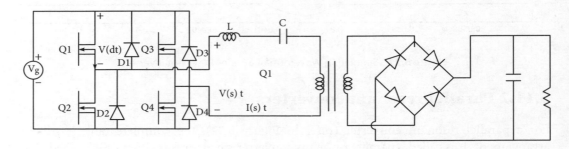

Figure 5.23 ZVS and ZCS-PWM Converters.

Zero Voltage Switching

In case of transistors, turn-on happens at zero voltage. Diodes may likewise work with zero-voltage exchanging. Zero-voltage exchanging removes the exchanging issues caused by diode external charge and device output capacitances. Zero-voltage exchanging is generally favoured in cutting-edge converters. Zero-voltage converters are PWM converters, in which an inductor charges and releases device capacitances, thus acquiring the ability to switch zero-voltage.

5.12 Applications of Inverters

Inverters can be used for the following applications.

- Variable rate AC engine drives
- Induction heating
- Aircraft power suppliers
- Uninterruptible Power Supplies (UPS)

- High-voltage DC transmission lines
- Battery vehicle drives
- Regulated voltage and frequency power supplies

Summary

By the end of this chapter, readers will be able to simulate the basic circuits of single-phase and three-phase inverters. Modified methods and their various applications have also been explained. The users will be able to design inverters with various load conditions required for various applications.

Solved Problems

1. For a single-phase half-bridge inverter with R load of 3 Ω and DC input voltage 24 V, calculate the rms output voltage, output power, average and peak current values, THD, and the HF and the DF of lower order harmonics.

    Solution

    Data: $V = 24$ V, $R = 3$ Ω

    (a) The rms output voltage at fundamental frequency,

    $$V_{1rms} = \frac{2V}{\sqrt{2}\pi} = 10.8 \text{ V}$$

    (b) The output power

    $$P_0 = \frac{V_0^2}{R} \text{ and } V_0 = \frac{V}{2}$$

    where, V_0 = rms output voltage

    $$\therefore P_0 = \frac{12^2}{3} = 48 \text{ W}$$

    (c) The average and peak current of each transistor.

    The average current

    $$I_{T(av)} = \frac{1}{T}\int_0^{T/2} \frac{V}{2R} dt = \frac{V}{2RT}(T/2) = \frac{V}{4R}$$

    ∴ Average transistor current

    $$I_{r(av)} = \frac{24}{3 \times 4} = 12 \text{ Å}$$

The transistor peak current

$$I_{T(peak)} = \frac{v/2}{R} = 4 \text{ Å}$$

(d) Peak reverse-blocking voltage V_{BR} for each transistor

$$V_{BR} = 2 \times \frac{v}{2} = 24 \text{ V}$$

(e) THD

$$\text{THD} = \frac{1}{V_{1rms}} \left(\sum_{n=2,3}^{\infty} V_{nrms}^2 \right)^{1/2}$$

$V_{01rms} = 10.8 \text{ V}$ as already calculated in (a)

The rms harmonic voltage

$$= \left[\sum_{n=3,5,7,\ldots}^{\infty} V_{n\,rms}^2 \right]^{1/2}$$

$$= \left(V_0^2 - V_{01rms}^2 \right)^{1/2} = \left[12^2 - (10.8)^2 \right]^{1/2} = 5.23 \text{ V}$$

$$\text{THD} = \frac{5.23}{10.8} = 0.484 = 48.4\%$$

(f) The DF

$$\text{DF} = \frac{1}{V_{01rms}} \left[\sum_{n=3,5,7}^{\infty} \left(\frac{V_{on\,rms}}{n^2} \right)^2 \right]^{1/2}$$

To find $\frac{V_{on\,rms}}{n^2}$, we have to find $V_{n\,rms}$ first

$$v_0 = \sum_{n=1,3,5}^{\infty} \frac{2V}{n\pi} \sin n\omega t = 0, \text{for}, n = 2, .4, .6;$$

$$\therefore v_0 = \frac{2V}{\pi} \sin \omega t + \frac{2V}{3\pi} \sin 3\omega t + \frac{2V}{5\pi} \sin 5\omega t + \frac{2V}{7\pi} \sin 7\omega t + \ldots$$

$$V_{3\,rms} = \frac{2V}{3\pi \sqrt{2}} = 3.6 \text{ V} \qquad V_{5\,rms} = \frac{2V}{5\pi \sqrt{2}} = 2.16 \text{ V}$$

$$V_{7\,rms} = 1.54 \text{ V} \qquad V_{9\,rms} = 1.2 \text{ V}$$

$$V_{11\,rms} = 0.982 \text{ V} \qquad V_{13\,rms} = 0.83 \text{ V}$$

$$\therefore \left[\sum_{n=3,5,7}^{\infty} \left(\frac{V_{n\,rms}^2}{n^2} \right)^2 \right]^{1/2} = \left[\left(\frac{V_3}{3^2} \right)^2 + \left(\frac{V_5}{5^2} \right)^2 + \left(\frac{V_7}{7^2} \right)^2 + \ldots \right]^{1/2}$$

$$= \left[0.16 + 0.0348 + 2.3 \times 10^{-3} + \ldots\right]^{1/2} = 0.44 \text{ V}$$

$$\therefore \quad DF = \frac{0.44}{10.8} = 0.041 = 4.1\%$$

(g) The lowest harmonic is the third harmonic

$$\therefore \text{ HF for the third harmonic} = HF_3 = \frac{V_{3rms}}{V_{1rms}} = \frac{3.6}{10.8} = 33.33\%$$

$$\text{DF the third harmonic } DF_3 = \frac{(V_{3rms}/3^2)}{V_{1rms}} = (3.6/9)/10.8 = 0.037 \text{ or } 3.7\%$$

2. The single-phase half-bridge inverter using transistors has a resistive load of 2 Ω. The DC supply is 24 V. Calculate:

   (a) RMS output voltage at fundamental frequency
   (b) Output power
   (c) Average and peak current
   (d) Peak reverse-blocking voltage of each transistor

Solution

(a) RMS output voltage at fundamental frequency is given by

$$V_{01rms} = 0.45 \times V = 0.45 \times 24 = 10.8 \text{ V}$$

(b) Output power $= \frac{V_{0rms}^2}{R}$. But $V_{0rms} = \frac{V}{2} = 12 \text{ V}$

$$\therefore \text{ Output power} = \frac{12^2}{2} = 72 \text{ W}$$

(c) Peak load current $= \frac{V}{2R} = \frac{24}{4} = 6 \text{ Å}$

Average load current = 0

(d) Peak reverse-blocking voltage of each transistor = V = 24 V.

Objective Type Questions

1. For a square-wave inverter of single-phase VSI feeding inductive load, the load current waveform will be
   (a) Sinusoidal (b) Rectangular
   (c) Trapezoidal (d) Triangular

2. The ratio of inverter gain is
   (a) $\dfrac{\text{DC output voltage}}{\text{AC input voltage}}$ (b) $\dfrac{\text{AC O/P voltage}}{\text{AC input voltage}}$
   (c) $\dfrac{\text{DC O/P voltage}}{\text{AC I/P voltage}}$ (d) $\dfrac{\text{AC O/P voltage}}{\text{AC I/P voltage}}$

3. In case of a three-phase inverter, the PWM control is used to
   (a) Reduce the THD with modest filtering
   (b) Minimize the load on DC side
   (c) Increase the life of the batteries
   (d) Reduce low-order harmonics and increase high-order harmonics

4. For the figure shown here, determine the rms voltage value of pole–pole voltage V_{12}, when the switches are operated at V_{10} and V_{20}, respectively.

   (a) $\dfrac{V_s * \emptyset}{\pi\sqrt{2}}$ (b) $V_s * \sqrt{\dfrac{\emptyset}{\pi}}$ (c) $V_s * \sqrt{\dfrac{\emptyset}{2\pi}}$ (d) $\dfrac{V_s}{\pi}$

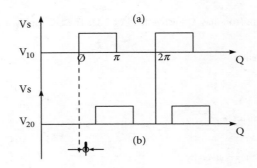
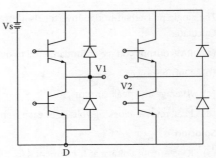

5. For a VSI operated with inductive load at 50 Hz, what is the duration of conduction of a feedback diode in a cycle, if the average load current is zero?
   (a) 5 ms
   (b) 10 ms
   (c) 20 s
   (d) 2.5 ms

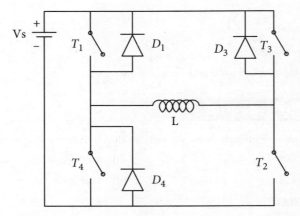

6. An inductive load, which is fed by a three-phase voltage source, produces an output voltage, which has hth order harmonic with magnitude a_h times of the fundamental frequency component. Therefore, the load current would have Hth harmonic of

(a) Zero

(b) a_h times the fundamental frequency component

(c) $h \cdot a_h$ times the fundamental frequency component

(d) a_{hm} times the fundamental frequency component

7. For a McMurray inverter, identify the correct statement
   1. Large withstanding ability
   2. Ability to carry commutating current
   3. Provide reverse bias
   4. Reactive current feedback

   (a) 1, 2 and 3 are correct
   (b) 1, 3 and 4 are correct
   (c) 2, 3 and 4 are correct
   (d) 1, 2 and 4 are correct

8. In the inverter circuit shown in the figure, if the SCRs are fired at delayed angles, the frequency of the output waveform will

   (a) Increase
   (b) Remain the same
   (c) Decrease
   (d) Depend on which SCR is fired first

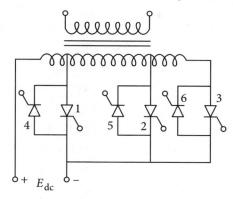

9. In the circuit of Problem 8 if SCR_i is on and then SCR_j is fired, the anode voltage of SCR_i will become nearly equal to

   (a) $+V_{DC}$
   (b) $-V_{DC}$
   (c) 1–2 V
   (d) Zero

10. Consider any inverter circuit with a capacitor C with 1 and 2 as two input terminals. When the SCR is turned on, the capacitor C will

    (a) Charge with terminal 2 as positive
    (b) Charge with terminal 1 as positive
    (c) Not charge at all unless SCR_2 is also turned on
    (d) Make SCR_2 on

11. In the SCR tap-switch inverter, when SCR_i is fired

    (a) Positive peak of the AC O/P is obtained
    (b) Negative peak of the O/P is obtained

(c) Two-third to peak value is obtained

(d) One-third of the peak value is obtained

12. In a single-phase VSI, the amplitude value is V_s, output power is P, the corresponding single-phase full-bridge inverter is

   (a) $V_s \cdot P$
   (b) $V_s/2, P/2$
   (c) $2V_s, 2P$
   (d) None of these

13. In case of VSIs,

   (a) Load voltage waveform V_o depends on load impedance Z, whereas load current waveform i_o does not depend on Z
   (b) Both V_o and i_o depend on Z
   (c) V_o does not depend on Z, whereas i_o depends on Z
   (d) None of these

14. A single-phase VSI is operated in load commutation mode when the operated load is

   (a) RLC overdamped
   (b) RLC underdamped
   (c) RLC critically damped
   (d) None of these

15. A single-phase bridge inverter delivers power to a series-connected RLC load with $R = 2\,\Omega, L = 8\,\Omega$. For this inverter load combination, load commutation is possible when the magnitude of 1/WC in ohms is

   (a) 10
   (b) 8
   (c) 6
   (d) zero

16. For single PWM inverters, the pulse width is equal to what value when the third harmonic is eliminated?

   (a) 30°
   (b) 60°
   (c) 120°
   (d) None of these

17. For single PWM inverters, the pulse width is equal to what value when the fifth harmonic is eliminated?

   (a) 30°
   (b) 72°
   (c) 36°
   (d) 108°

18. For single PWM inverters with pulse width 120°, the rms output voltage for 220 V DC is

   (a) 179.63 V
   (b) 254.04 V
   (c) 127.02 V
   (d) None of these

19. VSI is basically used when

   (a) Source inductance is large and load inductance is small
   (b) Source inductance is small and load inductance is small
   (c) Both source and load inductance are small
   (d) Both source and load inductances are large

20. In case of resonant pulse inverters,

   (a) DC output voltage variation is wide
   (b) The frequency is low

(c) Output voltage is never sinusoidal

(d) DC saturation of transformer core is minimized

21. In multiple-pulse modulation used in PWM inverters, the amplitudes of reference square wave and triangular carrier wave are, respectively, 1 V and 2 V. For generating 5 pulses per half-cycle, the pulse width should be

(a) 36° (b) 24°
(c) 18° (d) 12°

22. In sinusoidal-pulse modulation, used in PWM inverters, amplitude and frequency for triangular carrier and sinusoidal reference signals are, respectively, 5 V, 1 kHz and 1 V, 50 Hz. If zeros of the triangular carrier and reference sinusoid coincide, then the modulation index and order of significant harmonics are respectively

(a) 0.2, 9 and 11 (b) 0.4, 9 and 11
(c) 0.2, 17 and 19 (d) None of these

23. Which of the following statements is correct in connection with inverters?

(a) VSI and CSI, both require feedback diode

(b) Only CSI requires feedback diodes

(c) Gate turn off thyristors can be used in CSI

(d) Only VSI requires feedback diodes

24. In a constant source inverter, if frequency of output voltage is f Hz, then frequency of voltage input to constant source inverter is

(a) f (b) $2f$
(c) $3f$ (d) $4f$

25. By eliminating the third harmonic in an inverter with output frequency of 50 Hz, the frequency of other components in voltage output would be

(a) 250, 350, 500, high frequencies (b) 50, 250, 350, 500
(c) 50, 50, 350, 550 (d) None of these

26. For a CSI with capacitor C load, the voltage across the capacitor for constant source current is

(a) Square wave (b) Triangular wave
(c) Step function (d) None of these

27. In sinusoidal PWM, there are m cycles of the triangular carrier wave in the half-cycle of the reference sinusoidal signal. If zero of the reference sinusoid coincides with zero/peak of the triangular carrier waves, then number of pulses generated in each half-cycle is respectively

(a) $(m-1)/m$ (b) $(m-1)/(m-1)$
(c) m/m (d) None of these

28. When triangular PWM is applied in three-phase circuits, the VSI of BJT introduces

(a) Low-order harmonic voltages on the DC side

(b) Very high-order harmonic voltages on the DC side

(c) Low-order harmonic voltages on the AC side

(d) Very high-order harmonic voltages on the AC side

Review Questions

1. What is meant by the term inverter in power electronics?
2. List some of the applications of an inverter.
3. Classify the various types of inverters.
4. State why thyristors are not much preferred for inverters?
5. How is output frequency varied in case of a thyristor?
6. Give two advantages of CSI.
7. What is the main drawback of a single-phase half-bridge inverter?
8. Why should diodes be connected antiparallel with thyristors in inverter circuits?
9. What types of inverters require feedback diodes?
10. What is meant by a series inverter?
11. In a series inverter, what are the conditions to be used for selecting L and C?
12. What is meant by a parallel inverter?
13. What are the applications of a series inverter?
14. How is the inverter circuit classified based on commutation circuitry?
15. What is meant by the term McMurray inverter in power electronics?
16. List some of the applications of CSI.
17. What do you understand by the term PWM technique?
18. List the merits of PWM control.
19. What are the disadvantages of the harmonics present in the inverter system?
20. How can harmonic content be reduced?
21. Describe the working of a single-phase full-bridge inverter supplying R, RL load with relevant circuits and waveforms.
22. Describe the operation of a single-phase half-bridge inverter supplying R load with relevant waveforms.
23. Explain with necessary diagrams the working of a VSI.
24. Explain the various methods available for voltage control.
25. Discuss the function of a three-phase VSI supplying a balanced star connected load in 180° operating mode.
26. Describe the sinusoidal PWM technique with necessary diagram.
27. Explain the concept of current-commutated CSI using R load with a neat sketch.
28. Discuss the working of a single-phase series resonant inverter with appropriate circuits and waveforms.

Practice Questions

1. Explain the principle of operation of a three-phase VSI in 180° conduction mode with necessary waveforms and circuits. Also obtain the expression for line to line voltage.
2. Discuss the functioning of a three-phase VSI in 120° operating mode with relevant waveforms and obtain the expression for voltages.

3. Explain the following PWM techniques used in inverters.
   (i) Sinusoidal PWM
   (ii) Multiple PWM
4. Explain the operation of a single-phase capacitor-commutated CSI with R load.
5. Explain harmonic reduction by transformer corner lines and stepped-wave inverters.
6. Explain the different methods of voltage control adopted in an inverter with suitable waveforms.
7. Explain the working of a series inverter with the help of circuit diagram and relevant waveforms.
8. Draw the circuit diagram of a CSI and explain its operation with relevant waveforms.
9. Describe the working of a single-phase full-bridge inverter supplying R, RL loads with relevant circuit and waveforms.
10. What is the need for controlling the output voltage of inverters? Classify the various techniques adopted to vary the inverter gain. Explain briefly sinusoidal PWM.
11. Design a single-phase half-bridge inverter for $R = 5\,\Omega$ and input voltage $= 50$ V. Calculate the rms voltage, output power, and the average current of the thyristor using MATLAB simulation.
12. A three-phase inverter is supplied with 600 V. For a star connected load, calculate the rms load current, load power for 120° and 180° conduction for 20 Ω/phase using MATLAB.
13. Using MATLAB, for an R load $= 10\,\Omega$, with $V_s = 100$ V, calculate the rms output voltage, output power, and the average values of current and voltage of a single-phase full-bridge inverter.
14. Implement a modified McMurray inverter for half- and full-bridge inverter for $R = 5\,\Omega$, $L = 3$ mH, and input voltage $= 50$ V. Calculate the rms voltage, output power and the average current of the thyristor using MATLAB simulation.
15. Design a single-phase half-bridge inverter, which holds a resistive load of 2.5 Ω. Let the input DC voltage be 45 V. Using MATLAB, derive the rms output voltage and the output power at the operating frequency.

6

Controllers Using MATLAB (AC–AC Converters)

> **Learning Objectives**
> - ✓ To understand the concepts of AC voltage regulators
> - ✓ To classify the various types of voltage regulators
> - ✓ To understand the concepts of unidirectional and bidirectional operations
> - ✓ To study the operations of single-phase and three-phase circuits
> - ✓ To understand the concepts of cycloconverters and their types
> - ✓ To classify the difference between single-phase to single-phase, three-phase to single-phase, and three-phase to three-phase circuits

6.1 Introduction

The voltage controller, also called an AC voltage controller or AC regulator, is an electronic module with either thyristors, triode alternating currents (TRIACs), silicon controlled rectifiers (SCRs), or insulated gate bipolar transistors (IGBTs). The module changes a constant voltage, to get a variable output voltage conveyed to the load. AC controllers are basically employed to change the root mean square (rms) value of the changing voltage, which is connected to the load terminal of the circuit, by making thyristors lead in between the AC supply source and the load terminal. Fluctuating voltage is used for diminishing road lights, fluctuations, warming temperatures in homes or industry, regulation and control of fans and winding machines, and numerous other applications, in a similar manner to an autotransformer. Voltage controller modules are power devices. Since they need less maintenance and are extremely productive, voltage controllers have, to a great extent, supplanted such modules as enhancers and saturable reactors. There are two distinct methods of thyristor control utilized to control the AC power stream:

- On–off control or integral cycle control
- Phase control

6.1.1 ON–OFF control

Power thyristors can be used to fluctuate between on and off for a few cycles and then disengage for the other few cycles. Thus, thyristors act like rapid AC power semiconductor switches.

6.1.2 Phase control

In the phase angle control concept, thyristors are engaged to fluctuate the input AC supply to the load circuit for a few period of cycles, that is, the input AC supply is reduced by engaging the thyristors during a certain period of the cycle. For each half-cycle, the thyristor switch is turned on so that the supply voltage leads to the load; in the remaining period of the input half-cycle, the thyristor switch is turned off to disengage the AC supply from the load. By controlling the trigger point α, the output rms voltage over the load can be controlled. The trigger angle α is characterized as a result; the load current starts streaming at the point at which the thyristor turns on. AC voltage thyristor controllers are devices that utilize AC line switch, as they are line- or phase-commutated circuits in sequence with the input AC supply. When the voltage supply switches to the negative side, the load current moving through the leading thyristor decreases and finally reduces to zero. Phase-controlled thyristors operate slower than fast transformation inverters. They are moderately economical and used regularly. TRIACs are the most commonly switches as they meet the current and voltage regulations up to 400 Hz. Because of AC line compensation or common recompense, there is no need to add substitution hardware or parts and the circuits for AC voltage controllers are exceptionally basic. In case of phase-controlled AC voltage controllers with RL load, the response of the waveforms and the inferences of the expressions are however not very straightforward. Since, majority of the loads used in these controllers are of the RL type, RL loads ought to be considered in the investigation and configuration of AC voltage controller circuits.

6.2 Classification of AC Voltage Controllers

AC voltage controllers can be classified based on applications into the following types:

- Indirect AC–AC converters
- Cycloconverters

- Hybrid converters
- Matrix converters

Depending on the input supply configurations to the circuit, voltage regulators can be classified into the following types

- Single-phase AC voltage controllers
- Three-phase AC voltage controllers

In India, single-phase AC voltage controllers work with 230 V rms supply at 50 Hz frequency, and three-phase AC voltage controllers work with 400 V rms supply at 50 Hz supply. Based on the direction, the AC controllers are classified into the following types.

- Unidirectional controller
- Bidirectional controller

6.3 Single-Phase AC Voltage Controllers

The operation of AC voltage converters along with appropriate MATLAB circuits will be explained in this section. A two-quadrant converter can be considered as a substituting voltage source, a voltage which relates to the voltage got at the output terminals. The diodes joined in arrangement with every voltage source demonstrate the unidirectional conduction of every converter, whose output voltage can be either positive or negative, being a two-quadrant one. The leading of current is as shown in Figure 6.1 because thyristors – unidirectional exchanging devices – are utilized as part of the converters. Generally, the leakage element in the output voltage is neglected.

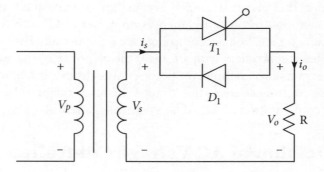

Figure 6.1 Circuit of a Single-Phase Half-Wave AC Voltage Controller with R Load.

The principle behind this ideal voltage converter is to modulate the individual converters' firing angles in a continuous manner to produce an AC sinusoidal voltage at its output terminals. Therefore, both the sources are supported with the same amplitude, frequency, and phase; besides, the converter's voltage is equal to the voltage of either of the two sources. The average power hence flows to and fro from either of the terminals, and the converter is inherently capable of operating with loads of either inductive or capacitive phase angle. Regardless of the phase of the current values with respect to the voltage values, the unidirectional property of the converter implies that a half-cycle of the load current will be carried by the positive converter and the negative half by the negative converter. As a result, the two quadrants operate in rectifying and inverting mode during the associated cycle of current.

6.3.1 Single-phase half-wave AC voltage controller with R load

The single-phase half-wave AC voltage controller circuit is shown in Figure 6.1.

The half-wave AC controller uses one thyristor and one diode in parallel with each other in inverse heading, that is, the anode of thyristor T1 is joined with the cathode of diode D1, and the cathode of T1 is connected with the anode of D1. The output voltage over the load resistor R and consequently the AC power stream to the load is controlled by fluctuating the trigger edge α. The trigger angle or the deferral point α alludes to the estimation of t or the moment at which the thyristor T1 is turned on, by applying a suitable gateway trigger pulse between the gate and cathode lead.

Thyristor T1 is forward one-sided among the positive half-cycle of input AC supply. It can be activated and made to lead by applying a suitable gate trigger pulse during the positive half-cycle of the input supply. When T1 is activated, it conducts and load current moves through thyristor T1, the load and through the transformer auxiliary winding.

Considering the thyristor T1 to be ideal, the thyristor switch can act as a shut switch when it is in the on state during π radians. In this state, the output voltage takes over the load, conducting from ω to π radians. When the input supply voltage reduces to zero, during ω to π radians, for a resistive load, the load current additionally reduces to zero at t and thus, the thyristor T1 turns off at 2π radians. Between the time period ω to 2π, when the supply voltage inverts and gets to be negative, the diode D1 gets to be forward one-sided, and thus turns on and conducts. The load current streams the other way during ω to 2π radians, when D1 is on and the output voltage takes after the negative half-cycle of the input supply (Refer to Figure 6.2).

The performance parameters for the AC voltage controller is given by

$$\text{Circuit turn off time} = \frac{\pi}{\omega} \text{ s} \tag{6.1}$$

$$V_C = \sum_{n=1,3,5}^{\omega} A_n \sin n\omega t + \sum_{n=1,3,5}^{\omega} B_n \cos n\omega t \; d(\omega t)$$

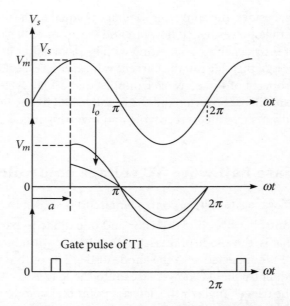

Figure 6.2 Waveform of a Single-Phase Half-Wave AC Voltage Controller.

where $A_n = \dfrac{V_m}{\pi}\left[\dfrac{\sin(n+1)\alpha}{(n+1)} - \dfrac{\sin(n-1)\alpha}{(n-1)}\right]$

$B_n = \dfrac{V_m}{\pi}\left[\dfrac{\cos(n+1)\alpha - 1}{(n+1)} - \dfrac{\sin(n-1)\alpha - 1}{(n-1)}\right]$

$V_{nm} = \sqrt{A_n^2 + B_n^2}$

$\phi_n = \tan^{-1}\dfrac{B_n}{A_n}$

For $n = 1$, V_{or} of output voltage is given by

$$V_{or} = \dfrac{V_m}{\sqrt{2\pi}}\left[(\pi - \alpha) + \dfrac{1}{2}\sin 2\alpha\right]^{1/2} \qquad (6.2)$$

$$I_{or} = \dfrac{V_{or}}{R}$$

$$P = I_{or}^2 R = \dfrac{V_{or}^2}{R} = \dfrac{V_m^2}{2\pi R}(\pi - \alpha) + \dfrac{1}{2}\sin 2\alpha \qquad (6.3)$$

$$= \frac{V_s^2}{\pi R}\left[(\pi-\alpha)+\frac{1}{2}\sin 2\alpha\right]$$

$$\text{Power factor (PF)} = \frac{\text{Real power}}{\text{Apparent power}} = \frac{V_{or}^2/R}{V_s V_{or}/R} = \frac{V_{or}}{V_s}$$

$$= \left\{\frac{1}{\pi}\left((\pi-\alpha)+\frac{1}{2}\sin 2\alpha\right)\right\}^{1/2} \tag{6.4}$$

The principle of operation of a single-phase half-wave AC voltage controller circuit is shown in Figure 6.3. In the MATLAB circuit model, the half-controlled circuit uses only one thyristor and one diode, which are connected parallel to each other, but in the opposite direction, where the anode of the thyristor switch is connected to the cathode terminal of the diode switch D and vice versa. The role of the triggering angle α is to control the voltage across the load resistor R and the AC power, which flows to the load.

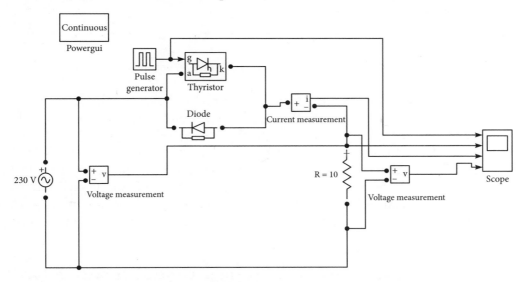

Figure 6.3 MATLAB Model of a Single-Phase Half-Wave AC Voltage Controller with R Load.

Solved MATLAB Example 1

The triggering angle α alludes to the time period and when the thyristor T is turned on, using the triggering pulse, the thyristor, is forward biased during the positive half-cycle of the supply. When the active thyristor T is directed in the forward direction, the forward current heads towards the load through the thyristor, load terminal, and through the transformer optional winding. During the period $\omega t = \alpha$ to π radians, the thyristor is set to shut down.

In this simulation model, the amplitude is set for $V_s = 230$ V operated at 50 Hz. The phase delay is calculated for $\alpha = 30°$ for thyristors and the load is considered as $R = 10\ \Omega$.

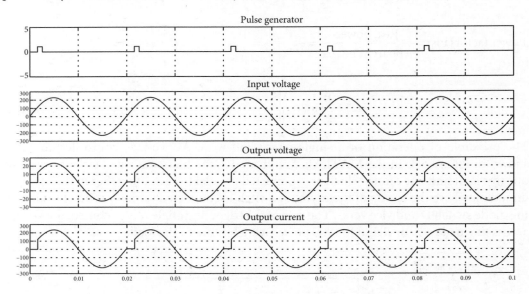

Figure 6.4 Simulation Waveform of a Single-Phase Half-Wave AC Voltage Controller with R Load.

The simulation waveform of a single-phase half-wave AC voltage controller with R load is demonstrated in Figure 6.4. When the thyristor turns on from the period $\omega t = \alpha$ to π radians, the input supply courses through the thyristor and the output voltage takes up over the load. When the supply source declines to zero value at $\omega t = \pi$ where the load is resistive, the load current reduces to zero and, the thyristor turns off. When $\omega t = \pi$ to 2π, the source voltage gets inverted and tends to be negative, as a result, the diode D turns out to be forward biased and subsequently turns on and conducts. The load current streams the other way when $\omega t = \pi$ to 2π radians and D is on; the output voltage takes after the negative half-cycle of the input.

6.3.2 Single-phase full-wave AC voltage controller with R load

Single-phase full-wave AC voltage controller circuit with two SCRs or a solitary TRIAC is by and large utilized as part of the vast majority of the AC control applications. The AC power stream to the load can be controlled in both the half-cycles by shifting the trigger edge.

The rms value of the load voltage can be shifted by changing the trigger point α. The input supply current fluctuates on account of the full-wave AC voltage controller and because of the symmetrical input supply current waveform, there is no DC segment of input supply current, that is, the normal estimation of the input supply current is zero. A single phase full-wave AC voltage controller with a resistive load is shown in Figure 6.5. We can control the AC power stream to the load in both the half-cycles by altering the trigger point α. Hence, the full-wave AC voltage controller is also called a bidirectional controller.

The circuit shown in Figure 6.5 utilizes two SCRs and is largely used as part of higher trend AC circuit applications. By fluctuating the firing angle α, the AC power and the rms estimation of the load voltage and current streaming towards the load can be controlled throughout the cycle. The source current fluctuates on account of a fully controlled AC regulator, whereas the symmetrical model of the source current do not allow any DC segment in the circuit, that is, the basic estimation of the source current is generally zero.

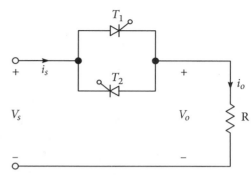

Figure 6.5 Circuit of a Single-Phase Full-Wave AC Voltage Controller with R Load.

Thyristor T1 is forward one-sided during the positive half-cycle of the input supply voltage. It is activated at a delay angle of α at 0 radians. Considering the on thyristor T1 as a perfect shut switch, the input supply voltage passes over the load resistor RL and the output voltage changes from V_o to V_s during $t = 0$ to π radians. The load current courses through the on thyristor T1 and through the load resistor RL in the descending heading during the conduction time of T1 from $t = 0$ to π radians.

Refer to Figure 6.6 for the waveform of a single-phase full-wave AC voltage controller with R load. At $t = \pi$, when the input voltage reduces to zero, the thyristor current, which is streaming through the load resistor R, also to zero and thus, T1 naturally turns off. No current streams in the circuit during $t = 0$ to π. Thyristor T2 is forward one-sided during the negative cycle of input supply and when thyristor T2 is activated at a delay angle α, the output voltage takes after the negative half-cycle of input from $t = 0$ to 2π. When T2 is on, the load current streams in the reverse course (upward bearing) through T2 during $t = 0$ to 2π radians. The time interim between the gate trigger of T1 and T2 is kept at π radians or 180°. At $t = 2\pi$, the input supply voltage reduces to zero and consequently, the load current also reduces to zero and thyristor T2 turn off normally.

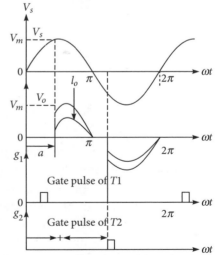

Figure 6.6 Waveform of a Single-Phase Full-Wave AC Voltage Controller with R Load.

Refer to Figure 6.7 that shows the single-phase fully controlled AC voltage controller with R load using MATLAB. It is advisable to control the power stream, the rms voltage and current by controlling the triggering angle α.

Solved MATLAB Example 2

The MATLAB model of a single-phase full-wave AC voltage controller with R load is shown in Figure 6.7. When thyristor T1 is activated at a trigger point, $0 \le \alpha \le \pi$, the switch is forward biased in the positive half-cycle. Considering thyristor switch T1 to be ideally on, the input voltage is seen over the resistive load R, where the output voltage is equal to the supply voltage during $\omega t = \alpha$ to π radians. The output voltage value flows through the descending line during the conduction period from $\omega t = \alpha$ to π radians. The supply voltage reduces to zero at $\omega t = \pi$, which in turns results in the thyristor current streaming through the load R tending to zero, subsequently making the conducting thyristor turn off, also resulting in the zero value of the current during the rest of the time period.

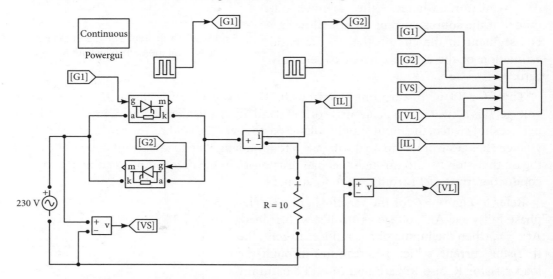

Figure 6.7 MATLAB Model of Single-Phase Full-Wave AC Voltage Controller with R Load.

In this simulation model, the amplitude is set for V_s = 230 V operated at 50 Hz. The phase delay is calculated for $\alpha = 30°$ for thyristors and the load is considered as $R = 10\ \Omega$.

Figure 6.8 shows the simulation waveform of a single-phase full-wave AC voltage controller with R load. Thyristor switch T2 is initially forward biased during the positive cycle of the supply. When thyristor T2 is activated at a triggering pulse, the output voltage starts towards the negative half of the input supply, i.e., during the period $\omega t = (\pi + \alpha)$ to 2π. When the thyristor T2 is on, the resistive load value streams in the opposite direction, that is, towards the upward bearing during $\omega t = (\pi + \alpha)$ to 2π radians. The interim of time between the firing pulses are kept at π radians.

When the time reaches $\omega t = 2\pi$, the supply voltage reduces to zero and pulls down the load current value to zero, which in turn turns off the thyristor.

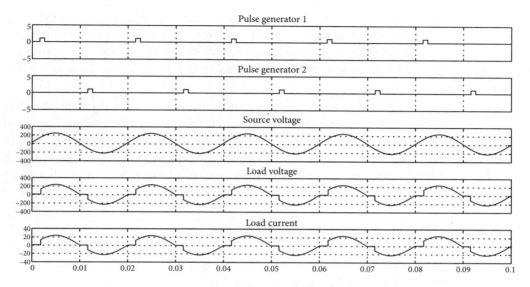

Figure 6.8 Simulation Waveform of a Single-Phase Full-Wave AC Voltage Controller with R Load.

6.3.3 Single-phase full-wave AC voltage controller with *RL* load

In this section, the operation and execution of a single-phase full-wave AC voltage controller with *RL* load is discussed. The majority of the loads in the voltage controller are of *RL* sort. The R communicates to the engine winding resistance and L speaks to the engine winding inductance.

A single-phase full-wave AC voltage controller circuit (bidirectional controller) with an *RL* load utilizing two thyristors T1 and T2, where T1 and T2 are two SCRs joined in parallel is shown in Figure 6.9. This set up can be utilized to actualize a full-wave AC controller if there are suitable rms load current and the rms output voltage values.

During the positive cycle of the input supply, thyristor T1 is forward biased. When T1 is activated at $t = \alpha$, the output voltage passes over the load after thyristor T1 is switched on due to the input supply voltage. The load current moves through T1 in the descending bearing and can be considered as the positive current pulse. Owing to the presence of inductance in the load terminal, the load current courses through T1, which would not reduce to zero when $t = \pi$; hence, the input voltage starts to be negative.

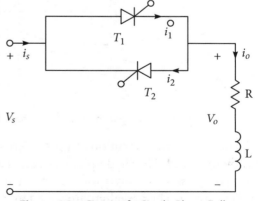

Figure 6.9 Circuit of a Single-Phase Full-Wave AC Voltage Controller with *RL* Load.

Thyristor T1 starts leading the load current until the energy is utilized from the inductor; then, the load current falls to zero at $t = \pi$ to β, where β represents the extinction angle, at which the load current reduces to zero. The elimination angle is measured from the starting of the positive half-cycle of the input supply to the point where the load current reduces to zero.

Thyristor T1 conducts from $t = \pi$ to β. The conduction angle of β for thyristor T1 is 2π, which relies on the postponement angle and the load impedance angle. The waveforms of the input supply voltage, the gate triggers T1 and T2, the thyristor current, the load current and the load voltage waveforms are shown in Figure 6.10. β is the extinction angle, which depends on the load inductance value. The output current of the voltage controller using RL load is given by

$$i_o = \frac{V_m}{2}\sin(\omega t - \emptyset) - \frac{V_m}{2}\sin(\alpha - \emptyset).\exp\left[\frac{R\alpha}{\omega L}\right].e^{-RT/L}$$

$$i_o = \frac{V_m}{2}\sin(\omega t - \emptyset) - \frac{V_m}{2}\sin(\alpha - \emptyset).\exp\left[\frac{-R}{\omega L}(\omega t - \alpha)\right] \quad (6.5)$$

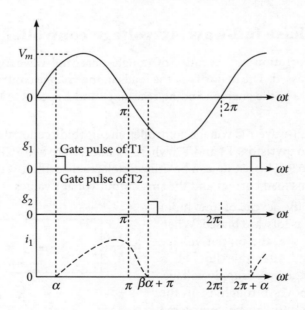

Figure 6.10 Waveform of a Single-Phase Full-Wave AC Voltage Controller with RL Load.

The operation and execution of a single-phase full-wave AC voltage controller with RL load is now analyzed. The single-phase circuit of a fully-controlled AC controller circuit with RL load comprising two thyristor switches, namely, T1 and T2 associated in parallel is shown in Figure 6.11. The best alternative for a thyristor could be a TRIAC, which is best suited for rms voltage and current values in real-time applications. However, the simulation analysis using TRIAC switch is not possible in MATLAB/SIMULINK.

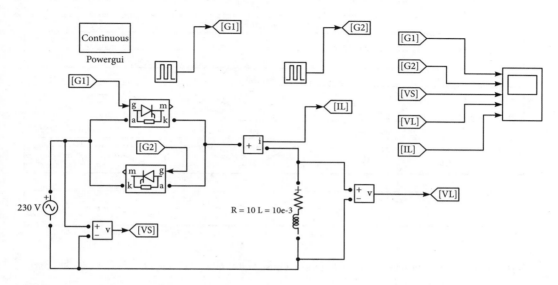

Figure 6.11 MATLAB Model of a Single-Phase Full-Wave AC Voltage Controller with *RL* Load.

Solved MATLAB Example 3

To initiate the operation of the circuit, during the positive supply of the source voltage, thyristor T1 is said to be forward biased and activated during the time period of $\omega t = \alpha$, using the trigger pulse. At this instant, the load voltage takes up after the supply source voltage, during the switch in the active state. The load current is said to course downwards and due to the presence of the inductance in the load, the load current does not reduce to zero at $\omega t = \alpha$. When the input supply begins negative, thyristor T1 will start to lead until the inductive energy is used completely, after which the load current reduces to zero at the point when $\omega t = \beta$. β is the extinction angle, which is measured from the start of the positive cycle of the supply source till the point at which the load current reduces to zero.

When the thyristor continues its conduction from $\omega t = \alpha$ to β, the conduction angle at this point is referred to as $\delta = (\beta - \alpha)$. This point completely relies on the triggering point α. Figure 6.12 shows the waveforms for firing pulses, supply voltage, load voltage, and load current.

Note AC–AC controllers can also be simulated using a tap-changing transformer based on the requirement. Tap-changing transformers have already been explained in Chapter 3.

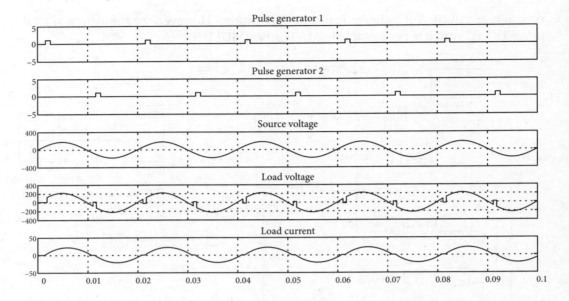

Figure 6.12 Simulation Waveform of a Single-Phase Full-Wave AC Voltage Controller with *RL* Load.

6.4 Cycloconverters and Its Types

A cycloconverter or a cycloinverter converts a consistent voltage, steady frequency AC waveform to another AC waveform of a lower frequency by arranging the output waveform from the AC supply without a transitional DC join. Generally, AC–AC conversion using semiconductor switches is done by two unique routes: (1) in two stages (AC–DC and afterwards DC–AC) as in DC join converters or (2) in one stage (AC–AC) as in cycloconverters. Cycloconverters are utilized as part of high-power applications driving and synchronous engines. They are normally stage controlled and they customarily utilize thyristors because of their simplicity of phase recompense. There are other more modern types of cycloconversion, for example, AC–AC grid converters and high-frequency AC–AC converters and these use self-controlled switches. These converters, in any case, are not common.

While phase-controlled SCR exchanging devices can be used for all types of cycloconverters requiring less effort and low control, TRIAC-based cycloconverters are naturally used for resistive load applications. The voltage and frequency of converters' output are both variable. The output to the input supply frequency ratio of a three-phase cycloconverter must be around 33 per cent for coursing current-mode cycloconverters or 50 per cent for blocking mode cycloconverters. Cycloconverters can be grouped with diverse output arrangements like the following

- Single-phase–single-phase step-up cycloconverters
- Single-phase–single-phase step-down cycloconverters
- Three-phase–single-phase cycloconverters
- Three-phase–three-phase cycloconverters

The cycloconverter has generally been utilized as drivers of high-power circuits, generally more than 1 MW, where none of the components of the circuit can be replaced further. The best examples are power plant drives over 5 MW and 13 MW wind passage fan drive and factory drives. The conventional set of thyristors requires an extensive number of thyristors, not less than 36, best suited for execution of complex control circuits. The converter also has some of these execution limits, the most noticeable of which is an output frequency restricted to around 33 percent of the input supply.

Converters are constructed using four set of thyristors, which are grouped into positive and negative banks. The output voltage is controlled using the phase control of the positive side when the current streams through the load, whereas the negative part thyristors are kept off; when negative current streams in the load, the positive thyristors are kept off. The thyristor is best kept in the off state, which is not leading, because the general mains may then be short circuited through the thyristor banks, bringing out a safer result. One of the major issues related to control of the circuit is to find the best way to swap between the banks in the best conceivable time to maintain a strategic distance from bending while guaranteeing that the two banks do not conduct in the meantime. A typical method by which the circuit always meets this necessity is to place an inside, tapped inductor in between the output terminals of the circuit terminals. Both banks can now lead together without shorting the mains. Similarly, the coursing current in the inductor keeps both banks working constantly, bringing about enhanced output waveforms. This procedure is not frequently utilized as the flowing current inductor has a tendency to be extravagant and cumbersome and the circling current decreases the power on the input supply.

6.4.1 Single-phase cycloconverters

To understand the operation standards of cycloconverters, the single-phase to single-phase cycloconverter is first analyzed. This converter comprises consecutive association of two full-wave rectifier circuits. The circuit of a single-phase cycloconverter is shown in Figure 6.13. For simplicity, let all thyristors go at $\alpha = 0$ terminating point, that is, let thyristors act like diodes. Note that the terminating points are a for the positive converter and b for the negative converter.

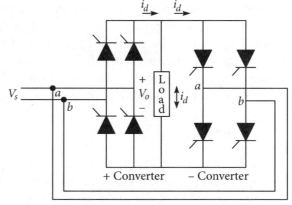

Figure 6.13 Single-Phase Cycloconverter.

Consider the operation of the cycloconverter to get one-fourth of the input frequency at the output. For the initial two cycles, the positive converter works supplying current to the load. It redresses the input voltage; thus, the load sees four positive half-cycles as shown in Figure 6.14. In the following two cycles, the negative converter works supplying current to the load in the converse bearing. The current waveforms do not appear in the figure as the resistive load current will have the same waveform as the voltage scaled by the resistance. Note that when one of the converters works, the other one is incapacitated, so that there is no current flowing between the two rectifiers.

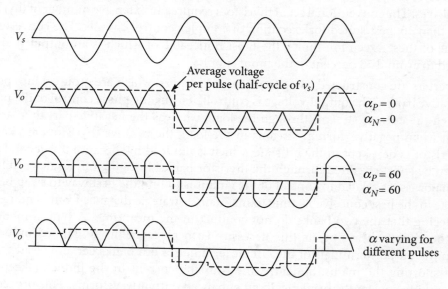

Figure 6.14 Waveform of a Single-Phase to Single-Phase Cycloconverter.

In a single-phase cycloconverter, the output frequency is not exactly the supply frequency. These converters help regular recompense, which is given by AC supply. During the positive half-cycle of the supply, thyristors T1 and T6 are forward one-sided. To begin with, the activating pulse is connected to T1 and consequently it begins directing. As the supply goes negative, T1 switches off, and in the negative half-cycle of the supply, T2 and T5 are forward one-sided. T2 is activated and subsequently it leads. In the following cycle of supply, T6 in the positive half-cycle and T1 in the negative half-cycle are activated. Subsequently, the output frequency is 1/2 times the supply frequency.

The frequency of the output voltage V_o in Figure 6.14 is four times as that of the input voltage,

That is, $f_o/f_i = 1/4$. (6.6)

Thus, this is a step-down cycloconverter. Then again, cycloconverters that have $f_o/f_i > 1$ are step-up cycloconverters. Note that progression down cycloconverters are more generally utilized than the progression up ones. The frequency of V_o can be changed by differing the quantity of positive and negative cycles. It can change as whole number products of f_i in 1Φ–1Φ cycloconverters. With the aformentioned operation, the 1Φ–1Φ cycloconverter can supply a specific voltage at a specific terminating edge α. The DC output of every rectifier is:

$$V_o = 2\sqrt{2}V_{rms}\cos\alpha \qquad (6.7)$$

Steady α operation gives a rough output waveform with a lot of harmonics. The dotted lines in Figure 6.14 demonstrate a square wave. If the square wave can be altered to look more like a sine wave, the sounds would be diminished. Thus, α is adjusted as shown in Figure 6.14. Currently, the six-ventured dotted line is more like a sine wave with less sound. The more pulse there are with various αs, the less are the harmonics.

The single-phase–single-phase cycloconverter comprises consecutive association of two full-wave rectifier circuits. Figure 6.15 shows a model of this converter with a resistive load. The input supply voltage is an AC voltage at a frequency. For simplicity, assume that the thyristors gets terminated when the condition is set to be $\alpha = 0°$. Thus at this instant, the thyristor behaves like a diode switch irrespective of the terminals.

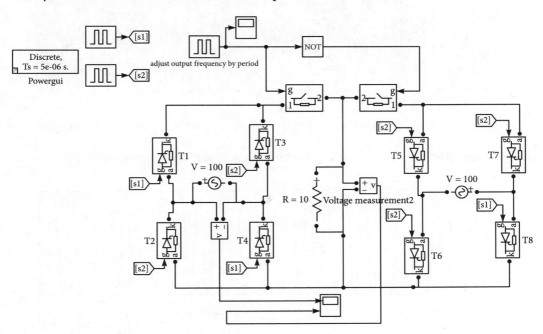

Figure 6.15 MATLAB Model of a Single-Phase to Single-Phase Cycloconverter.

MATLAB Example 4

The cycloconverter is set to one-fourth of the input frequency with V_s as the supply voltage for the initial two cycles, the converter acting under the positive region supplies current to the load terminal. The input supply is redressed consequently as shown in Figure 6.16.

Similarly, during the negative half of the cycle, the current heads in the opposite direction. When one converter is set to on, the other converter is debilitated, when no current circulates between the two set of converters, the waveform is as shown in Figure 6.17. It is to be noted that the waveforms of the negative cycle are not demonstrated because the current is the same.

6.4.1.1 Single-phase–single-phase step-up cycloconverters

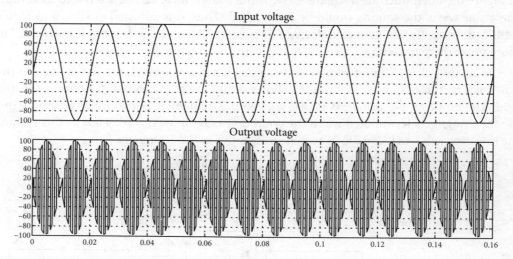

Figure 6.16 Simulation Waveform of a Single-Phase to Single-Phase Step-Up Cycloconverter.

6.4.1.2 Single-phase–single-phase step-down cycloconverters

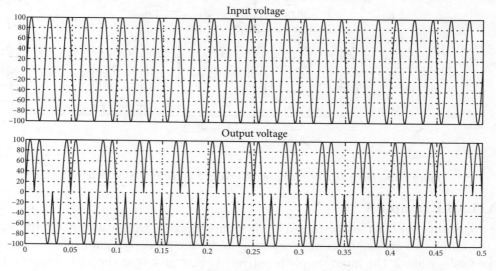

Figure 6.17 Simulation Waveform of a Single-Phase to Single-Phase Step-Down Cycloconverter.

Accordingly, by shifting α, the key output voltage can be controlled. Steady α operation gives a rough output waveform with lots of harmonics as shown in Figure 6.16. If the square wave can be adjusted to look more like a sine wave, the harmonics can be reduced. Currently, six-ventured applied sine waves with resonances are very less as diverse as result in less harmonics.

6.4.2 Three-phase cycloconverters

Three-phase cycloconverters are chiefly utilized as part of systems that involve AC machine drives. They are utilized with synchronous machines due to their output response qualities. The cycloconverter supplies a driving and solitary power element to the load during the input operation. The highly efficient synchronous machine, which draws any power and variable current from the converter, marks the exchange. Then again, adjacent machines can draw slacking current, so the cycloconverter does not have an angle contrasted with alternate converters in this perspective for running a prompting machine. Besides this, cycloconverters are utilized as a part of Scherbius drives for velocity control purposes driving wound rotor driving engines. Cycloconverters are compatible with rich output voltages, which will be discussed in later sections. When cycloconverters are utilized to run an AC machine, the leakage inductance of the machine channels a large portion of the higher frequency. The responses of the 3φ–1φ converters are similar to wye and delta sort of connections, where the converters are referred to as (3φ–3φ) cycloconverters; here, the output voltages are $2\pi/3$ radians phase moved from each other. The output waveforms of three 3φ–1φ converters of the same kind are joined in wye or delta and if the output voltages are $2\pi/3$ radians phase moved from one another, the subsequent converter is a three-phase to three-phase (3φ–3φ) cycloconverter.

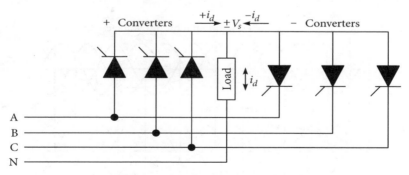

Figure 6.18 3Φ–1Φ Half-Wave Cycloconverter.

The converters in Figures 6.18 and 6.19 with wye associations will show the consequent operation of the converters. If the three converters associated are half-wave converters, then the new converter is known as a 3φ–3φ half-wave cycloconverter. Converters widely range themselves as 3φ–3φ bridge cycloconverters or 6-pulse cycloconverters or 36-thyristor cycloconverters, 3φ–3φ half-wave cycloconverters or 3-pulse cycloconverters or 18-thyristor cycloconverters.

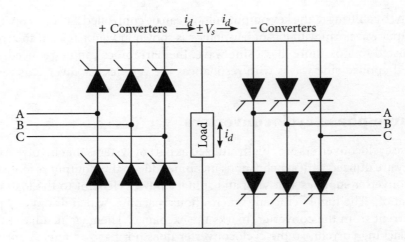

Figure 6.19 3Φ–1Φ Bridge Cycloconverter.

6.4.2.1 Three-phase–single-phase cycloconverters

There are two types of three-stage to single-stage (3Φ–1Φ) cycloconverters: 3Φ–1Φ half-wave cycloconverter and 3Φ–1Φ bridge cycloconverter. Like the 1Φ–1Φ type, the 3Φ–1Φ cycloconverter applies redressed voltage to the load. Both positive and negative converters can create voltages at either extremity; however, the positive converter can supply positive current and the negative converter can supply negative current. Subsequently, the cycloconverter can work in four quadrants: $(+v, +i)$ and $(-v, -i)$ correction modes and $(+v, -i)$ and $(-v, +i)$ reversal modes. The waveform of the output voltage and the major output voltage are shown in Figure 6.20.

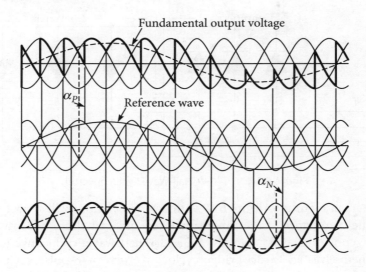

Figure 6.20 Waveforms of 3Φ–1Φ Cycloconverter.

The extremity of the current determines whether the positive or negative converter supplies energy to the load. As usual, the terminating plot for the positive converter is α_P, and that of the negative converter is α_N. When the extremity of the current changes, the converter already supplying the current is handicapped and the other one is empowered. The load dependably requires the major voltage to be constant. In this manner, during the current extremity inversion, the normal voltage supplied by both the converters should be equivalent. Changing from one converter to the next would bring about an undesirable voltage level. Therefore, converters are compelled to deliver the same normal voltage at all times. The accompanying condition for the terminating edges ought to be met.

$$\alpha_P + \alpha_N = V \tag{6.8}$$

The fundamental output voltage is given by

$$V(t) = 2V_0 \sin \omega t \tag{6.9}$$

where V_0 is the rms value of the fundamental voltage.

The fundamental voltage is also given by

$$V_o = \sqrt{2} V_o \sin t\omega \tag{6.10}$$

The converter is utilized to change three-phase AC waveform into single-phase AC waveform. To change three-phase–single-phase supply to a lower frequency, the fundamental rule is to shift continuously the terminating point of the three thyristors of the three-phase half-wave circuit. Three single-phase AC sources are utilized with 120° phase shift from one another to provide three-phase AC source. Down the middle-wave converter, six thyristors are utilized. These six converters are parted into two gatherings: positive gathering and negative gathering for bidirectional stream of current. In the positive gathering, the thyristors are operated at cathode mode. In the negative gathering, the thyristors are operated at anode mode. Since it works in the circulating the current-mode, the reactor is utilized.

MATLAB Example 5

The three-phase to single-phase cycloconverter, utilizing two three-phase half-wave converters, is shown in Figure 6.21. The rule of operation is same here as described earlier. Each thyristor is activated in succession, whether it is of P type or N type. It might be noticed that the thyristors are associated with the input phase supply individually, in arrangement with the load impedance, as shown in Figure 6.21. The frequency is 150 Hz, three times the input frequency of 50 Hz, as this converter is a three-pulse one. Along these lines, the inductance in the inductive RL load must be high, compared with one utilized as part of the prior case, to make the current constant. This inductance is used as the channel for the output current.

310 *Power Electronics with MATLAB*

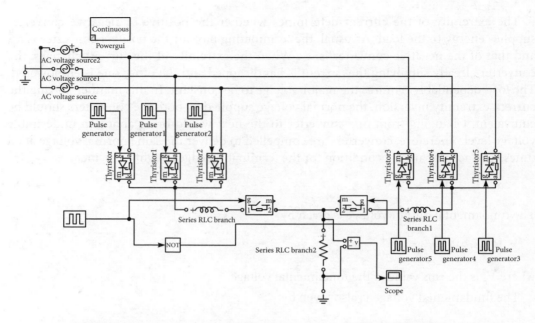

Figure 6.21 MATLAB Model of a Three-Phase to Single-Phase Cycloconverter.

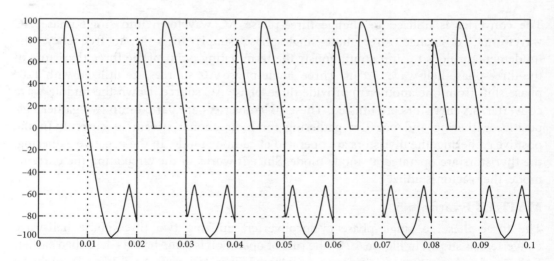

Figure 6.22 Simulation Waveform of a Three-Phase to Single-Phase Cycloconverter.

The simulation waveform of a three-phase to single-phase cycloconverter is shown in Figure 6.22. The method of operation here is a noncircling current one. It might be noticed that the harmonics in the output voltage is higher than those in the case utilizing two three-phase full-wave span converters. This is because six pulses are utilized as a part of a cycle for the earlier case, the frequency being 300 Hz. Additionally, three thyristors are utilized as a part of every converter, that is, a sum of just six devices for two converters are required, while in the earlier case, six thyristors are utilized for every extension converter, requiring a sum of 12 devices. This implies that the expense is much lower, as additionally the control circuit for this situation is much more direct and less expensive, as only three pulses for every converter are required. This may be favoured, as only the resonant substance is more, which may not be a disadvantage in the majority of the applications. If it is utilized to drive AC engines, the high impedance at the frequency is relied on to make the output current a close sinusoidal one, with the outcome that no extra sifting segment is required.

6.4.2.2 Three-phase to three-phase cycloconverters

A converter is said to be a three-phase to three-phase (3Φ–3Φ) cycloconverter, when the output response of three 3Φ–1Φ is in star or delta connection, and if the output voltage response are $2\Phi/3$ radians phase shifted with each other. The cycloconverter with star connections is shown in Figures 6.23 and 6.24.. If the three converters are connected in half-wave converter method, then the converter is said to be a 3Φ–3Φ half-wave; if the converters are connected in bridge method, then the converter is said to be a bridge converter. The three-pulse half-wave converter is also called an 18-thyristor cycloconverter and the bridge converter is said to be a six-pulse cycloconverter or a 36-thyristor cycloconverter.

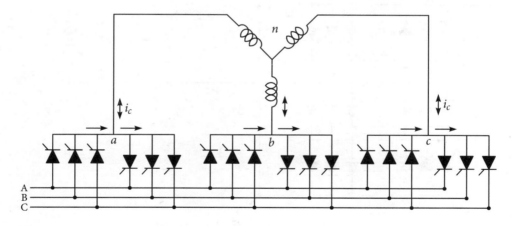

Figure 6.23 3Φ–3Φ Half-Wave Cycloconverter.

The circuit of a three-phase to three-phase cycloconverter is shown in Figure 6.23. Two three-phase half-wave converters associated consecutive for every phase, with three thyristors for every extension, are required here. The aggregate number of thyristors utilized is 18, in this manner decreasing the expense of power parts furthermore, less control circuits are expected to create the terminating pulse for the thyristors, as will be explained later. This may be compared with the six three-phase full-wave six-pulse span converters, having six thyristors for every converter, with the aggregate devices utilized being 36. Although this will diminish the harmonics in both output voltage and current waveforms, they are expensive. The frequency is 150 Hz, three times the frequency of 50 Hz. In Figures 6.23 and 6.24, the flowing current method of operation is utilized, in which both positive and negative converters in every phase conduct. Between gatherings, a reactor in every phase is required. However, if noncoursing current method of operation is utilized, where one and only converter is positive or negative in every phase, and the circuit conducts at once, reactors are not required.

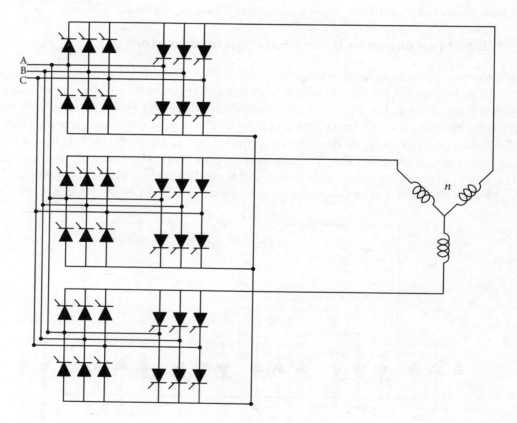

Figure 6.24 3Φ–3Φ Bridge Cycloconverter.

Three-phase cycloconverters are for the most part utilized as part of AC machine drive systems running a three-phase synchronous machine. Synchronous machines are more effective due to their output power variable attributes. Thus, a cycloconverter can supply solitary power component loads when its input is continually decreasing. Power can be drawn from any current element from the converter in case of cycloconverters. This makes the cycloconverter useful in the synchronous machine.

However, driving machines can draw current, so the cycloconverter does not have control compared with alternate converters for running an affectation machine. Despite this, cycloconverters are used as part of Scherbius drives for rate control purposes driving wound rotor excitation engines.

Cycloconverters produce resonant rich output voltages, which we will consider in later sections. When cycloconverters are used to run an AC machine, the leakage inductance of the machine channels the majority of the higher frequency sounds and lessens the extent of the lower order harmonics.

6.4.2.2.1 Noncirculating mode and circulating current mode

The operation of cycloconverters has already been explained earlier. When the load current is sure, the positive converter supplies the required voltage and the negative converter is impaired. During negative load current, the converter supplies the voltage at the negative end, while the positive converter is blocked. This process is known as blocked or noncirculating mode. Cycloconverters that use this methodology are called blocking-mode cycloconverters.

When both the converters are empowered, the supply is short circuited. To avoid this, an inter group reactor (IGR) can be added between the converters if required. Rather than obstructing the converters during current inversion, if they are both empowered, then a circling current is created. This current is known as the flowing current. It is unidirectional as thyristors permit the current to stream in one and only heading. Some cycloconverters permit this flowing current at all times. These are called coursing current cycloconverters.

Noncirculating mode

Noncirculating mode cycloconverters do not let circling current stream, and accordingly, they do not need IGRs. When the current reduces to zero, the converters are blocked on both the positive and negative ends. The load current stops in turn when the converter stays off for a short while. At that point, contingent upon the maximum current, one of the converters is empowered. With every zero intersection of the current, the converter, which was incapacitated before the zero intersection, is empowered. A switch flip-flop, which flips when the current goes to zero, can be utilized for this. The operation waveforms for a three-phase blocking-mode or noncirculating mode cycloconverter are shown in Figure 6.25.

The blocking-mode operation has a few advantages and disadvantages compared to the flowing-mode operation. During the delay time, the current stays at zero twisting the voltage and current waveforms. This contortion implies more harmonics contrasted with the circulating-mode cycloconverters. Notwithstanding this, the current inversion issue brings more control for these cycloconverters. No massive IGRs are utilized, so the size and cost is not as much as that of the circling current case. Another advantage is that one and only converter is in conduction at all times as opposed to two. This implies less problems and higher effectiveness.

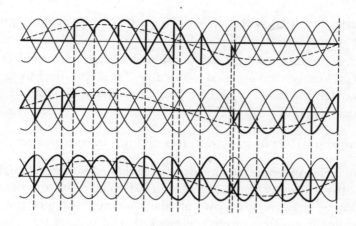

Figure 6.25 Noncirculating Mode.

Circulating mode

For this situation, both the converters work at all times delivering the same major output voltage. The terminating points of the converters fulfil the terminating edge condition; in this way when one converter is in correction mode, the other one is in reversal mode and vice versa. If both the converters deliver unmodified sine waves, then there would not be any coursing current as the quick potential contrast between the outputs of the converters would be zero. As a general rule, an IGR is associated between the outputs of two-phase-controlled converters in either amendment or reversal mode. The voltage waveform over the IGR is shown in Figure 6.26. This shows the difference in the momentary output voltages delivered by the two converters. Note that it is zero when both the converters create the same quick voltage. The inside tap voltage of an IGR is the voltage connected to the load and it is the mean of the voltages connected to the terminals of IGR; accordingly the load voltage value is diminished.

The coursing current cycloconverter applies a smoother load voltage with less sounds contrasted with the blocking-mode case. Additionally, the control is straightforward as there is no current inversion issue. However, the cumbersome IGR is a major load for this converter. Also, the quantity of devices leading is twice that of the blocking-mode converter. Because of these inconveniences, this cycloconverter is not much used.

The blocked-mode cycloconverter converter and the circling-current cycloconverter can be consolidated to give a half-cycle system, which has the advantages of both. The resulting cycloconverter resembles a flowing-mode cycloconverter circuit; however, relying on the maximum of the output current, one and only converter is empowered; the other is incapacitated as with the blocking-mode cycloconverters. When the load current reduces below a point, both the converters are empowered. In this way, the current has a smooth inversion. At the point when the current increases over a point in the other heading, the active converter is debilitated. This cycloconverter works in the blocking mode more often than not, so a small IGR can be utilized. The productivity is marginally higher than that of the flowing current cycloconverter yet significantly less than the blocking-mode cycloconverter. However, the contortion created by the blocking-mode operation vanishes because of the coursing current operation around zero current. Also, the control of the converter is less intricate than that of the blocking-mode cycloconverter.

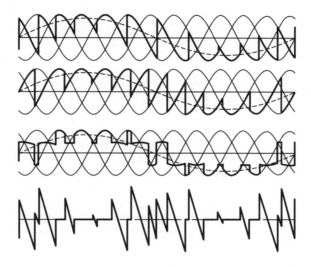

Figure 6.26 Circulating Mode.

AC–AC conversions in industry mostly involve cycloconverters. Innovative cycloconversion are of great research interest.

MATLAB Example 6

It might be noticed that the circuit in each of the three phases is like the cycloconverter circuit shown in Figure 6.27. The terminating succession of the thyristors for the phase B and C are the same as that for phase A, yet they are slack by the angle and separate. Subsequently, an adjusted three-phase voltage is acquired at the output terminals, to be boosted to the three-phase load. The normal estimation of the output voltage is changed by fluctuating the terminating points of the thyristors, though its frequency is shifted by changing the time interim, after which the following thyristor is activated. With an adjusted load, the unbiased association is slightly more and may be precluded, in this manner. Typically, the output frequency of the cycloconverter is lower than the supply frequency, constrained to around 33 per cent of it. This is essential for acquiring reasonable force output, proficiency and harmonics. If the output frequency is to be expanded, the harmonic content in the output voltage builds up as its waveform is made out of less fragments of the supply voltage. In this manner, the problems in cycloconverters in AC engines increase. By utilizing more complex converter circuits with higher pulse numbers, the output voltage waveform is enhanced, with the ratio of output to input frequency expanded to around fifty per cent. Refer to Figure 6.28 for simulation waveform of the three phases.

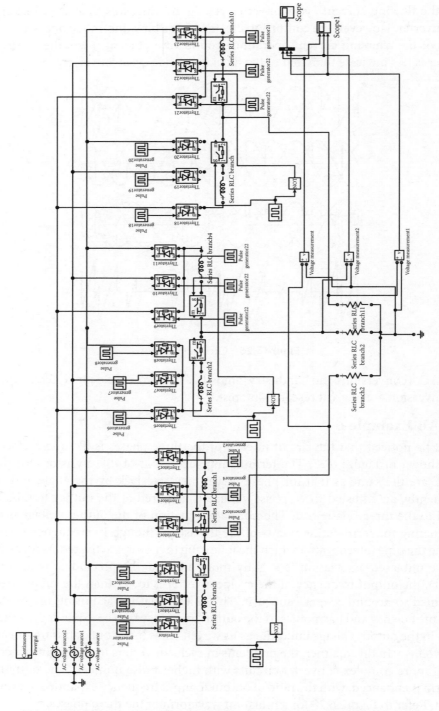

Figure 6.27 MATLAB model of a Three-Phase to Three-Phase Cycloconverter.

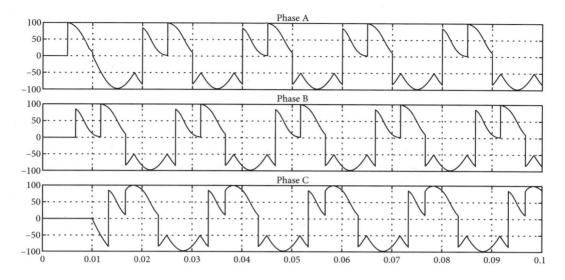

Figure 6.28 Simulation Waveform of a Three-Phase to Three-Phase Cycloconverter.

6.5 Load-Commutated Cycloconverter

The concept of forced commutation is applied in a step-up cycloconverter where additional circuitry is essential for force commutation; however, it does not depend on the source or load voltage. A step-down cycloconverter relies on natural commutation. The natural or line commutation is provided by the supply voltage. In such cases, the output frequency is less than the input supply frequency.

Here, the concept of load commutation differs from natural and line commutation. In load commutation, thyristors are commutated by reversal of the load voltage. This implies that the load circuit must have a generated electromagnetic field that should be independent of the source voltage. The best-suited example is a permanent magnet synchronous machine. In case of such loads, the load frequency may be equal to or greater than the source frequency for both these cases; thyristors will be naturally commutated by the reversal of load circuit emf.

6.6 Matrix Converter

The matrix converter is a conventional rectifier–inverter power frequency converter. It gives sinusoidal output and output waveforms, with negligible higher sounds and no subharmonics; it has inborn bidirectional energy stream ability; the input power element can be completely controlled. It has insignificant energy storing prerequisites, which disposal of massive and lifetime-constrained energy-consuming capacitors. However, the lattice converter has a few disadvantages. It has the greatest output voltage exchange proportion constrained to 87 per cent for sinusoidal input and output waveforms. It requires more semiconductor devices than an ordinary AC–AC circuitous power frequency converter. Since no solid bidirectional switches exist, discrete unidirectional devices, differently created, must be utilized for every bidirectional switch. It is especially sensitive to the unsettling influences of the input voltage system.

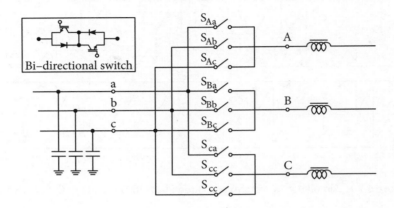

Figure 6.29 Schematic of Matrix Converter.

The matrix converter comprises nine bidirectional switches that permit any output to be joined with any input. The circuit plan is as shown in Figure 6.29. The output terminals of the converter are joined with a three-phase voltage-sustained system, more often than not a matrix, while the output terminals are associated with a three-phase current-encouraged system. The capacitive channel on the voltage-boosted side and the inductive channel on the current-sustained side are characteristically vital. Their size is reverse corresponding to the system converter exchanging frequency. It is because of its natural bidirectionality and symmetry that a double association is additionally possible for the matrix converter: a current-encouraged system at the input and a voltage-fed system at the output.

With nine bidirectional switches, the system converter can hypothetically accept distinctive exchanging state combinations. However, not all the switches can be helpfully utilized. For control system utilization, the decision of the system converter changing states mixes to be utilized depends on two fundamental tenets – considering that the converter is supplied by a voltage source and normally boosts an inductive load, the input stages should never be short-circuited and the output streams should not be interfered. From a pragmatic perspective, these guidelines suggest that unrivaled one bidirectional switch per output stage must be exchanged at any moment. By this requirement, in a three-phase to three-phase system converter, 27 exchanging combinations are allowed.

6.7 Applications of Voltage Controllers

The various applications of voltage controllers include the following:
- Illumination and lightening control in AC power circuits
- Induction heating appliances
- Tap changing transformer
- Speed control of AC engine

> **Summary**
>
> At the end of this chapter, students will be able to understand the concepts of a AC voltage regulator and its types. Students can get a clear view about the industrial applications of these regulators. The cycloconverter and its types have been explained and students can understand the use of various applications in power electronics era.

Solved Problems

1. Determine the on–off time intervals, rms output voltage, input PF, average thyristor current and rms thyristor current for a full-wave AC voltage controller having 230 V at 50 Hz supply with load = 50. T_{on} = 30 cycles and T_{off} = 40 cycles, respectively.

   Solution

   $$V_{in\,rms} = 230 \text{ V}, \quad V_m = \sqrt{2} \times 230 \text{ V} = 325.269 \text{ V}, \quad V_m = 325.269 \text{ V}$$

   $$T = \frac{1}{f} = \frac{1}{50 \text{ Hz}} = 0.02 \text{ s}, \quad T = 20 \text{ ms}.$$

   N = number of input cycles when controller is on; n = 30.

   m = number of input cycles when controller is off; m = 40.

   $$t_{ON} = n \times T = 30 \times 20 \text{ ms} = 600 \text{ ms} = 0.6 \text{ s}$$

   $$t_{ON} = n \times T = 0.6 \text{s} = \text{controller on time}$$

   $$t_{OFF} = m \times T = 40 \times 20 \text{ ms} = 800 \text{ ms} = 0.8 \text{ s}$$

   $$t_{OFF} = m \times T = 0.8 \text{s} = \text{controller off time}$$

   $$\text{Duty cycle } k = \frac{n}{m+n} = \frac{30}{40+30} = 0.4285$$

RMS output voltage

$$V_{O\,rms} = V_{i\,rms} \times \sqrt{\frac{n}{m+n}}$$

$$V_{O\,rms} = 230\,V \times \sqrt{\frac{30}{30+40}} = 230\sqrt{\frac{3}{7}}$$

$$V_{O\,rms} = 230\,V \times \sqrt{0.42857} = 230 \times 0.65465$$

$$Vo_{rms} = 150.570\,V$$

$$I_{O\,rms} = \frac{V_{O\,rms}}{Z} = \frac{V_{O\,rms}}{R_L} = \frac{150.570\,V}{50\ 3} = 3.0114\,A$$

$$P_O = I_{O\,rms}^2 \times R_L = 3.0114^2 \times 50 = 453.426498\,W$$

Input PF $= \sqrt{k}$

$$PF = \sqrt{\frac{n}{m+n}} = \sqrt{\frac{30}{70}} = \sqrt{0.4285}$$

PF = 0.654653

Average thyristor current rating

$$I_{T\,Avg} = \frac{I_m}{Ð} \times \left(\frac{n}{m+n}\right) = \frac{k \times I_m}{Ð}$$

where $I_m = \frac{V_m}{R_L} \times \frac{\sqrt{2} \times 230}{50} = \frac{325.269}{50}$

$I_m = 6.505382\,A =$ Peak (maximum) thyristor current

$$I_{T\,Avg} = \frac{6.505382}{\pi} \times \left(\frac{3}{7}\right)$$

$I_{T\,Avg} = 0.88745\,A$

RMS current rating of thyristor

$$I_{T\,rms} = \frac{I_m}{2}\sqrt{\frac{n}{m+n}} = \frac{I_m}{2}\sqrt{k} = \frac{6.505382}{2} \times \sqrt{\frac{3}{7}}$$

$I_{T\,rms} = 2.129386\,A$

2. For a single-phase half-wave controller operated at 230 V rms at 50 Hz having load resistance $R = 50$, calculate rms output voltage, output power, rms load current, average load current, input PF, average thyristor current and rms thyristor current, by assuming T_1 is triggered at 60° and that the transformer turns ratio is 1:1.

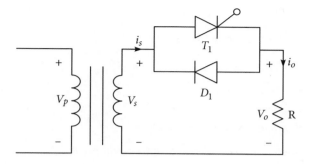

Solution

V_p = 230 V, rms primary supply voltage

f = Input supply frequency = 50 Hz

$R_L = 50$

$$\alpha = 60° = \frac{\pi}{3} \text{ radians}$$

V_s = rms secondary voltage

$$\frac{V_p}{V_s} = \frac{N_P}{N_S} = \frac{1}{1} = 1$$

Therefore, $V_p = V_s = 230$ V

where, N_P = Number of turns in the primary winding.

N_S = Number of turns in the secondary winding.

rms value of output (load) voltage $V_{o\,rms}$

$$V_{O\,rms} = \sqrt{\frac{1}{2\pi} \int V_m^2 \sin^2 \omega t \cdot d\omega t}$$

We have obtained the expression for $V_{o\,rms}$ as

$$V_{O\,rms} = V_s \sqrt{\frac{1}{2\pi}[2\pi - \alpha] + \frac{\sin 2\alpha}{2}}$$

$$V_{O\,rms} = 230 \sqrt{\frac{1}{2\pi}\left[2\pi - \frac{\pi}{3}\right] + \frac{\sin 120°}{2}}$$

$$V_{O\,rms} = 230 \sqrt{\frac{1}{2\pi} \cdot 5.669} = 230 \times 0.94986$$

$$V_{O\,rms} = 218.4696 \text{ V} \approx 218.47$$

rms load current $I_{o\,rms}$

$$I_{O\,rms} = \frac{V_{O\,rms}}{R_L} = \frac{218.46966}{50} = 4.3639 \text{ A}$$

Output load power P_o

$$P_o = I_{O\,rms}^2 \times R_L = 4.36939^2 \times 50 = 954.5799 \text{ W}$$

$$P_o = 0.9545799 \text{ kW}$$

Input PF $\quad PF = \dfrac{P_o}{V_s \times I_s}$

V_s = rms secondary supply voltage = 230 V.

I_s = rms secondary supply current = rms load current

$I_s = I_{O\,rms} = 4.36969 \text{ A}$

$$PF = \dfrac{954.5799 \text{ W}}{230 \times 4.36939 \text{ W}} = 0.9498$$

Average output (load) voltage

$$V_{ODC} = \dfrac{1}{2\pi}\left[\int V_m \sin\omega t.d\omega t\right]$$

We have obtained the expression for the average/DC output voltage as

$$V_{ODC} = \dfrac{V_m}{2\pi}\cos\alpha - 1$$

$$V_{ODC} = \dfrac{\sqrt{2} \times 230}{2\pi}[\cos 60° - 1] = \dfrac{325.2691193}{2\pi}0.5 - 1$$

$$V_{ODC} = \dfrac{325.2691193}{2\pi}0.5 = -25.88409 \text{ V}$$

Average DC load current

$$I_{ODC} = \dfrac{V_{ODC}}{R_L} = \dfrac{-25.88409}{50} = -0.51768 \text{ A}$$

Average and rms thyristor currents

$$I_{T\,Avg} = \dfrac{1}{2\pi}\left[\int_\alpha^\pi I_m \sin\omega t.d\omega t\right]$$

$$I_{T\,Avg} = \dfrac{I_m}{2\pi}\left[\int_\alpha^\pi \sin\omega t.d\omega t\right]$$

$$I_{T\,Avg} = \dfrac{I_m}{2\pi}\left[-\cos\omega t\,/_\alpha^\pi\right]$$

$$I_{T\,Avg} = \dfrac{I_m}{2\pi}\left[-\cos\pi - \cos\alpha\right]$$

$$I_{T\,Avg} = \dfrac{I_m}{2\pi}\left[1+\cos\alpha\right]$$

where $I_m = \dfrac{V_m}{R_L} =$ Peak thyristor current $=$ Peak load current

$$I_m = \dfrac{\sqrt{2} \times 230}{50}$$

$$I_m = 6.505382 \text{ A}$$

$$I_{T\,Avg} = \dfrac{V_m}{2\pi R_L}[1 + \cos\alpha]$$

$$I_{T\,Avg} = \dfrac{\sqrt{2} \times 230}{50 \times 2\pi}[1 + \cos 60°]$$

$$I_{T\,Avg} = \dfrac{\sqrt{2} \times 230}{100\pi}[1 + 0.5]$$

$$I_{T\,Avg} = 1.5530 \text{ A}$$

rms thyristor current $I_{T\,rms}$ can be calculated using the expression

$$I_{T\,rms} = \sqrt{\dfrac{1}{2\pi}\left[\int_\alpha^\pi I_m^2 \sin^2\omega t \cdot d\omega t\right]}$$

$$I_{T\,rms} = \sqrt{\dfrac{I_m^2}{2\pi}\left[\int_\alpha^\pi \dfrac{1 - \cos 2\omega t}{2} d\omega t\right]}$$

$$I_{T\,rms} = I_m\sqrt{\dfrac{1}{4\pi}\left[\omega t \Big/_\alpha^\pi - \left(\dfrac{\sin 2\omega t}{2}\right)\Big/_\alpha^\pi\right]}$$

$$I_{T\,rms} = \sqrt{\dfrac{1}{4\pi}\left[\pi - \alpha - \left(\dfrac{\sin 2\pi - \sin 2\alpha}{2}\right)\right]}$$

$$I_{T\,rms} = \sqrt{\dfrac{1}{4\pi}\left[\pi - \alpha + \left(\dfrac{\sin 2\alpha}{2}\right)\right]}$$

$$I_{T\,rms} = \dfrac{1}{\sqrt{2}}\sqrt{\dfrac{1}{2\pi}\left[\left(\pi - \dfrac{\pi}{3}\right) + \left(\dfrac{\sin 120°}{2}\right)\right]}$$

$$I_{T\,rms} = 4.6\sqrt{\dfrac{1}{2\pi}\left[\left(\dfrac{2\pi}{3}\right) + \left(\dfrac{0.8660254}{2}\right)\right]}$$

$$I_{T\,rms} = 4.6 \times 0.6342 = 2.91746 \text{ A}$$

$$I_{T\,rms} = 2.91746 \text{ A}$$

3. For a single-phase full-wave controller using RL load with input supply of 230 V at 50 Hz, the load value is L = 10 mH, R = 10 Ω. The delay angles are equal where $\alpha_1 = \alpha_2 = \dfrac{\pi}{3}$. Calculate the conduction angle of T1, rms output voltage and input PF. Comment on the result.

Solution

$$V_S = 230 \text{ V}, \ f = 50 \text{ Hz}, \ L = 10 \text{ mH}, \ R = 10 \ \Omega, \ \alpha = 60°, \ \alpha = \alpha_1 = \alpha_2 = \dfrac{\pi}{3} \text{ radians}$$

$$V_m = \sqrt{2}V_S = \sqrt{2} \times 230 = 325.2691193 \text{ V}$$

$$Z = \text{Load impedance} = \sqrt{R^2 + \omega L^2} = \sqrt{10^2 + \omega L^2}$$

$$\omega L = 2\pi f L = 2\pi \times 50 \times 10 \times 10^{-3} = \pi = 3.14159 \ \Omega$$

$$Z = \sqrt{10^2 + 3.14159^2} = \sqrt{109.8696} = 10.4818 \ \Omega$$

$$I_m = \dfrac{V_m}{Z} = \dfrac{\sqrt{2} \times 230}{10.4818} = 31.03179 \text{ A}$$

Load impedance angle $\phi = \tan^{-1}\left(\dfrac{\omega L}{R}\right)$

$$\phi = \tan^{-1}\left(\dfrac{\pi}{10}\right) = \tan^{-1}(0.314159) = 17.44059°$$

Trigger angle $\alpha > \beta$. Hence, the type of operation will be a discontinuous load current operation. We get

$$\beta < \pi + \alpha$$

$$\beta < 180° + 60° \ ; \ \beta < 240°$$

Therefore, the range of β is from 180° to 240°. $180° < \beta < 240°$

Extinction angle β is calculated using the equation

$$\sin(\beta - \phi) = \sin(\alpha - \phi)e^{\dfrac{-R}{\omega L}(\beta - \alpha)}$$

In the exponential term, the value of α and β should be substituted in radians. Hence,

$$\sin(\beta - \phi) = \sin(\alpha - \phi)e^{\dfrac{-R}{\omega L}(\beta_{Rad} - \alpha_{Rad})} \ ; \ \alpha_{Rad} = \left(\dfrac{\pi}{3}\right)$$

$$\alpha - \phi = 60 - 17.44059 = 42.5594°$$

$$\sin(\beta - 17.44°) = \sin(42.5594°)e^{\dfrac{-10}{\pi}(\beta - \alpha)}$$

$$\sin(\beta - 17.44°) = 0.676354 e^{-3.183(\beta - \alpha)}$$

$$180° \rightarrow \pi \text{ radians}, \quad \beta_{Rad} = \frac{\beta° \times \pi}{180°}$$

Assuming $\beta = 190°$

$$\beta_{Rad} = \frac{\beta° \times \pi}{180°} = \frac{190° \times \pi}{180°} = 3.3161$$

L.H.S: $\sin(190° - 17.44°) = \sin 172.56 = 0.129487$

R.H.S: $0.676354 \times e^{-3.183\left(3.3161 \times \frac{\pi}{3}\right)} = 4.94 \times 10^{-4}$

Assuming $\beta = 183°$

$$\beta_{Rad} = \frac{\beta° \times \pi}{180°} = \frac{183° \times \pi}{180°} = 3.19395$$

$$\beta - \alpha = \left(3.19395 - \frac{\pi}{3}\right) = 2.14675$$

L.H.S: $\sin(\beta - \phi) = \sin(183 - 17.44) = \sin 165.56° = 0.24936$

R.H.S: $0.676354 \times e^{-3.183 \times 2.14675} = 7.2876 \times 10^{-4}$

Assuming $\beta = 180°$

$$\beta_{Rad} = \frac{\beta° \times \pi}{180°} = \frac{180° \times \pi}{180°} = \pi$$

$$\beta - \alpha = \left(\pi - \frac{\pi}{3}\right) = \frac{2\pi}{3}$$

L.H.S: $\sin(\beta - \phi) = \sin(180 - 17.44) = 0.2997$

R.H.S: $0.676354 e^{-3.183\left(\pi - \frac{\pi}{3}\right)} = 8.6092 \times 10^{-4}$

Assuming $\beta = 196°$

$$\beta_{Rad} = \frac{\beta° - \pi}{180°} = \frac{196° - \pi}{180°} = 3.420845$$

L.H.S: $\sin(\beta - \phi) = \sin(196 - 17.44) = 0.02513$

R.H.S: $0.676354 e^{-3.183\left(3.420845 \times \frac{\pi}{3}\right)} = 3.5394 \times 10^{-4}$

Assuming $\beta = 197°$

$$\beta_{Rad} = \frac{\beta° - \pi}{180°} = \frac{197° - \pi}{180°} = 3.43829$$

L.H.S: $\sin(\beta-\phi) = \sin(197-17.44) = 7.69 = 7.67937 \times 10^{-3}$

R.H.S: $0.676354 e^{-3.183\left(3.43829 \times \frac{\pi}{3}\right)} = 4.950386476 \times 10^{-4}$

Assuming $\beta = 197.42°$

$$\beta_{Rad} = \frac{\beta° - \pi}{180°} = \frac{197.42° - \pi}{180°} = 3.4456$$

L.H.S: $\sin(\beta - \phi) = \sin(197.42 - 17.44) = 3.4906 \times 10^{-4}$

R.H.S: $0.676354 e^{-3.183\left(3.4456 \times \frac{\pi}{3}\right)} = 3.2709 \times 10^{-4}$

Conduction angle $\delta = \beta - \alpha = 197.42° - 60° = 137.42°$

RMS output voltage

$$V_{o\,rms} = V_S \sqrt{\frac{1}{\pi}\left[(\beta - \alpha) + \frac{\sin 2\alpha}{2} - \frac{\sin 2\beta}{2}\right]}$$

$$V_{o\,rms} = 230 \sqrt{\frac{1}{\pi}\left[\left(3.4456 - \frac{\pi}{3}\right) + \frac{\sin 2 \times 60°}{2} - \frac{\sin 2 \times 197.42°}{2}\right]}$$

$$V_{o\,rms} = 230 \sqrt{\frac{1}{\pi}[2.39843 + 0.4330 - 0.285640]}$$

$$V_{o\,rms} = 230 \times 0.9 = 20.7.0445 \text{ V}$$

Input PF

$$PF = \frac{P_o}{V_S \times I_S}$$

$$I_{o\,rms} = \frac{V_{o\,rms}}{Z} = \frac{207.04445}{10.4818} = 19.7527 \text{ A}$$

$$P_o = I_{o\,rms}^2 \times R_L = 19.7527^2 \times 10 = 3901.716 \text{ W}$$

$$V_S = 230 \text{ V}, \quad I_S = I_{o\,rms} = 19.7527 \text{ A}$$

$$PF = \frac{P_o}{V_S \times I_S} = \frac{3901.716}{230 \times 19.7527} = 0.858$$

4. For a 120 V single-phase full-wave controller operated with load resistance of 6 Ω, determine the rms output voltage, average thyristor current, power and PF for triggering angle π/2.

Solution

$$\alpha = \frac{\pi}{2} = 90°, \quad V_s = 120 \text{ V}, \quad R = 6 \, \Omega$$

RMS value of output voltage

$$V_o = V_s \left[\frac{1}{\pi} \left(\pi - \alpha + \frac{\sin 2\alpha}{2} \right) \right]^{\frac{1}{2}}$$

$$V_o = 120 \left[\frac{1}{\pi} \left(\pi - \frac{\pi}{2} + \frac{\sin 180}{2} \right) \right]^{\frac{1}{2}}$$

$$V_o = 84.85 \text{ V}$$

RMS output current

$$I_o = \frac{V_o}{R} = \frac{84.85}{6} = 14.14 \text{ A}$$

Load power

$$P_0 = I_o^2 \times R$$

$$P_0 = 14.14^2 \times 6 = 1200 \text{ W}$$

Input current is the same as load current

Therefore, $I_s = I_o = 14.14$ A

Input supply volt–ampere (VA) = $V_s I_s$ = 120 × 14.14 = 1696.8 VA

Therefore,

$$\text{Input PF} = \frac{\text{Load power}}{\text{Input volt-amp}} = \frac{1200}{1696.8} = 0.707 \text{ lag}$$

Each thyristor conducts only for a half-cycle

Average thyristor current $I_{T \text{ Avg}}$

$$I_{T \text{ Avg}} = \frac{1}{2\pi R} \int_{\alpha}^{\pi} V_m \sin \omega t \cdot d\omega t$$

$$= \frac{V_m}{2\pi R}(1 + \cos \alpha); \quad V_m = \sqrt{2} V_s$$

$$= \frac{\sqrt{2} \times 120}{2\pi \times 6}(1 + \cos 90) = 4.5 \text{ A}$$

rms thyristor current $I_{T\,rms}$

$$I_{T\,rms} = \sqrt{\frac{1}{2\pi}\int_{\alpha}^{\pi}\frac{V_m^2\sin^2\omega t}{R^2}d\omega t}$$

$$= \sqrt{\frac{V_m^2}{2\pi R^2}\int_{\alpha}^{\pi}\frac{1-\cos 2\omega t}{2}d\omega t}$$

$$= \frac{V_m}{2R}\left[\frac{1}{\pi}\left(\pi - \alpha + \frac{\sin 2\alpha}{2}\right)\right]^{\frac{1}{2}}$$

$$= \frac{\sqrt{2}V_s}{2R}\left[\frac{1}{\pi}\left(\pi - \alpha + \frac{\sin 2\alpha}{2}\right)\right]^{\frac{1}{2}}$$

$$= \frac{\sqrt{2}\times 120}{2\times 6}\left[\frac{1}{\pi}\left(\pi - \frac{\pi}{2} + \frac{\sin 180}{2}\right)\right]^{\frac{1}{2}}$$

$$= 10\text{ A}$$

5. A single-phase half-wave AC regulator using one SCR in antiparallel with a diode feeds 1 kW at 230 V heater. Find load power for a firing angle of 45°.

Solution

$$\alpha = 45° = \frac{\pi}{4},\ V_s = 230\text{ V},\ P_0 = 1\text{ kW} = 1000\text{ W}$$

At standard rms supply voltage of 230 V, the heater dissipates 1 kW of output power
Therefore,

$$P_0 = V_0 \times I_0 = \frac{V_0 \times V_0}{R} = \frac{V_0^2}{R}$$

Resistance of heater

$$R = \frac{V_0^2}{P_0} = \frac{230^2}{1000} = 52.9\ \Omega$$

rms value of output voltage

$$V_0 = V_s\left[\frac{1}{2\pi}\left(2\pi - \alpha + \frac{\sin 2\alpha}{2}\right)\right]^{\frac{1}{2}};\text{ for firing angle } \alpha = 45°$$

$$V_0 = 230\left[\frac{1}{2\pi}\left(2\pi - \frac{\pi}{4} + \frac{\sin 90}{2}\right)\right]^{\frac{1}{2}} = 224.7157\text{ V}$$

rms value of output current

$$I_0 = \frac{V_0}{R} = \frac{224.9}{52.9} = 4.2479\text{ A}$$

Load power

$$P_o = I_o^2 \times R = 4.25^2 \times 52.9 = 954.56 \text{ W}$$

6. For an SCR to be operated at 45°, determine the rms and average current value flowing through the heater shown in the figure.

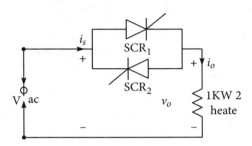

Solution

$$\alpha = 45° = \frac{\pi}{4}, \quad V_S = 220 \text{ V}$$

Resistance of heater

$$R = \frac{V^2}{R} = \frac{220^2}{1000} = 48.4 \text{ }\Omega$$

Resistance value of output voltage

$$V_o = V_S \sqrt{\frac{1}{\pi}\left(\pi - \alpha + \frac{\sin 2\alpha}{2}\right)}$$

$$V_o = 220 \sqrt{\frac{1}{\pi}\left(\pi - \frac{\pi}{4} + \frac{\sin 90}{2}\right)}$$

$$V_o = 220 \sqrt{\frac{1}{\pi}\left(\pi - \frac{\pi}{4} + \frac{1}{2}\right)} = 209.769 \text{ V}$$

rms current flowing through heater $= \frac{V_o}{R} = \frac{209.769}{48.4} = 4.334 \text{ A}$

Average current flowing through the heater $I_{Avg} = 0$

7. A 220 V at 50 Hz operated single-phase voltage controller consists of R = 4 Ω and L = 6 mH load. Determine the firing angle control range, rms load current value, power and PF, thyristor current on average and rms values.

Solution

For control of output power, minimum angle of firing angle α is equal to the load impedance angle θ

$\alpha = \theta$ Load angle

$$\theta = \tan^{-1}\left(\frac{\omega L}{R}\right) = \tan^{-1}\left(\frac{6}{4}\right) = 56.3°$$

Maximum possible value of α is 180°

Therefore, control range of firing angle is $56.3° < \alpha < 180°$

Maximum value of rms load current occurs when $\alpha = \theta = 56.3°$. At this value of α, the maximum value of rms load current

$$I_0 = \frac{V_s}{Z} = \frac{220}{\sqrt{4^2 + 6^2}} = 30.5085 \text{ A}$$

Maximum power $P_0 = I_0^2 R = 30.5085^2 \times 4 = 3723.077 \text{ W}$

Input VA $= V_s I_0 = 220 \times 30.5085 = 6711.87 \text{ W}$

$$PF = \frac{P_0}{\text{Input VA}} = \frac{3723.077}{6711.87} = 0.5547$$

Average thyristor current will be maximum when $\alpha = \theta$ and conduction angle $\gamma = 180°$

Therefore, maximum value of average thyristor current

$$I_{T\,Avg} = \frac{1}{2\pi} \int_{\alpha}^{\pi+\alpha} \frac{V_m}{Z} \sin\omega t - \theta \, d\omega t$$

Note: $i_0 = i_{T_1} = \frac{V_m}{Z}\left[\sin(\omega t - \theta) - \sin(\alpha - \theta)e^{\frac{-R}{\omega L}(\omega t - \alpha)}\right]$

At $\alpha = 0$,

$$i_{T_1} = i_0 = \frac{V_m}{Z}\sin(\omega t - \theta)$$

$$I_{T\,Avg} = \frac{V_m}{2\pi Z}\left[-\cos(\omega t - \theta)\right]_{\alpha}^{\pi+\alpha}$$

$$I_{T\,Avg} = \frac{V_m}{2\pi Z}\left[-\cos(\pi + \alpha - \theta) + \cos(\alpha - \theta)\right]$$

But $\alpha = 0$,

$$I_{T\,Avg} = \frac{V_m}{2\pi Z}\left[-\cos\pi + \cos 0\right] = \frac{V_m}{2\pi Z} \times 2 = \frac{V_m}{\pi Z}$$

$$\therefore I_{T\,Avg} = \frac{V_m}{\pi Z} = \frac{\sqrt{2} \times 220}{\pi \sqrt{4^2 + 6^2}} = 13.7336 \text{ A}$$

Similarly, maximum rms value occurs when $\alpha = 0$ and $\gamma = \pi$.

Therefore, maximum value of rms thyristor current

$$I_{TM} = \sqrt{\frac{1}{2\pi} \int_{\alpha}^{\pi+\alpha} \left\{\frac{V_m}{Z} \sin(\omega t - \theta)\right\}^2 d\omega t}$$

$$I_{TM} = \sqrt{\frac{V_m^2}{(2\pi Z)^2} \int_{\alpha}^{\pi+\alpha} \left[\frac{1 - \cos 2\omega t - 2\theta}{2}\right] d\omega t}$$

$$I_{TM} = \sqrt{\frac{V_m^2}{4\pi Z^2} \left[\omega t - \frac{\sin 2\omega t - 2\theta}{2}\right]_{\alpha}^{\pi+\alpha}}$$

$$I_{TM} = \sqrt{\frac{V_m^2}{4\pi Z^2}(\pi + \alpha - \alpha - 0)}$$

$$I_{TM} = \frac{V_m}{2Z} = \frac{\sqrt{2} \times 220}{2\sqrt{4^2 + 6^2}} = 21.57277 \text{ A}$$

8. The single-phase AC voltage controller has a 120-V rms at 60-Hz source. The load resistance is 15 Ω. Determine (a) the delay angle required to deliver 500 W to the load, (b) the rms source current, (c) the rms and average currents in the SCRs and (d) the PF.

Solution

(a) The required rms voltage to deliver 500 W to a 15 Ω load is

$$P = \frac{V_{orms}^2}{R}$$

$$V_{orms} = \sqrt{PR} = \sqrt{(500)(15)} = 86.6 \text{ V}$$

The rms load voltage is determined by

$$V_{orms} = \frac{V_m}{\sqrt{2}} \sqrt{\left(1 - \frac{\alpha}{\pi} + \frac{\sin 2\alpha}{2\pi}\right)}$$

$$86.6 = 120 \sqrt{\left(1 - \frac{\alpha}{\pi} + \frac{\sin 2\alpha}{2\pi}\right)}$$

Which yields $\alpha = 1.54$ rad $= 88.1°$

(b) The source current is given by

$$I_{orms} = \frac{V_{orms}}{R} = \frac{86.6}{15} = 5.77 \text{ A}$$

(c) The SCR currents are given by

$$I_{SCR,rms} = \frac{I_{orms}}{\sqrt{2}} = 4.08 \text{ A}$$

$$I_{SCR,avg} = \frac{\sqrt{2} \, 120}{2\pi (15)} [1 + \cos(88.1°)] = 1.86 \text{ A}$$

(d) The PF is

PF = P/S = 500/(120 × 5.77) = 0.72

9. A resistance heating element fed from an AC bus is controlled by an AC switch with integral half-cycle control. The base period is 20 half-cycles of the AC supply. Determine the number of half-cycles for which the AC switch should be gated to bring down the heating power to 30 per cent of the maximum.

Solution

The average power in the resistance element is proportional to the square of the rms voltage:

$$P = k\left(V\sqrt{D}\right)^2$$

$$P = kDV^2$$

Therefore, for 30 per cent power, D = 0.30. The number of on half-cycles, 0.30 × 20 = 6 half-cycles.

10. A single-phase resistance load is supplied from a 120 V AC bus in series with a phase-controlled static AC switch. The firing delay angle is 60°. Determine the rms value of the load voltage.

Solution

$$V_{rms} = V(1 - \alpha/\pi + \sin 2\alpha/2\pi)^{1/2}$$

Substituting $\alpha = \frac{1}{3}\pi$ in this equation gives

$$V_{rms} = 0.8969 \, V$$

The required duty cycle to give the same rms load voltage would be given by

$$\sqrt{D} = 0.8969$$

Therefore, D = 0.8969² = 0.8045

The number of on half-cycles would be 0.8045 × 20 = 16, to the nearest integer.

11. The TRIAC light dimmer circuit is used to adjust the intensity of a 120 V, 100 W incandescent lamp working from 120 V at 60 Hz mains. C = 0.33 μF and R = 3.33 kΩ (a 5 kΩ potentiometer set at 3.33 kΩ). The breakover voltage of the diode AC (DIAC) is 40 V. Determine the firing delay angle α under these conditions.

Solution

We have $\omega CR = 2\pi \times 60 CR = 0.4143$

$$\Phi = \tan^{-1} 0.4143 = 0.3928 \text{ rad}$$

Numerical substitution in the following equation gives

$$\frac{V_b}{V_m} = \frac{1}{\sqrt{1 + \omega^2 C^2 R^2}} [\sin(\alpha - \Phi) + \frac{e^{-\alpha}}{1} \omega CR \sin\Phi]$$

$$0.2551 = \sin(\alpha - 0.3928) + 0.3828 e^{-2.4137\alpha}$$

Computed solution of this equation gives

$$\alpha = 0.546 \text{ rad} = 31.3°$$

12. An AC load circuit consists of a resistance of 30 Ω in series with an inductance of 0.2 H. It is fed from a 120 V, at 60 Hz AC source. Determine the firing angle range in which the load voltage can be controlled.

    Solution

    We have $R = 30\ \Omega$, $X = 2\pi \times 60 \times 0.2 = 75.4\ \Omega$ and $X/R = 2.5133$, so that

    $\Phi = \tan^{-1} 2.5133 = 68.3°$

    The control range for the firing angle is $68.3°–180°$

13. An AC voltage controller has a resistive load of $R = 10\ \Omega$. The rms input voltage is $V_s = 120$ V at 60 Hz. The thyristors switch is on for $n = 25$ cycles and off for $m = 75$ cycles. Determine (a) the rms output voltage V_o, (b) the input PF and (c) the average and rms current of thyristors.

    Solution

    $R = 10\ \Omega$, $V_s = 120$ V, $V_m = \sqrt{2} \times 120 = 169.7$ V and $k = \dfrac{n}{n+m} = \dfrac{25}{100} = 0.25$.

    The rms value of the output voltage is

    $V_o = V_s \sqrt{k} = V_s \sqrt{\dfrac{n}{n+m}} = V_s \sqrt{\dfrac{25}{100}} = 60$ V

    And the rms load current is $I_o = V_o/R = 60/10 = 6.0$ A.

    The load power is $P_o = I_o^2 R = 6^2 \times 10 = 360$ W. Because the input current is the same as the load current, the VA input is

    $VA = V_s I_s = V_s I_o = 120 \times 6 = 720$ W.

    The input PF is

    $PF = \dfrac{P_o}{VA} = \sqrt{\dfrac{n}{n+m}} = \sqrt{k}$

    $= \sqrt{0.25} = \dfrac{360}{720} = 0.5$ (lagging)

    The peak thyristor current is $I_m = V_m/R = 169.7/10 = 16.97$ A. The average current of the thyristors is

    $I_A = \sqrt{\dfrac{n}{2\pi(n+m)}} \displaystyle\int_0^\pi I_m \sin\omega t\, d(\omega t) = \dfrac{I_m n}{\pi(n+m)} = \dfrac{k I_m}{\pi}$

    $= \dfrac{16.97}{\pi} \times 0.25 = 1.33$ A

    The rms current of thyristors is

    $I_R = \left[\sqrt{\dfrac{n}{2\pi(n+m)}} \displaystyle\int_0^\pi I_m^2 \sin^2 \omega t\, d(\omega t)\right]^{\frac{1}{2}} = \dfrac{I_m}{2} \sqrt{\dfrac{n}{n+m}} = \dfrac{I_m \sqrt{k}}{2}$

    $= \dfrac{16.97}{2} \times 0.25 = 4.24$ A

14. A single-phase AC voltage controller has a resistive load of $R = 10\ \Omega$ and the input voltage is $V_s = 120$ V at 60 Hz. The delay angle of thyristor T_1 is $\alpha = \pi/2$. Determine (a) the rms value of output voltage V_o, (b) the input PF, and (c) the average input current.

Solution

$R = 10\ \Omega,\ V_s = 120\ \text{V},\ V_m = \sqrt{2} \times 120 = 169.7\ \text{V},\ \alpha = \pi/2$

The rms value of the output voltage is

$$V_o = 120\sqrt{\frac{3}{4}} = 103.92\ \text{V}$$

The rms load current

$$I_o = V_o / R = \frac{103.92}{10} = 10.392\ \text{A}.$$

The load power

$$P_o = I_o^2 R = 10.392^2 \times 10 = 1079.94\ \text{W}$$

Because the input current is the same as the load current, the input VA rating is

$$\text{VA} = V_s I_s = V_s I_o = 120 \times 10.392 = 1247.04\ \text{VA}.$$

The input PF is

$$\text{PF} = \frac{P_o}{\text{VA}} = \frac{V_o}{V_s}\left[\frac{1}{2\pi}\left(2\pi - \alpha + \frac{\sin 2\alpha}{2}\right)\right]^{\frac{1}{2}}$$

$$= \sqrt{\frac{3}{4}} = \frac{1079.94}{1247.04} = 0.866\ (\text{lagging})$$

The average output voltage

$$V_{DC} = -120 \times \frac{\sqrt{2}}{2\pi} = -27\ \text{V}$$

And the average input current

$$I_D = \frac{V_{DC}}{R} = -\frac{27}{10} = -2.7\ \text{A}.$$

Objective Type Questions

1. AC voltage regulators converters convert
   (a) Fixed mains voltage to fixed AC voltage
   (b) Fixed mains voltage directly to variable AC voltage without change in frequency
   (c) Fixed mains voltage directly to variable AC voltage with change in frequency

2. Sequence control of AC regulators is employed for
   (a) The improvement of PF and reduction of harmonics
   (b) The reduction of PF only
   (c) The reduction of harmonics only
   (d) The improvement of PF and increase in harmonics
3. In a single-phase full-wave AC regulator, varying the delay angle a from 0 to p can vary the rms output voltage from
   (a) V_s to $V_s/4$
   (b) V_s to $V_s/2$
   (c) V_s to $3V_s/2$
   (d) V_s to 0
4. The conduction angle (d) of SCR T_1 in a single-phase full-wave controller is
   (a) $d = b - a$
   (b) $d = b + a$
   (c) $d = a - b$
   (d) $d = b + 0$
5. A single-phase half-wave AC voltage regulator using one SCR antiparallel with a diode, feeds 1 kW at 230 V heater. For a firing angle of 180°, the load power is
   (a) 5 W
   (b) 300 W
   (c) 400 W
   (d) 500 W
6. Cycloconverter converts
   (a) AC voltage to DC voltage
   (b) DC voltage to DC voltage
   (c) AC voltage to AC voltage at same frequency
   (d) AC voltage at supply frequency to AC voltage at load frequency
7. A six-pulse cycloconverter is fed from a 415 V three-phase supply with reactance of 0.3 W/phase. The output load voltage for firing angle of 45° and load current 40 A is given by
   (a) 272 V
   (b) 549 V
   (c) 200 V
   (d) 180 V
8. A three-pulse cycloconverter feeds a single-phase load of 200 V at 50 A at a PF of 0.8 lagging. PF of the supply current is given by
   (a) 0.48
   (b) 0.9
   (c) 0.1
   (d) 0.38
9. A cycloconverter can be considered to be composed of two converters
   (a) Back–back
   (b) Series connected
   (c) Parallel connected
   (d) Series–parallel connected
10. In a three-pulse cycloconverter with an intergroup reactor operating in circulating current mode, both P and N converter groups synthesize the
    (a) Same fundamental sine wave
    (b) Different fundamental sine wave
    (c) Same fundamental cosine wave
    (d) Different fundamental cosine wave
11. A six-pulse cycloconverter is fed from a 415 V three-phase supply with reactance of 0.3 cycles/phase. The output load voltage for firing angle of 30° and load current 30 A is given by
    (a) 272 V
    (b) 549 V
    (c) 200 V
    (d) 180 V

12. A three-pulse cycloconverter feeds a single-phase load of 200 V at 50 A at a PF of unity. PF of the supply current is given by
    (a) 0.48 (b) 0.9 (c) 0.1 (d) 0.38

13. A cycloconverter can be considered to be composed of two converters
    (a) Connected back to back
    (b) Series connected
    (c) Parallel connected
    (d) Series–parallel connected

14. For a P-pulse cycloconverter, the peak value of the output voltage with maximum value of supply voltage (V_{smax}) is given by
    (a) $V_{0(max)} = \dfrac{P}{\pi} \sin \dfrac{\pi}{P} V_{s(max)}$
    (b) $V_{0(max)} = \dfrac{2P}{\pi} \sin \dfrac{\pi}{P} V_{s(max)}$
    (c) $V_{0(max)} = \dfrac{P}{\pi} \cos \dfrac{\pi}{P} V_{s(max)}$
    (d) $V_{0(max)} = \dfrac{3P}{\pi} \sin \dfrac{3\pi}{P} V_{s(max)}$

15. For a P-pulse cycloconverter, when the output voltage is reduced in magnitude by the firing delay angle, then
    (a) $V_{0(max)} = \dfrac{P}{\pi} \sin \dfrac{\pi}{P} V_{s(max)} \cos\alpha$
    (b) $V_{0(max)} = \dfrac{P}{\pi} \cos \dfrac{\pi}{P} V_{s(max)} \sin\alpha$
    (c) $V_{0(max)} = \dfrac{2P}{\pi} \sin \dfrac{2\pi}{P} V_{s(max)} \cos\alpha$
    (d) $V_{0(max)} = \dfrac{2P}{\pi} \cos \dfrac{2\pi}{P} V_{s(max)} \sin\alpha$

16. In a three-pulse cycloconverter with an intergroup reactor operating in circulating current mode, both P and N converter groups synthesize the
    (a) Same fundamental sine wave
    (b) Different fundamental sine wave
    (c) Same fundamental cosine wave
    (d) Different fundamental cosine wave

Review Questions

1. Compare on–off control and phase control?
2. What is the advantage of on–off control?
3. What is the disadvantage of on–off control?
4. Explain duty cycle in on–off control method?
5. Write short notes on unidirectional or half-wave AC voltage controller?
6. Mention some of the demerits of unidirectional or half-wave AC voltage controller?
7. Write short notes on unidirectional or bidirectional or half-wave AC voltage controller?
8. Discuss the control range of a firing angle in an AC voltage controller with RL load?
9. What type of gating signal is used in a single-phase AC voltage controller with RL load?
10. State the demerits of continuous gating signal.
11. Explain high-frequency carrier gating.
12. Explain sequence control of AC voltage regulators.

13. Mention some of the demerits of sequence control of AC voltage regulators.
14. Define the term cycloconverter.
15. Classify the types of cycloconverters.
16. What is meant by step-up cycloconverters?
17. What is meant by step-down cycloconverters?
18. Mention some of the applications of cycloconverter.
19. What is meant by positive converter group in a cycloconverter?
20. What is meant by negative converter group in a cycloconverter?
21. Explain the various control strategies of voltage controller in detail.
22. Explain the working principle of a single-phase half-wave voltage controller with R load with neat diagrams.
23. Explain the concept of a cycloconverter in detail with relevant diagrams.
24. Explain the concept of a single-phase to single-phase step-up cycloconverter in detail with relevant diagrams.
25. Explain the concept of a single-phase to single-phase step-down cycloconverter in detail with relevant diagrams.
26. Explain the concept of a three-phase to single-phase cycloconverter in detail with relevant diagrams.
27. Explain the concept of a three-phase to three-phase cycloconverter in detail with relevant diagrams.
28. Mention few applications of voltage controllers used in real time.

Practice Questions

1. Explain with waveforms the operation and draw the circuit diagram of a three-phase to single-phase cycloconverter.
2. With suitable diagrams, explain the concept of a single-phase AC voltage controller with only one thyristor feeding resistive load by phase and on–off control method. Also derive the expressions for rms voltages for the aforementioned cases.
3. Describe the operation of a single-phase full-wave AC voltage controller with the help of voltage and current waveform. Also derive the expression for average value of output voltage.
4. Explain sinusoidal and multiple pulse width modulation (PWM) techniques used in inverters.
5. Explain the operation of the step-down cycloconverter of both bridge and midpoint configuration with necessary waveforms.
6. With the aid of a circuit diagram, explain the operation of three-phase to three-phase cycloconverter employing three-phase half-wave circuits and list a few of its applications.
7. With a neat circuit and waveforms, explain the operation of a single-phase half-wave phase controller and single-phase full-wave phase controller.
8. Discuss the principle of working of a single-phase to single-phase step-up cycloconverter. Also discuss the factors that affect the performance of cycloconverters.
9. Explain the working of a single-phase AC voltage controller with RL load when its firing angle is more than the load PF angle. Illustrate with waveforms.

10. For a single-phase voltage, a controller feeds power to a resistive load of 3 Ω from 230 V at 50 Hz source. Calculate (a) the maximum values of average and rms thyristor currents for any firing angle and (b) the minimum circuit turn-off time for any firing angle.

11. A single-phase AC voltage controller with RL load is supplied with an input supply of 230 V at 50 Hz. The load values are $L = 10$ mH, $R = 10$ Ω. Determine the triggering angle, rms output voltage and PF of the circuit using MATLAB simulation.

12. Design a MATLAB circuit of a single-phase half-wave AC voltage regulator using one single SCR antiparallel with the diode feeds of 1 kW and 230 V heater circuit. For a triggering angle of 45°, determine the load power.

13. A single-phase voltage controller is engaged for controlling the power flow of the circuit from 220 V and 50 Hz supply source into the load circuit of $R = 4$ Ω and $L = 6$ mH. Using MATLAB simulation, calculate the maximum value of the rms load current, power and PF, and maximum value of average and rms thyristor currents. Also determine the triggering angle of the circuit for the given values.

14. A three-phase full-wave converter of input rms voltage of 120 V with 120° phase shift has a load resistance of 6 Ω. The triggering angle is said to be 180°. Using MATLAB, calculate

    (a) Output rms voltage
    (b) Power output
    (c) Input PF

    Comment on your results.

15. For a single-phase full-wave AC voltage controller supplying R load, the input supply voltage is 230 V, rms at 50 Hz. Design a MATLAB circuit for $R = 100$ Ω, the firing angle for thyristors T1 and T2 being the same. Determine

    (a) Triggering angle of the thyristors
    (b) Output rms voltage
    (c) PF

7
Simulation and Digital Control Using MATLAB

> **Learning Objectives**
> - ✓ To understand the basic concepts of digital control using MATLAB
> - ✓ To understand the basic concepts of fuzzy logic principles
> - ✓ To understand the basic concepts of neural network
> - ✓ To understand real-time implementation using MATLAB

7.1 Introduction

The traditional methods of analysis may not be useful in the practical analysis of sensors. Traditional calculations may also not be satisfactory to simulate real life situation. Fuzzy-logic-based systems may be used to overcome traditional calculation issues. Fuzzy logic liberates us from the rigid thinking of ideal arrangements of components that are utilized as part of typical systems. Fuzzy logic models guarantee a limited subjective analysis of a quantitative element system while being pertinent to any system that can be portrayed in technical terms. Fuzzy models give a compact and dynamic representation of systems that do not have a complete quantitative model or of indeterminate systems. Static thinking is not encouraged in fuzzy semantic representation of a quantitative system.

Fuzzy logic models use fuzzy sets to make a limited number of inputs, output, and conditions of a quantitative system. As of now, most fuzzy models are executed as if–then standards, where the system input is compared against certain rules and the model's output is the combined output of considerable number of principles compared in parallel. This basic consistent system, a fuzzy inference system (FIS), does not really provide a fixed value but can just assess an improved subjective model of a plant.

Later work has extended the usefulness of this structure by providing machine learning systems that adjust and tune fuzzy semantic models and consequently produce new models through self-association. The conventional input and output relations have been supplanted by artificial neural networks (ANNs) that take in synaptic weights in its preparation process. ANNs are versatile and can adjust weights according to the environment. ANNs can bargain normally with relevant data, since the data is obtained from the general structure and the initiation condition of the system. Each neuron is possibly influenced by the movement of every single other neuron. The ANN can be prepared to settle on choices and are likewise tolerant if a neuron or uniting connection is impaired or an example is of bad quality; thus, because of appropriation of data in the system, destruction must be major to cause problems. Since neurons are the elemental units for all ANN, it is possible to share the calculation and structures among diverse applications. Consistent coordination of modules is thus possible.

Learning or tuning is the first step in building technical fuzzy models created from heuristic area information that are constantly upgraded. Learning is accomplished by utilizing a neuro-fuzzy structure and changing the directed learning systems initially produced for neural systems. These systems incorporate angle plunge back-proliferation, minimum mean-squares, and a half-breed philosophy that joins slightest squares to improve direct parameters and uses back-engendering to streamline the nonlinear parameters. These same learning systems can consequently realize any discretionary nonlinear mapping of input and output without a starting etymological fuzzy model. The subsequent self-composed fuzzy models do not essentially have an etymological translation that would be perceived by a human. Generally, systems grown through self-association are never translated technically, yet are used adequately, for example, for coordinating and bend fitting. Fuzzy systems are regularly favoured for bend fitting on the grounds that fuzzy tenets utilized by the system have a nearby effect, in actuality giving a versatile component to actualizing B-splines. It is possible to coordinate the fuzzy logic controller (FLC) with ANN so that the expression for the information utilized as part of the systems is seen by people. This diminishes the problem of depicting the ANN. The fuzzy controller determines how to enhance its execution utilizing ANN structure and adapts along these lines by experience. Neuro-figuring is quick compared to routine registering because of the gigantic parallel reckoning available in this method. In addition, it has the properties of adaptation to noncritical failure and fault tolerance. Neural systems are used as estimators. Neural-system-based control does not require a numerical model of a system like routine control does with the necessary exactness. This chapter describes the application of fuzzy logic and neural networks in power electronics.

7.2 Fuzzy Logic Principles

Fuzzy logic has become the go to method for making a refined control system. It is just the expansion of binary logic. The essential difference between fuzzy logic and binary logic is that in binary logic, we take only two cases, either 0 or 1, which suggests low or high states. In fuzzy logic, we take each and every state into thought. Fuzzy logic is a system for identifying with the information in a way that humans do. It then controls the information like humans. Fuzzy logic can be used in applications that are either difficult to handle logically or where the use of fuzzy logic gives better execution.

Fuzzy logic has two kinds of effects. In the general sense, fuzzy logic is an intelligent system with an extension of logical values in multivalues. Laterally, fuzzy logic is used for all intents and purposes along with fuzzy logic sets, which will relate it to the various classes of constraints, where investment is the matter to be considered. In this point of view, fuzzy logic contrasts both in thought and substance with standard multivalued predictable structures. Fuzzy logic incorporates the accompanying elements:

- Fuzzy logic is theoretically straightforward.
- Fuzzy logic is adaptable.
- Fuzzy logic is tolerant of uncertain information.
- Fuzzy logic can demonstrate nonlinear elements of discretionary intricacy.
- Fuzzy logic can be based on the experience of specialists.
- Fuzzy logic can be mixed with customary control methods.
- Fuzzy logic can be used in lieu of characteristic language.

7.2.1 Fuzzy logic tool box

The fuzzy logic tool box is an accumulation of capacities based on MATLAB. It provides devices to make and alter the FIS (fuzzy interference system). Fuzzy systems can be modelled using Simulink model. To design simulink model systems that uses fuzzy logic, we essentially need to duplicate the FLC (fuzzy logic control) by logging into the library of the fuzzy logic toolbox present in the Simulink Library Browser by typing

>>fuzblock

The library shown in Figure 7.1 appears.

Figure 7.1 presents the fuzzy logic toolbox that includes the fuzzy logic controller log with rule viewer. The fuzzy logic toolbox includes membership functions in the sublibrary that contains the Simulink blocks for built-in membership functions.

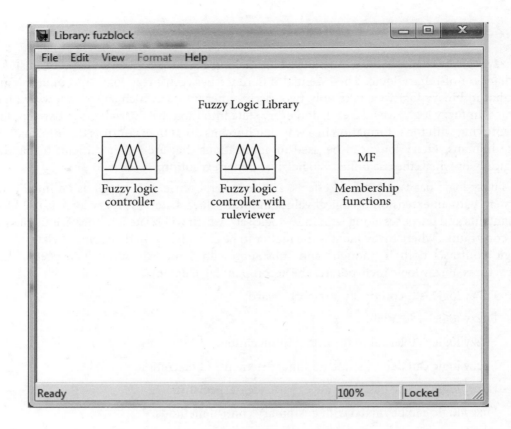

Figure 7.1 Fuzzy Logic Toolbox.

The FIS details are displayed using the FIS editor. To initiate the FIS editor, use the following command in MATLAB prompt:

>>*fuzzy*

Figure 7.2 shows the FIS editor that includes the names of the input and the output variables. The functions indicated do not depict the genuine states of the systems; they are just symbols.

The FIS editor deals with the unusual states of the systems using the name of the variables, the type and the number of variables. The fuzzy toolboxes are not confined by the input quality; they may be constrained by the internal memory of the machine. To obtain an input of really good quality, the quantity may be too large, which creates the complexity that needs to be investigated by the FIS. Refer to Figure 7.3 to define the shapes of all the membership functions associated with each variable.

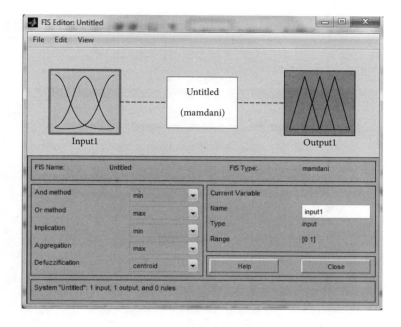

Figure 7.2 FIS Editor.

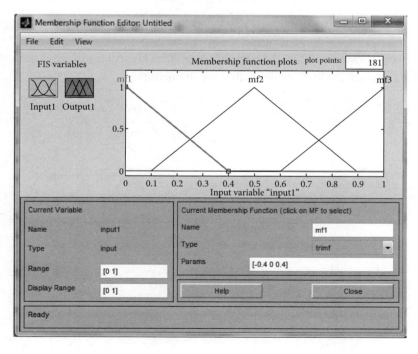

Figure 7.3 Membership Function.

Rule editor is used to edit the list of rules that defines the behaviour of the system (refer to Figure 7.4).

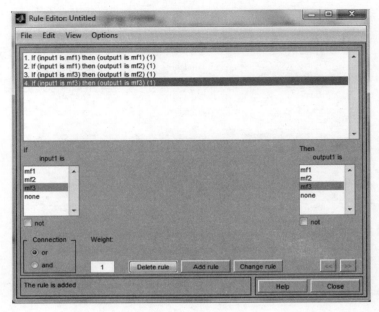

Figure 7.4 Rule Editor.

The rule viewer (shown in Figure 7.5) displays the FISs. The ruler indicates the influence of membership functions, which would influence the results.

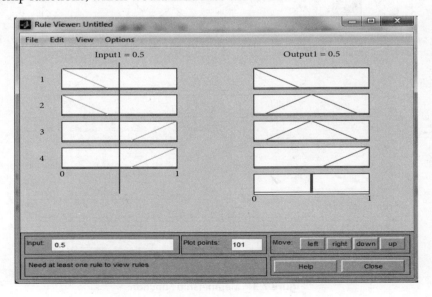

Figure 7.5 Rule Viewer.

Surface viewer (Figure 7.6) which is a representation of the dependency of the outputs with the variables, displays and plots the output surface map of the system.

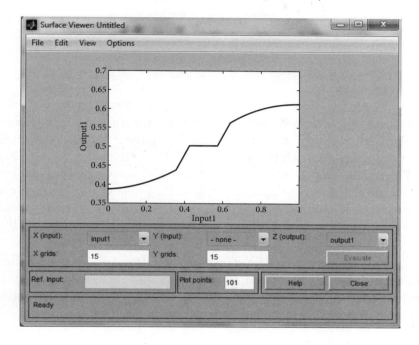

Figure 7.6 Surface Viewer.

The fuzzy logic system has become the most popular criteria in microcontrollers in recent years.

Fuzzy logic controllers deal with very normal input values, where complex mathematical models are not needed; nonlinearities may however be present. The three stages of modelling include fuzzification, rule base and defuzzification. The fuzzification models are subjected to regeneration into linguistic variables functions, which operate almost like that shown in Figure 7.7. The different types of fuzzy levels used are NB (negative big), NS (negative small), ZE (zero), PS (positive small) and PB (positive big). In Figure 7.7, a and b are supported by the values of the numerical variable.

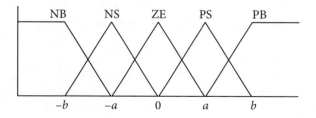

Figure 7.7 Membership Function for Input and Output of FLC.

In FLC, the user has the flexibility to choose and compute the input variables say, with an error E and a change in error ΔE. Since dP/dV vanishes, the following approximation is used.

$$E(n) = P(n) - P(n-1) / V(n) - V(n-1) \tag{7.1}$$

$$\Delta E(n) = E(n) - E(n-1) \tag{7.2}$$

When E and ΔE are calculated and reformed by the variable to be linguistic, the response obtained is often said to be an amendment of the dual relationship with ΔD of the converter, which can be listed in a table of rule base as given in Table 7.1. The variable, which is linguistic, is set to ΔD for various set of combinations of E and area unit, which is supported by the data of the available user. For an instance, the operation purpose is left to the Maximum Power Point (MPP), where E is the metal, ΔE is ZE. As a result, the quantitative relation is basically accumulated; thus, ΔE is the metal to achieve the output response.

In FLC, the numerical value is derived from the linguistic variable in the defuzzification stage as shown in Table 7.1. As a result, an analogue signal is produced, where the normal controllers have to perform better in various conditions. As a result, effectiveness is based on error computation and merging with the rule base table.

Table 7.1 Fuzzy Rule Base Table

E\ ΔE	NB	NS	ZE	PS	PB
NB	ZE	ZE	NB	NB	NB
NS	ZE	ZE	NS	NS	NS
ZE	NS	ZE	ZE	ZE	PS
PS	PS	PS	PS	ZE	ZE
PB	PB	PB	PB	ZE	ZE

7.2.2 Implementation

To overcome the problems present in past strategies, the FLC is mostly used in recent power systems. The main distinction from the routines of the past is that exact depiction, of the system is not required. Hence, fuzzy logic permits the determination of various principles based on etymology and consequently, controller tuning is essential. Controller tuning results in values which are qualitatively not the same as the existing configuration systems. Above all, the fuzzy logic control is nonlinear and versatile in character; it provides a strong variety of parameters, supply and load voltage influences. The current trend is to extensively use fuzzy logic as a helpful device to model control systems that are nonlinear in nature, in the same way as the sun-based PV(photovoltaic) exhibits to track the high force. The control signals of the fuzzy controller could lapse and also may fail, while the output control uses pulse width modulation (PWM). The metal–oxide–semiconductor field-effect transistor (MOSFET) is trained by the PWM technique and the MOSFET switch is inclined to the leading state when the voltage value is high and into the nonleading state when the voltage value is low.

The current stream to the storage unit is controlled by modifying the conducting cycles and the best-suited controller circuit could be a DC–DC controller with PWM control. Refer to Figure 7.8, which depicts the buck converter.

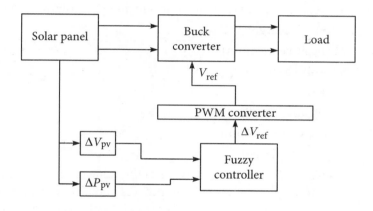

Figure 7.8 Block Diagram of the Buck Converter.

The input parameters set to the FLC causes change in the exhibit power ΔP_{pv} and change in PV voltage ΔV_{pv} relating to the moments of inspecting time. The response from the FLC is the voltage taken for reference ΔV_{ref}, which implies $I_{pv} \times V_{pv}$ completely. The response is fed to the PWM generator, which results in obtaining the reference voltage of the buck converter. The fuzzy-based technique yields results of output with incremental voltage of extremity and variable effectiveness. The transient conditions of the system are result in incremental voltage.

7.2.3 Description and design of FLC

Fuzzy model (refer to Figure 7.9) of the system is such that it is dependent on the earlier master information of the system. The FLC is categorized into four set of segments, namely, fuzzification, rule base, inference and defuzzification. The input segments include (ΔP_{pv}) the PV array power, (ΔV_{pv}), the PV array voltage and (ΔV_{ref}) representing the change in reference voltage.

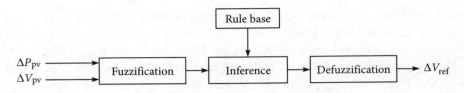

Figure 7.9 Block Diagram of the FLC.

Fuzzification

The fuzzy model is developed on a trial-and-error basis to meet the desired performance criteria.

As already mentioned before, the fuzzy logic sets include the following set of variables:

PL (positive large)
PM (positive medium)
PS (positive small)
Z (zero)
NS (negative small)
NM (negative medium)
NL (negative large)

Consider a system in which the fuzzy set PS assumes the value of membership greater than zero at the initial state to respond in a speed up process and also to prevent the reference voltage variation. Along with this, PM and NM are additional fuzzy sets added to improve the control surface.

Refer to Figure 7.10, which represents the membership functions where they are designed to model the unsymmetrical nature of the circuit. The functions are dense and of high sensitivity. They are normalized with suitable gain points, which are used to match the inputs to the respective universes of discourse.

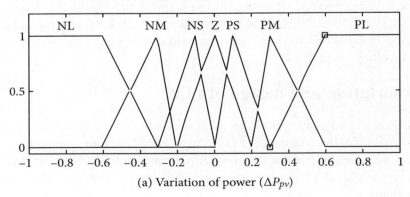

(a) Variation of power (ΔP_{pv})

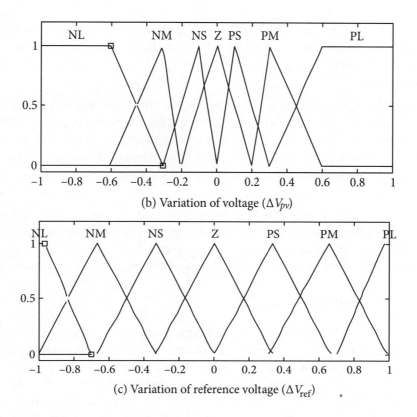

Figure 7.10 Membership Functions for the Fuzzy Model (a) Input ΔP_{pv}, (b) Input ΔV_{pv}, and (c) Output ΔV_{ref}.

Rule Base

Table 7.2 Rule Base for the Fuzzy Model

$\Delta V_{pv}/\Delta P_{pv}$	NL	NM	NS	Z	PS	PM	PL
NL	PL	PL	PL	PL	NM	Z	Z
NM	PL	PL	PL	PM	PS	Z	Z
NS	PL	PM	PS	PS	PS	Z	Z
Z	PL	PM	PS	Z	NS	NM	NL
PS	Z	Z	NM	NS	NS	NM	NL
PM	Z	Z	NS	NM	NL	NL	NL
PL	Z	Z	NM	NL	NL	NL	NL

The master rule or the fuzzy algorithm always tracks the maximum power, that is, the last change in the voltage value of the reference V_{ref} would cause the power value to increase by altering the reference voltage in a unique direction. Conversely, if the power drops, the voltage should be directed in the opposite direction. Refer to Table 7.2 for the rule base, which includes 49 rules.

Inference Method

The output of the fuzzy controller is decided by the inference technique. The deduction system of Mamdani, which is a max–min strategy is used within the system. The system is said to be more proficient in computation and preferably, has more interpolative properties over techniques depending on suggestion capabilities. Hence, Mamdani deduction system is one of the most prominent systems for in control building.

Defuzzification

The response of the fuzzy controller is fuzzy operated, where a fine-qualified result is very essential for defuzzification. The centroidal strategy sees to it that the normally exploited routines of defuzzification are the ones which are opted for this system. The effects and characteristics, which are demonstrated to give the best outcome, are said to be unique.

7.2.4 Simulation and results

Figure 7.11 represents the Simulink model of fuzzy-logic-based maximum power tracking for a solar panel with 60 W. The PV panel is validated using a buck converter and a resistive load. The main objective of this fuzzy model is to trace the maximum power irrespective of variations in the panel voltage. According to fuzzy principles, there are two input variables used, namely, change in PV power array ΔP_{pv} and change in PV array voltage ΔV_{pv} corresponding to the timing instants of the samples. The change in the reference voltage represents the output variable ΔV_{ref} and is integrated to achieve the desired V_{ref} value. The duty cycle of the buck converter is determined using the reference voltage V_{ref}. Equations (7.3), (7.4) and (7.5) represent the power required, change in voltage array and the reference variable.

$$\Delta P_{pv} = \left[P_{pv}(k) - P_{pv}(k-1) \right] \times K1 \tag{7.3}$$

$$\Delta V_{pv} = \left[V_{pv}(k) - V_{pv}(k-1) \right] \times K2 \tag{7.4}$$

$$\Delta V_{ref} = \left[V_{ref}(k) - V_{ref}(k-1) \right] \times K3 \tag{7.5}$$

where K1, K2 and K3 represent the gain coefficients and time index is represented as k. The inputs and outputs are converted to linguistic variables to obtain the FLC output value; also to perform this kind of operation, the fuzzy conversion membership functions can be used. The membership functions are fixed between −1 and 1 using input scaling factors, namely, K1 and K2, and represent the output scaling factors K3. Here, simple numbers are processed using the controller after scaling is done using fuzzy computation. The input and output terms representing the linguistic terms are characterized by seven membership functions as shown in Figure 7.10.

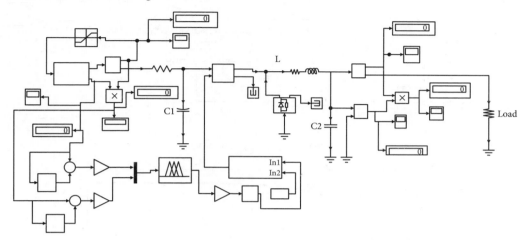

Figure 7.11 FLC Simulink Model.

The input variables ΔV_{pv} and ΔP_{pv} are processed in such a way that the output variables ΔV_{ref} are based on the control rules shown in Table 7.2. The output value is fed to the pulse generator, which makes the reference voltage respond to the buck converter. The basic principle of operation of the pulse generator is built using the comparison of the signals, where one form of the signal refers to the triangular waveform and the other refers to fixed linear signals, representing the time equivalent of the triggering voltage. The FLC Simulink model is shown in Figure 7.11.

U1 and U2 are the two variables representing reference voltage time signal and triangular signal of the if block used in simulation as shown in Figures 7.12 and 7.13. The output response of the modulator is utilized to feed the pulse generator of the MOSFET power semiconductor switch. The power device is switched into the conducting state whenever the voltage value is high and into the nonconducting state whenever the voltage value is low. Thus as a result, the fuzzy logic algorithm tracks the maximum power that is based on the rules defined. The simulation results are shown in Figure 7.14.

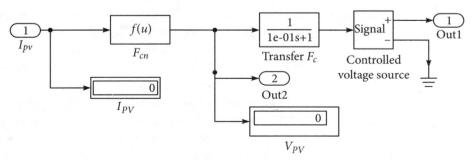

Figure 7.12 PV Source Block.

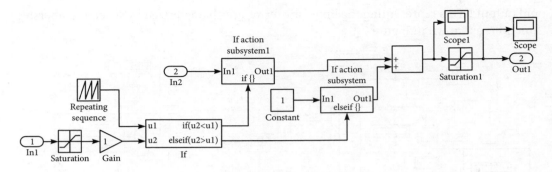

Figure 7.13 Subsystem Block.

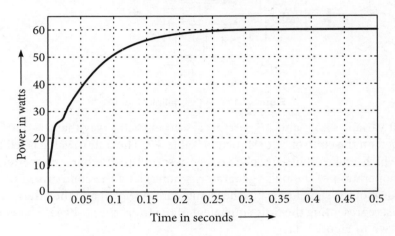

Figure 7.14 Power Characteristic Curve with FLC.

The Simulink model of FLC shown earlier in Figure 7.11 is developed in an environment to track the maximum power point and the value obtained is 59.9 W (refer Figure 7.14).

7.3 Neural Network Principles

Neural systems are made out of basic components working in parallel. These components are modelled on natural sensory systems. As in nature, the associations between components to a great extent focuses on the system capacity. A neural system can be prepared to perform a specific capacity by changing the estimations of the associations (weights) between components.

Ordinarily, neural systems are balanced, or prepared, so that a specific input prompts a particular target output. Figure 7.15 shows such a circumstance. There, the system is balanced, in view of a correlation of the output and the objective, until the system output coordinates the objective. Generally, numerous data/target sets are expected to prepare a system.

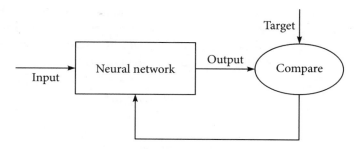

Figure 7.15 Overview of Neural Network.

Neural systems are prepared to perform complex capacities in different fields, including for example, response, distinguishing proof, arrangement, discourse, vision and control systems. They can likewise be prepared to take care of issues that are troublesome for customary PCs. The tool kit encourages the utilization of neural system ideal models for development and they are utilized as a part of building, budgetary and other useful applications.

The human mind predominantly motivates fake neural systems. This does not imply that ANNs are perfect recreations of the natural neural systems inside our cerebrum as the workings of the human mind is still a mystery. The machine, which has the intention to establish the best methodology in the same way as the mind performs specific tasks, is referred to as neural network. The execution is done utilizing the electronic parts integrated with programming on a computerized PC. Neural systems perform common calculations through a procedure of learning.

7.3.1 Background of neural networks

A massive processor, which is made up of simple units relatively matching the natural property of storing enormous data and making use of it for future analysis, is referred to as neural networks. The neural network system basically resembles the activity of the human brain in two aspects namely:

- Acquiring knowledge through the learning process
- Storage of acquired knowledge

The ultimate objective of a neural network is to acquire knowledge from the learning environment and to improve the end results in a better way through the interaction of various modifications applied to synaptic weights and biases. The learning process requires various iteration levels and includes the following set of events:

- The environment stimulates the neural network
- The parameters are frequently changed
- The network uniquely responds to the environment

The neuron represents the fundamental processing structure of the environment. The basic building block of the human brain, the neuron is similar to the neural network. In general, the biological neuron of the human body impacts the responses of multiple inputs with nonlinear operation. Figure 7.16 shows the relationship of the four parts of the neuron.

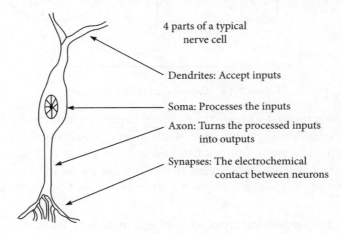

Figure 7.16 A Simple Neuron.

As seen in the figure, the neuron consists of dendrites, soma, axon and the synapses. The dendrites are the extensions of the soma, which acts as an input channel where the inputs are received using the synapses. The soma courses the input signals, which becomes a processed output, through the axons and the synapses.

Researchers are trying to recreate the brain in an artificial way to replicate the natural capabilities of the brain, so that users can engineer solutions to any kind of sophisticated problems. To perform such operations, the ANN simulates the basic functions of the natural neurons. Refer to Figure 7.17, which depicts an artificial neuron.

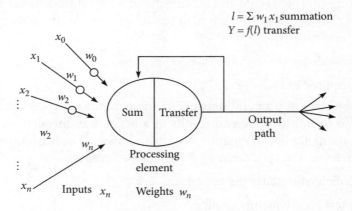

Figure 7.17 A Basic Artificial Neuron.

The input variables are represented in terms of $x(n)$. These input variables are multiplied using the connection weights $w(n)$. The input variables are summed up and transferred to generate the results as outputs. The small package lends itself to large-scale implementation. Further implementation is possible with other structures having different transformations.

7.3.2 Implementation

Let us consider a simple neural network. The feedforward network is three layered. It is created using *newff* command. The input layer consists of two neurons; 50 neurons are in the hidden layer; and one layer is present in the output layer. The inputs represent temperature and insulation. 23 input data samples are given to the network for training the sets. The target values represent the voltages that are obtained from modelling a particular input set among the training values. The active function used in the input and output layers are *tansig* and *purelin*.

The *tansig* activation function calculates its output according to the following equation:

$$\text{tansig}(n) = 2/(1 + \exp(-2 \times n)) - 1 \tag{7.6}$$

The *purelin* activation function calculates its output according to the following equation:

$$\text{purelin}(n) = n \tag{7.7}$$

The basic structure of the neural network is shown below in Figure 7.18.

Mean square error is the error criterion that is considered for training. *Trainlm* is the function considered, where trainlm is the network training function that updates the weight and bias values according to Levenberg–Marquardt (LM) optimization. This is a numerical optimization technique that has many advantages over its counterparts. The major merit is that it requires less number of data for training the network and achieves accurate results. The other advantage of using LM optimization method is that it produces accurate results even though the systems are not completely observable and controllable.

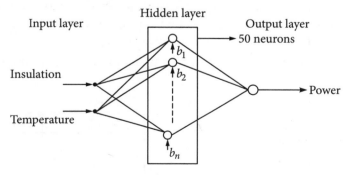

Figure 7.18 Basic Structure of a Neural Network.

The Hessian matrix can be computated where the performance function is approximated as

$$H = J^T J \tag{7.8}$$

The gradient component is computed as

$$g = J^T e \tag{7.9}$$

The Jacobian matrix (J) can be obtained using backpropagation (BP), which is less complex than the Hessian matrix concepts. The LM method provides approximation to the Hessian matrix in the following Newton-like updates.

$$x_{k+1} = x_k - [J^T J + \mu I]^{-1} J^T e \tag{7.10}$$

While modelling the system, the scalar value μ is zero, which represents the Newton's method using the Hessian matrix approximation. When the scalar value μ is set to be large, the gradient value is obtained with a small step size. When μ is decreased after reduction in performance function, the successive step is increased. In this way, the performance functions are always dealt with in various levels of iterations. Thus, Newton's method is a good and fast mechanism to give better performance results.

To update the weights of the network, error BP learning is implemented to minimize the error of the mean square. The algorithm holds two passes namely,

- Forward pass
- Backward pass

During the forward pass, 23 pairs of temperature and insulation are given to the network for training all the time. This type of training is called batch training. These input values propagate through the network layer by layer to generate the output voltage. The inputs are mainly used for gradient computation and updation of the network biases and weights. At the end, the output voltage is compared with the target value. The error value is the difference between the two values propagated through the backward pass, whereas the weights of the network are updated in a recursive manner.

The BP algorithm uses the following rule for updating the weights of the network.

$$w(k+1) = w(k) - \eta g(k) \tag{7.11}$$

where, η learning rate
       $g(k)$ gradient vector

The gradient vector value is computed in the backward mode by applying the chain rule base. The various parameters used for training are epochs, show, goal, time, min_grad, max_fail, mu, mu_dec, mu_inc and mu_max. The various sets of training parameters are listed here:

net.trainParam.epochs	Epochs in maximum to train
net.trainParam.goal	Efficiency target
net.trainParam.max_fail	Validation failures in maximum
net.trainParam.mem_reduc	Speed or memory factor
net.trainParam.min_grad	The value of minimum performance gradient
net.trainParam.mu	Initial Mu
net.trainParam.mu_dec	Decrease factor of Mu
net.trainParam.mu_inc	Increase factor of Mu
net.trainParam.mu_max	Mu-Maximum
net.trainParam.show	Progress between epochs
net.trainParam.time	Maximum training time in seconds

The goal value of the performance level is specified as 0.15. The number of epochs to the maximum is set as 30 on a trial basis. When the epoch is lower than the number, the training of the datasets is stopped before reaching the performance level. The maximum validation failure is set as 5. The mu value is decreased and is set as 0.01. This results in a training curve of every 50 epochs, where the other values are set with default values. The default values are listed here:

net.trainParam.mu_inc	10
net.trainParam.mu_max	1e−10
net.trainParam.mem_reduc	1
net.trainParam.min_grad	1e−25

The initial value of μ is set to be mu. The performance function is multiplied by mu_dec or mu_inc depending on whether there is increase and decrease of the step. If the value of mu is larger than mu_max, the process is terminated after the trained network is tested using nine sets of data.

7.3.3 Algorithm for ANN

Step-1 Construct the network and set the synaptic weights with random values

Step-2 The network is applied with input sets

Step-3 The network parameter can be initialized and the output values estimated to train the network

Step-4 Compare the desired output with the actual output to determine the error value

Step-5 Calculate the amount of each weight, which can be changed to make corrections of each individual weight

Step-6 Repeat Steps 3 and 4 until the error for the training sets is reduced to a small value

Step-7 Use testing sets to validate the network

The neural network Simulink model is obtained with *gensim* command. After the generation of the network, a total of nine sets of insulation and temperature values are given to the network for validation purpose. The dataset is given at different instances to the network in intervals of 1 s. After the simulation of the network, the values of current, voltage and power can be obtained. The overall Simulink model is shown in Figure 7.19.

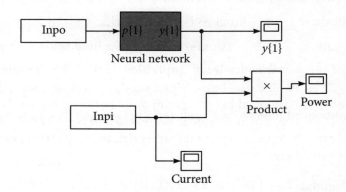

Figure 7.19 Overall Simulink Model.

The basic model of the neural network is shown in Figure 7.20. The primary and the secondary layers are shown in Figures 7.21 and 7.22, respectively.

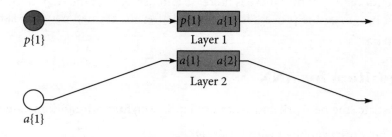

Figure 7.20 Structure of the Neural Network.

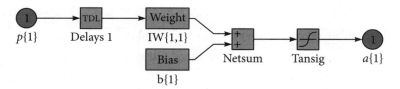

Figure 7.21 Structure of the First Layer of Neural Network.

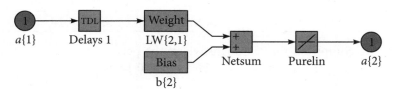

Figure 7.22 Structure of the Second Layer of Neural Network.

7.3.4 Simulation results

Training sets from Table 7.3 are chosen to uncover all the spaces of input to acquire the best performance of temperature ranging from −40 °C to 52 °C irradiation of solar power that has a wide range from 50 W/m² to 1000 W/m². The insulation value, temperature value and the ANN target voltage values are represented in the corresponding columns in Table 7.3.

Table 7.3 Training Data

50	50	16.15
100	−40	7.95
100	0	11.82
180	30	15.69
200	10	13.75
220	40	17.04
600	52	19.45
700	42	18.58
770	47	19.15
830	35	18
850	50	19.68
900	10	15.41
910	12	15.6
920	13	18.17
950	23.75	16.96
955	23.87	16.97

960	24	16.96
965	24.12	16.96
970	24.25	17.02
975	24.37	17.02
980	24.5	17.08
985	24.62	17.13
995	24.87	17.14

The trainings are completed for 25 epochs where the performance goal is met. After the completion of training, the mean square error value is 0.146852. Figure 7.23 represents the training curve.

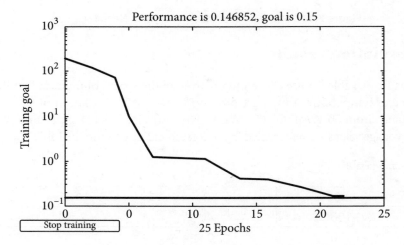

Figure 7.23 Training Curve.

For nine sets of insulation along with temperature conditions, the training network is validated. After training of the sets, the corresponding voltages and power values obtained are listed in Table 7.4.

Table 7.4 Results of ANN

The ANN Insulation Value (W/m²)	Temperature Value (°C)	The Target Modelling Voltage (V)	Modelling Power (W)	ANN Voltage (V)	ANN Power (W)
960	24	16.96	56.84	16.9698	56.8380
175	5	13.1	7.892	12.04	7.6394
1000	25	17.1	59.9159	17.0862	59.8885
700	42	18.52	44.82	19.2	44.8192

965	24.125	16.96	57.21	16.98	57.2030
930	12	16.6	48.64	16.5	48.6104
995	24.87	17.14	59.41	17.0674	59.4013
820	40	18.46	53.14	18.05	52.9013
980	24.5	17.08	58.31	17.02	58.3007

The curves for voltage, current and power are shown in Figures 7.24, 7.25 and 7.26, respectively.

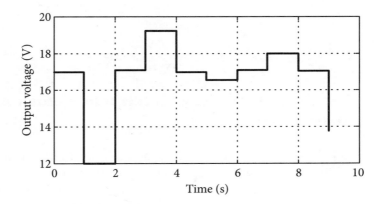

Figure 7.24 Voltage Curve for Neural Network.

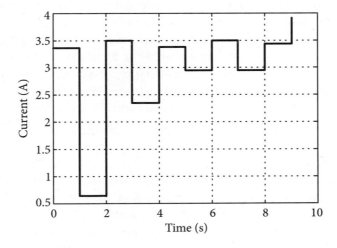

Figure 7.25 Current Curve for Neural Network.

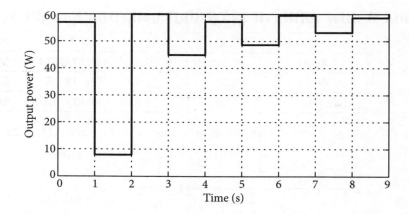

Figure 7.26 Power Curve for Neural Network.

7.4 Converter Control Using Microprocessors and Microcontrollers

A microprocessor is a chip that has the components of a CPU on a single integrated circuit (IC). An IC is a multipurpose, programmable device that acknowledges advanced data as information, systems it as demonstrated by rules set away in its memory and generates output. A chip is a generally valuable structure. Microcontrollers join a microprocessor with periphery devices in introducing structures. The chip is used in far-reaching and domestic units, machines, cars, auto keys, test instruments, toys, light switches/dimmers and electrical circuit breakers, smoke alerts, and batteries. Nonprogrammable controls would require variable, or uncontrolled execution to finish the results possible with a chip. A chip control program can be powerful, uniquely designed for the assorted needs of an element, allowing redesigns in execution with an insignificant overhaul of the element. Different components can be executed in particular models of an element offering at trivial creation cost. Microprocessor control of a structure can give control strategies that would be impossible to execute using electromechanical controls or electronic controls.

A microcontroller is an incorporated circuit containing a processor focus, memory, and programmable I/O peripherals. Microcontrollers are used as a piece of subsequently controlled devices. A microcontroller is also seen as an autonomous structure with a processor, memory and peripherals that can be used as an embedded system. Microcontrollers have been utilized today as a part of other equipment, for instance, vehicles, telephones, machines, and peripherals for PCs. While some embedded devices are incredibly exceptional, various others have tremendous requirements for memory and task length, do no live up to expectation devices, and have a low programming multifaceted nature. Consistent data and output devices fuse switches, exchanges, solenoids, LEDs, sensors for data, for instance, temperature. The PWM square makes it possible for the CPU to control power converters, resistive loads, motors, etc., without using loads of CPU ports.

Along these lines, MATLAB, gives serial data correspondence handiness on PCs. In addition, Simulink, MATLAB's insight-based programming environment, enables customers to reproduce and look at element models. The serial correspondence handiness of MATLAB enables PCs to respond with microcontrollers to transmit control commands and get substantial data. Besides, we utilize MATLAB Simulink along with the graphical user interface (GUI) environment for activities, allowing enhanced data operation and recognition.

Summary

At the end of this chapter, users will be able to understand the concepts of fuzzy logic, neural network concepts and its practical applications. The impact of integrated circuits in MATLAB is also explained.

Solved Examples

1. A realtor wants to explain to his clients his best offers. He wants to show comfortness of the various houses in terms of bedrooms. The available types of homes are represented by the following set.

    U = {1, 2, 3, 4, 5, 6, 7, 8, 9, 10}

    The realtor wants to explain a 'Best home for a 4 member family' using fuzzy set. U is the number of bedrooms explained for home set.

    Solution

    The fuzzy set named 'Best home for a 4 member family' is represented as

    House for Four =

    FuzzySet[{{1,2}, {2,5},{3,8},{4,1},{5,7},{6,3}}, Universal Space→{1,10}];

    FuzzyPlot[Housefor Four, showdots→True];

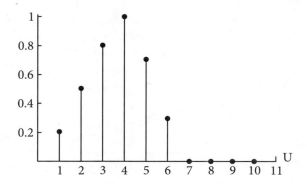

2. The given fuzzy set represents the age integral of human beings.

$$U = \{0, 1, 2, 3, ..., 100\}$$

Solution

The interval is represented by the universal space ranging from 0 to 100.

SetOptions[FuzzySet, UniversaalSpace → {0,100}];

Assuming 'Young' age group, whose membership function is given by

Young = FuzzyTrapezoid[0,0,25,40];

Assuming 'Old' age group, whose membership function is given by

Old = FuzzyTrapezoid[50,65,78,40];

Assuming 'MiddleAged' age group, whose membership function is given by

MiddleAged = Intersection[Complement[Young], Complement [Old]];

FuzzyPlot [Young, MiddleAged,Old,PlotJoined → True];

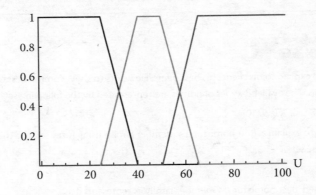

3. Design a linear neuron to predict the next value in a time series given the last five values.

Solution

time = 0:0.025:5;

signal = sin(time*4*pi);

Plot(time,signal)

xlabel('Time');

ylabel('Signal');

title('Signal to be Predicted');

signal = con2seq(signal);

Xi = signal(1:4);

X = signal(5:(end-1));

timex = time(5:(end-1));

T = signal(6:end);

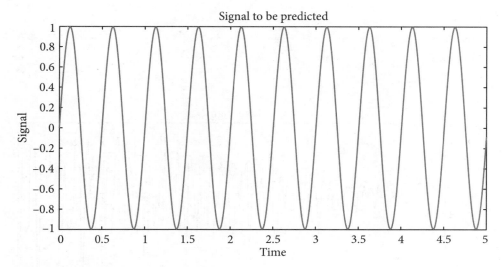

```
net = newlind(X,T,Xi);
view(net)
Y = net(X,Xi);
figure
plot(timex,cell2mat(Y),timex,cell2mat(T),'+')
xlabel('Time');
ylabel('Output - Target +');
title('Output and Target Signals');
```

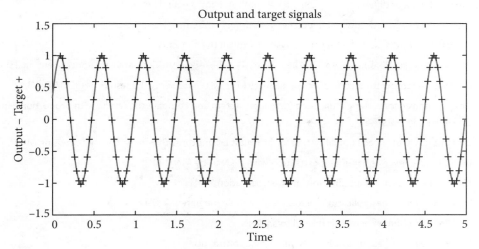

```
figure
E = cell2mat(T)-cell2mat(Y);
plot(timex,E,'r')
```

```
hold off
xlabel('Time');
ylabel('Error');
title('Error Signal');
```

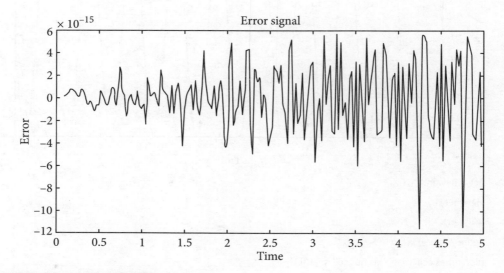

Practice Questions

1. Explain the importance of neural network systems in real time.
2. Explain the major difference between a serial computer and a neural network.
3. Explain the design procedure of neural network using competitive learning.
4. Draw the block diagram of a fuzzy logic system. Explain the concepts.
5. Explain the logical reasons why a fuzzy logic system can be used as an alternative for nonconventional systems.
6. Construct a temperature controller using a fuzzy logic system to verify the results using MATLAB.
7. Design a power converter circuit using fuzzy logic system and compare the results with proportional integral controller concepts.
8. Design a low-cost speed control system for brushless DC motor with a fuzzy logic system using MATLAB.
9. Construct a converter parameter extraction with ANN using MATLAB.
10. Explain the concepts of fuzzification and defuzzification.
11. Explain the real-time applications of fuzzy logic and neural network systems.
12. Explain how a human neuron is related to an ANN.
13. Discuss the home heating system with fuzzy logic control using MATLAB.
14. Explain the technique of fuzzy logic blood pressure during anesthesia in brief using MATLAB.
15. Explain the merits of fuzzy logic system and neural networks and vice versa.

Review Questions

1. Define artificial neural network (ANN).
2. List the differences between ANN and a biological network.
3. Define the activation function.
4. What are the applications of neural networks?
5. What are dendrites?
6. What are the different types of training?
7. Define back propagation network (BPN).
8. What is meant by batch training?
9. What is LM optimization?
10. Define feedback networks?
11. What is energy function or Lyapunov function?
12. Discuss the basic building blocks used in ANN?
13. To describe the state of a system, what are the variables required?
14. How does ANN resemble a brain?
15. What are fuzzy sets?
16. What is the procedure to determine fuzzy sets?
17. Define fuzzification.
18. What are fuzzy relations?
19. List the operations on fuzzy relations.
20. Differentiate fuzzification and defuzzification?
21. List the defuzzification methods.
22. Explain the method of center of sums in defuzzification technique.
23. What are the features of membership functions?
24. What is a normal fuzzy set?
25. Where do you apply a normal fuzzy set?
26. Define height of a fuzzy set.

Multiple Choice Questions

1. Fuzzy logic is a form of
   - (a) Two-valued logic
   - (b) Crisp set logic
   - (c) Many-valued logic
   - (d) Binary set logic
2. The truth values of traditional set theory is _____ and that of fuzzy set is_____
   - (a) Either 0 or 1, between 0 and 1
   - (b) Between 0 and 1, either 0 or 1
   - (c) Between 0 and 1, between 0 and 1
   - (d) Either 0 or 1, either 0 or 1

3. How many types of random variables are available?
   (a) 1
   (b) 2
   (c) 3
   (d) 4

4. The values of the set membership is represented by
   (a) Discrete set
   (b) Degree of truth
   (c) Probabilities
   (d) Both (b) and (c)

5. What is meant by probability density function?
   (a) Probability distributions
   (b) Continuous variable
   (c) Discrete variable
   (d) Probability distributions for continuous variables

6. Which of the following is used for probability theory sentences?
   (a) Conditional logic
   (b) Logic
   (c) Extension of propositional logic
   (d) None of the mentioned

7. Choose the set of fuzzy operators used in fuzzy set theory.
   (a) AND
   (b) OR
   (c) NOT
   (d) EX-OR

8. The more linguistic operators that can be applied to fuzzy set theory are called
   (a) Hedges
   (b) Lingual variables
   (c) Fuzzy variables
   (d) None of the mentioned

9. Where can the Bayes rule be used?
   (a) Solving queries
   (b) Increasing complexity
   (c) Decreasing complexity
   (d) Answering probabilistic query

10. What does the Bayesian network provide?
    (a) Complete description of the domain
    (b) Partial description of the domain
    (c) Complete description of the problem
    (d) None of the mentioned

11. Fuzzy logic is usually represented as
    (a) IF–THEN–ELSE rules
    (b) IF–THEN rules
    (c) Both (a) and (b)
    (d) None of the mentioned

12. _____ is/are the way/s to represent uncertainty.
    (a) Fuzzy logic
    (b) Probability
    (c) Entropy
    (d) All of the mentioned

13. _____ are algorithms that learn from their more complex environments (hence eco) to generalize, approximate and simplify solution logic.
    (a) Fuzzy relational database
    (b) Ecorithms
    (c) Fuzzy set
    (d) None of the mentioned

14. Which condition is used to influence a variable directly by all the others?
    (a) Partially connected
    (b) Fully connected
    (c) Local connected
    (d) None of the mentioned
15. What is the consequence between a node and its predecessors while creating a neural network?
    (a) Conditionally dependent
    (b) Dependent
    (c) Conditionally independent
    (d) Both (a) and (b)
16. To output a zero, a three-input neuron is trained when the input is 110. To output 1, the input is 111. Therefore, the output will be zero when the input is
    (a) 000 or 110 or 011 or 101
    (b) 010 or 100 or 110 or 101
    (c) 000 or 010 or 110 or 100
    (d) 100 or 111 or 101 or 001
17. For a single layer artificial network, the number of neurons will be
    (a) single
    (b) many
    (c) double
    (d) infinite
18. In general, an auto associative network is
    (a) Neural network without loop
    (b) Neural network with feedback
    (c) Neural network with single loop
    (d) Preprocessing single-layered feedforward network
19. For a four input neuron network, the transfer function is equal to 2, which is linear. The inputs given to the network are 4, 10, 5 and 20, respectively. Hence, the output produced will be
    (a) 238
    (b) 97
    (c) 167
    (d) 345
20. Identify the correct statement.
    (i) Neural networks have very high computational rate compared to conventional methods
    (ii) Neural networks can be learned only by example
    (iii) Neural network is the mimic of human brain
    (a) All of the mentioned are true
    (b) (ii) and (iii) are true
    (c) (i), (ii), and (iii) are true
    (d) None of the mentioned
21. Identify the correct statement.
    (i) The training time of the network is dependent on the size of the network
    (ii) Neural network can be replicated on a conventional system
    (iii) Artificial neurons are similar to biological neurons
    (a) All of the mentioned
    (b) (ii) is true
    (c) (i) and (ii) are true
    (d) None of the mentioned

22. Mention the merits of neural networks over conventional computers.
    (i) They can be better learnt with examples
    (ii) Fault tolerance is more
    (iii) They are more applicable for real-time applications
    (a) (i) and (ii) are true
    (b) (i) and (iii) are true
    (c) Only (i)
    (d) All of the mentioned

23. Indentify the correct statement.
    Neural networks with a single layer do not have the ability to
    (i) Recognize pattern performance
    (ii) Identify the parity of a picture
    (iii) Determine the connection of the picture
    (a) (ii) and (iii) are true
    (b) (ii) is true
    (c) All of the mentioned
    (d) None of the mentioned

24. Which is true for neural networks?
    (a) It has set of nodes and connections
    (b) Each node computes its weighted input
    (c) Node could be in excited state or nonexcited state
    (d) All of the mentioned

25. Neuro software is:
    (a) A software used to analyze neurons
    (b) A powerful and easy neural network
    (c) Designed to aid experts in real world
    (d) A software used by neurosurgeons

26. Why is the XOR problem exceptionally interesting to neural network researchers?
    (a) User-friendly expression
    (b) Binary operation is complex
    (c) Solution through single-layer perceptron
    (d) Simply linear

27. What is backpropagation?
    (a) Another name of curvy function used in perceptron
    (b) Spread of error back through the network to adjust the inputs
    (c) Spread of error back through the network to adjust the inputs to allow weights
    (d) None of the mentioned

28. The benefits of linearly separable neural networks are
    (a) Class of problems that can be solved easily
    (b) Class of problems with perceptron that can be solved easily
    (c) Continuous mathematical functions
    (d) Mathematical functions that can be drawn

29. Which of the following is not the promise of ANN?
    (a) It can explain results
    (b) It can survive the failure of some nodes
    (c) It has inherent parallelism
    (d) It can handle noise

30. Neural networks are complex _____ with many parameters.
    (a) Linear functions
    (b) Nonlinear functions
    (c) Discrete functions
    (d) Exponential functions
31. After all the weighted inputs of a perceptron are added, the output will be either 0 or 1, when it exceeds a certain value. Comment on this statement.
    (a) True
    (b) False
    (c) Sometimes it can also output intermediate values
    (d) Cannot say
32. Unit step function is also called
    (a) Step function
    (b) Heaviside function
    (c) Logistic function
    (d) Perceptron function
33. To solve the XOR problem with multiple perceptron, each perceptron can partition off a linear part of the space and they can give combined results.
    (a) True—Multiple perceptrons can classify complex parts also
    (b) False—Incapable of solving linearly incapable functions
    (c) True—They should be explicitly hand coded
    (d) False—Single perceptron is enough
34. The network that involves backward links from output to the input and hidden layers is called a _____.
    (a) Self-organizing map
    (b) Perceptron
    (c) Recurrent neural network
    (d) Multilayered perceptron
35. Which of the following is an application of neural network?
    (a) Sales forecasting
    (b) Data validation
    (c) Risk management
    (d) All of the mentioned

8
Power Electronics Applications

> **Learning Objectives**
> - To learn about the applications of power electronics in various fields
> - To understand uninterruptible power supply and its types
> - To understand switch-mode power supply in electronics era
> - To learn about HVDC transmission and VAR compensators
> - To understand RF heating, battery charger and switch-mode welding

8.1 Introduction

Power devices transform and control electrical energy at high productivity using exchanging mode power semiconductor devices. Some regular uses of power devices include DC- and AC-directed power supplies, uninterruptible power supply (UPS) systems, electrochemical procedures, heating and lighting control, power line static VAR, and resonant compensation, high-voltage, direct current (HVDC) systems, photo voltaic and energy unit converter systems, solid state circuit breakers, high-frequency heating, variable speed engine drives and so forth. Engine drive zones might include applications in PCs and peripherals, solid state starters, home apparatuses, paper and material factories, wind generation systems, aerating, cooling and heating pumps, rolling and concrete plants, machine devices and mechanical autonomy, pumps and compressors, ship impetus and so forth. It is no big surprise that the use of power devices is presently spreading quickly from the developed countries to the developing nations of the world.

8.2 Uninterruptible Power Supply (UPS)

A UPS is a device that maintains a ceaseless supply of electric energy to hardware by supplying power from a different source when the principle power supply is not accessible. The UPS is typically embedded between the business utility mains and the basic loads. When a power cut occurs, the UPS will quickly change from the utility mains energy to its own energy source. The UPS is utilized where constant supply of power is critical such as in clinical centres or escalated care units.

The various categories of UPS systems include the following:
- Static systems
- Dynamic systems
- Hybrid systems

8.2.1 Static systems

Static UPS systems depend on power electronic devices. The continuous improvement of devices such as insulated-gate bipolar transistors (IGBTs) permits high-frequency operation, which brings about a quick transient reaction and low total harmonic distortion (THD) in the output voltage. Static UPS system comprises a few noteworthy components such as rectifiers and battery chargers, inverters, static exchange switches, and storage banks. All UPS systems have no less than one extensive, low-frequency, attractive segment, for the most part a transformer. In the early days, these transformers were expensive and large. Presently, compact size, less costly high-frequency magnetics are utilized. All UPS systems utilize an inner battery that delivers AC power through an inverter. How and when this inverter starts is possibly the most important factor as this, to a great extent, decides the adequacy of the UPS. According to universal norms, static UPS systems can be characterized into three primary classifications:

- Offline UPS
- Online UPS
- Line interactive UPS

8.2.1.1 Offline UPS

As offline UPS are disconnected from the net system, in ordinary operation, power is supplied specifically from AC mains. When the mains shut, an exchange switch disengages the electrical cable and connects the inverter with the load. When the mains power is restored, the load is reconnected to the electrical cable. This is the most straightforward type of UPS system. In reality, this system cannot be considered as an UPS as the inverter is typically off. They might be more rightly called standby power sources. The offline UPS system comprises a battery set, a charger and a power semiconductor switch. The switch connects the fundamental power supply with the load and with the batteries. Hence, the battery set stays charged during ordinary operation as shown in Figure 8.1.

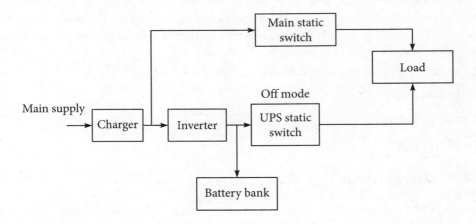

Figure 8.1 Block Diagram of Offline UPS.

In any case, when the fundamental power supply is not able to give the required power, then the static switch connects the load to the inverter to supply energy from the batteries. The exchange time from the ordinary operation to the external energy operation is by and large under 10 ms; this brief span when there is no power does not influence normal PC loads. With this arrangement, the UPS essentially exchanges utility power through the load when a power trough, or spike happens; in the meantime, UPS switches the load onto battery power and detaches the utility power until it comes back to an adequate level. Offline UPS systems are usually appraised at 600 VA for small PCs and home applications.

8.2.1.2 Online UPS

In online systems (as shown in Figure 8.2), the rectifier–inverter combination supplies the load power from the AC mains during ordinary operation. When the mains shut, the battery supplies DC current to the inverter; no time delay is involved. When the rectifier–inverter system comes up short, the load can be exchanged to AC mains utilizing an exchange switch. The online UPS systems have more applications than offline UPS systems; however, the value, weight and volume are higher.

The inverter of an online UPS supplies nonstop energy to the basic load. Under states of loads with high inrush streams, which are past the limit of the inverter, the static bypass switch gives the mains energy to the load. These UPS protects the load from surge voltage, transient voltage and noise. During ordinary or even unusual line conditions, the inverter supplies energy from the mains through the rectifier, which charges the batteries consistently. Notwithstanding this, it can remedy variable power.

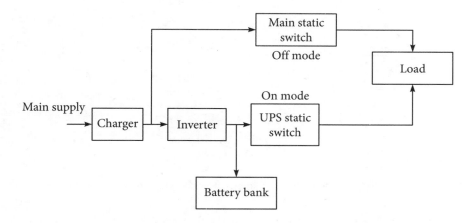

Figure 8.2 Block Diagram of Online UPS.

When the line comes up short, the inverter still supplies energy to the loads from the batteries. There is no exchange time between ordinary to no mains energy modes. When the connections are done, online UPS system is the most dependable UPS arrangement because of its directness, and the nonstop charge of the batteries, which implies that they are constantly prepared for power blackout. This sort of UPS provides freedom for input and output voltage adequacy and frequency. So high-output voltage quality can be acquired. When an overload happens, the bypass switch connects the load straightforwardly with the utility mains, ensuring the nonstop supply of the load, and maintaining a strategic distance so that it does not harm the UPS module.

In this circumstance, the output voltage must be synchronized with the utility stage; generally, the bypass operation will not be permitted. Proficiency of the online UPS systems is up to 94 per cent, which is constrained because of the twofold change impact. Online UPS systems are regularly utilized in delicate hardware or situations. All business UPS units of 5 kVA or more are online UPS systems.

8.2.1.3 Line interactive UPS

In line interactive UPS system, when the mains supply is available, the static switch is on. The static switch connects load with mains supply through inductor L. The batteries are charged through the charger block. When primary power supply is shut down, the mains static switch is open. Thus, the inverter turns on and gives energy to the load. When the primary power is accessible, the charger/inverter works as a charger and when the principle supply is not accessible, it acts as an inverter. It is considered as a device halfway between online and offline. Refer to Figure 8.3 for line interactive UPS.

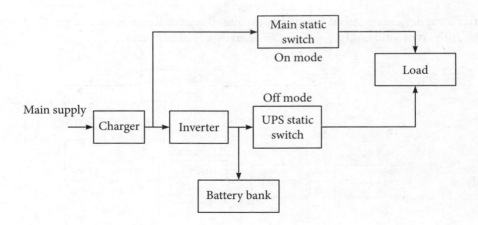

Figure 8.3 Block Diagram of Line Interactive UPS.

It comprises a solitary bidirectional converter that connects the batteries to the load. Line interactive UPS units regularly rate somewhere around 0.5 and 5 kVA for small server systems. The productivity of the line interactive UPS is around 97 per cent when there are no issues in the line. Under typical operation, the fundamental power supply gives supply to the load, and the batteries can be charged through the bidirectional inverter, that is, it becomes a DC/AC converter or charger. It might likewise have dynamic power shifting capacities. When a shut down happens in the fundamental power supply, the static switch separates the load from the line and the bidirectional converter works as an inverter, which supplies energy from the batteries. The principle favourable circumstances of the line interactive UPS are the direct and the lower expense compared to online and offline UPS.

Line interactive UPS normally have an automatic voltage regulator (AVR). The AVR permits the UPS to successfully go up or down the approaching line voltage without changing to battery power. Hence, this kind of UPS can adjust to most long haul over-voltages or under-voltages without using the batteries. Another point of interest is that it diminishes the quantity of exchanges to the battery. Hence, the lifetime of the batteries are augmented. The drawback of the line interactive UPS is that under ordinary operation, it is impractical to manage output voltage and frequency. A line interactive UPS system has all its components disconnected from the net UPS systems. Notwithstanding that, it permits the load to be supplied from the mains by using a low- and high-voltage method of operation. Without this component, the UPS would frequently be running on battery; but with this component, when a full-power shut down occurs, the battery would be halfway released, and the self-sufficiency time might be lacking for a legitimate shutdown. Line interactive UPS systems are ideally suitable when the supply is inclined to hangs and surges.

Both line interactive UPS and offline UPS systems regularly draw high-peak variable streams from the utility. It results in the load drawing power from the transformer. As opposed to this, online UPS systems typically draw low-peak component streams from the utility. This is because the system contains power factor corrector (PFC), which has the capacity to draw undistorted (low-peak element) streams, which minimize burden to the power lattice.

8.2.1.4 Merits

- It connects primary supply to load.
- Since the inverter is dependably on, the nature of the load voltage is free from twisting.
- Unsettling influences on the supply, for example, power outage, brownouts, spikes, and so on, does not affect the output.
- Voltage regulation is better.
- Exchange time is for all intents and purposes zero since the inverter is dependably on. No exchanging is included.
- 100 per cent line regulation.
- Good brownout protection.
- Generally, sinusoidal output.
- Power variable adjustment and higher unwavering quality.

8.2.1.5 Demerits

- General productivity of UPS is decreased since inverter is dependably on.
- The wattage of the rectifier is increased since it needs to supply energy to the inverter and also charge the battery.
- Online UPS is costlier than other UPS systems.

8.3 Switch-Mode Power Supply

The switch-mode power supply (SMPS), which is a black box with two input terminals and two output terminals, is indistinguishable from linear power supply. The direct controller manages a consistent stream of current from the input to the end to keep up a steady load voltage. The SMPS directs the current stream by slashing the input voltage and controlling the normal current duty cycle.

The pulse-width-balanced SMPS is classified into two types of fundamental standard of operation.

- Forward-mode SMPS
- Flyback-mode SMPS

8.3.1 Forward-mode SMPS

Forward-mode SMPS (Figure 8.4) include an expansive group of exchanging power supply components. They can be recognized by an *LC* channel soon after the power switch or after the output rectifier on the auxiliary of a transformer. A buck controller is a type of forward-mode controller. In this classification, the power switch is put between the input voltage and the inductor. In the middle of the power switch and the channel section (inductor), there might be a transformer for moving the input voltage up or down as in transformer-isolated forward controllers. When the switch is turned on, the load current goes from the information source, through the inductor to the load, and back again through the arrival lines to the data source. As of now, the diode is converse one-sided. When the switch is off, the inductor still anticipates that current will course through it. The previous course of current through the input source is open-circuited as of now. So the freewheeling diode (FWD) begins to lead and keeps up a close current circle through the load. When the switch turns on once more, the voltage put away in the inductor reverse biases the FWD. In short, the forward current continually courses through the inductor; hence, these supplies are called forward-mode exchanging controllers. The measure of energy conveyed to the load is controlled by the switch.

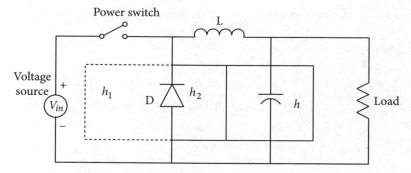

Figure 8.4 Forward-Mode SMPS.

8.3.2 Flyback-mode SMPS

This sort of SMPS (shown in Figure 8.5) utilizes the same four fundamental components as that of forward-mode exchanging controllers, yet they are modified. Here the inductor is put straightforwardly between the input voltage source and the power switch. When the switch is turned on, current is drawn through the inductor. It causes energy to be put away in the inductor. When the switch is closed, the current cannot alter the course immediately and it tries to stream in the same bearing for some time. Then, the inductor voltage switches. The diode turns on and the energy from the inductor stores in the capacitor. Since the inductor voltage goes back over the input voltage, the voltage that shows up on the output capacitor is higher than the input voltage. The main storage place for the load is the output channel capacitor. It makes the output swell voltage of flyback converters worse than their forward-mode controllers. Because of the limitation of the time required to discharge the inductor's flux into the output capacitor, the duty cycle is restricted.

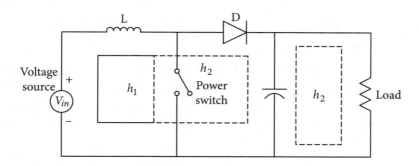

Figure 8.5 Flyback-Mode SMPS.

Based on the presence or absence of a transformer in the circuit, the SMPS is classified as follows: (1) nontransformer-isolated switching power supply topologies and (2) transformer-isolated switching power supply topologies.

The nontransformer-isolated kinds of SMPS are straightforward. They are utilized when some outer segments give DC separation or assurance of exchanging supply. These outer segments are generally 50–60 Hz transformers or separated mass power supplies. Generally, they are utilized as part of nearby board-level voltage regulation. In these topologies, only the semiconductors give the DC detachment from the input to the output.

The portions of the nontransformer-isolated topologies are as follows:
1. Buck regulator
2. Boost regulator
3. Buck–boost regulator

Power supplies that are expected to run straightforwardly from the AC source require a transformer to segregate the load side from AC lines. Transformers can likewise be utilized as part of power supplies where segregation is required for different reasons, for example, medicinal hardware use.

Components of the transformer-confined topologies include the following.
1. Flyback regulator
2. Push–pull regulator
3. Half-bridge regulator
4. Full-bridge regulator

8.4 High-Voltage DC Transmission

An HVDC electric power transmission system utilizes direct current for the mass transmission of electrical power and ac current for the more regular alternating current (AC) systems. For long-distance transmission, HVDC systems might be less costly and endure lower electrical issues. For submerged power links, HVDC stays away from the overwhelming streams required to charge and release link capacitance in every cycle. For shorter separations, the higher expense of DC transformation gear compared with an AC system might be defended because of the advantages of direct current connections.

HVDC permits power transmission between unsynchronized AC transmission systems. Since the power moving through an HVDC connection can be controlled autonomously of the phase point in the middle of source and load, it can balance out a system against upheavals because of fast changes in power. HVDC likewise permits exchange of power between lattice systems running at various frequencies, for example, 50 Hz and 60 Hz. This enhances the stability and economy of every matrix, by permitting trade of power between contradictory systems.

8.5 VAR Compensators

A static VAR compensator, which is shown in Figure 8.6, is an arrangement of electrical devices for giving quick-acting receptive power on high-voltage power transmission networks. Switched virtual circuits belong to the flexible AC transmission system device family; they control voltage, power element, sounds, and balance out the system. In contrast to a synchronous condenser, which is a turning electrical machine, a static VAR compensator has no noteworthy moving parts. Before the development of the SVC, power component remuneration protected extensive pivoting machines, for example, synchronous condensers or exchanged capacitor banks.

The SVC is a mechanized impedance coordinating device, intended to convey the system closer to the solidarity power variable. SVCs are utilized as part of two fundamental circumstances:

- Connected to the power system, to manage the transmission voltage—transmission SVC
- Connected close to huge mechanical loads, to enhance power quality—industrial SVC

In transmission applications, the SVC is utilized to direct the system voltage. If the power system's receptive load is capacitive, the SVC will utilize thyristor-controlled reactors (TCRs) to consume VARs from the system, bringing down the system voltage. Under inductive conditions, the capacitor banks are naturally exchanged along these lines, giving a higher system voltage. By interfacing the TCR, which is constantly variable alongside a capacitor bank step, the net result is a ceaselessly variable driving or slacking power.

In mechanical applications, SVCs are ordinarily set near high and quickly changing loads, for example, circular segment heaters, where they can smooth voltage.

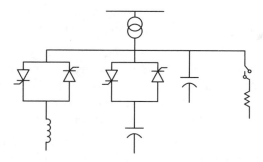

Figure 8.6 VAR Compensators.

Generally, an SVC contains one or more banks of settled or exchanged shunt capacitors or reactors, of which no less than one bank is exchanged by thyristors. Components that might be utilized to make an SVC commonly include the following:

- TCR, where the reactor might be air- or iron-cored
- Thyristor-switched capacitor (TSC)
- Harmonic filter
- Mechanically exchanged capacitors or reactors

By using phase balance exchanging thyristors, the reactor might be variably exchanged into the circuit thus giving a ceaselessly variable MVAR (Mega Unit of Reactive Power) infusion to the electrical network. In this design, coarse voltage control is given by the capacitors; the TCR gives smooth control. Smoother control and more adaptability can be provided by thyristor-controlled capacitor switching.

Thyristors are electronically controlled. They are similar to all semiconductors, produce heat, and deionized water is generally used to cool them. Chopping the receptive load into the circuit in this way infuses undesirable harmonics; thus, banks of high-power filters are normally given to smooth the waveform. Since the channels themselves are capacitive, they additionally send out MVARs to the power system. More unpredictable plans are possible where exact voltage regulation is required. Voltage regulation is provided by the closed-loop controller. Remote supervisory control and manual alteration of the voltage set-point are likewise regular.

For the most part, static VAR remuneration is not done at line voltage; a bank of transformers steps the transmission voltage of 230 kV down to a much lower level, for instance, 9.0 kV. This decreases the size and number of segments required in the SVC, in spite of the fact that the conductors must be substantial to handle the high streams connected with the lower voltage. In some static VAR compensators for mechanical applications, for example, electric heaters, where there might be a medium current–voltage bus bar current, for instance, at 33 kV or 34.5 kV, the static VAR compensator might be specifically connected to spare the expense of the transformer. Another normal connection point for SVC is on the delta tertiary winding of Y-associated autotransformers used to connect one transmission voltage to another voltage.

SVC is dynamic as it uses thyristors connected in arrangement and reverse parallel, shaping thyristor valves. Close loop semiconductors, typically a few inches in measurement, are generally found inside a valve house. The principle advantage of SVCs over basic mechanically exchanged pay plans is their fast momentary reaction to changes in the system voltage. Hence, they are used near their zero-indication and amplify the power as required.

They are less expensive, have higher limit, are quicker, and more dependable than element remuneration plans, for example, synchronous condensers. However, static VAR compensators are more costly than mechanically exchanged capacitors; certain system administrators use a blend of the two innovations now and then in the same establishment, utilizing the static VAR compensator to provide quick changes and the mechanically changed capacitors to provide constant state VARs.

8.6 Battery Charger

A device that plays a vital role on an optional cell or a rechargeable battery by building an electric current through it, is called a battery charger. The charging is completely based on the size and sort of battery that is being charged. Some types of battery have higher flexibility for damage and they are energized when they connected to a reliable voltage source and steady-state current. Some basic chargers of this type require manual separation near the terminal of the charge cycle or a clock to terminate the charging cycle at a settled time. Other categories cannot withstand long and high-rated overcharging where the charger might have temperature or voltage identification circuits and microcontrollers to alter the charging current, decide the condition of charge and to terminate the charge.

A stream charger generally gives a little measure of current, sufficient to check self-release of a battery that is not used for a while. Moderate battery chargers may take a few hours to complete a charge. Electric vehicles require high-rate chargers. Since the battery charger is expected to be connected with the battery, it might not have voltage regulation or be able to change the DC voltage output. Battery chargers outfitted with both voltage regulation and sifting are called battery eliminators. Some battery chargers include the following:

- Simple and fast charger
- Inductive charger
- Smart charger
- Motion power charger
- Pulse and drop charger
- Solar charger
- Universal battery

8.7 Switch-Mode Welding

A curve welding creates a circular segment between the terminals of two cathodes, where electrical disengagement is critical. Electrical disengagement is given by a high-frequency transformer in which current at the output decreases the welding current; maintaining a low swell current after the generation of a bend is also critical.

8.8 RF Heating

RF heating, also called electronic warming, is a method in which a high-frequency trading electric field, or radio wave or microwave electromagnetic radiation warms a dielectric material. At higher frequencies, this warming is achieved by a subnuclear dipole turn within the dielectric. RF dielectric warming at moderate frequencies, in view of its more conspicuous penetration compared to microwave warming, is used as a strategy for rapid warming and for reliably setting up certain heat-constant devices.

8.9 Electronic Ballast

An electrical ballast is a device proposed to confine the measure of current in an electric circuit. A recognizable and generally utilized example is the inductive counterweight used as a part of substituting current fluorescent lights, to constrain the current through the tube, which would otherwise ascent to dangerous levels because of the negative differential resistance in the tube's voltage–current characteristics.

8.9.1 Characteristics of fluorescent lamps

Counterweights can be as basic as an arrangement of resistors, inductors or capacitors, or a blend thereof as electronic ballast is used with fluorescent lights and high-power release lights.

8.9.1.1 Moment start

A moment start balance does not preheat the terminals, rather it utilizes a moderately high voltage to start the release circular segment. Although it is energy proficient, it outputs the least light start cycles, as material is impacted from the surface of the cool terminals every time the light is turned on. Moment start weights are most appropriate for applications with long-duty cycles, where the lights are not every now and again turned on and off.

8.9.1.2 Quick start

A quick start weight applies voltage and warms the cathodes at the same time. It gives long light life and more cycle life. However, it utilizes marginally more energy as the terminals at every end of the light keep on expending warming power as the light works.

8.9.1.3 Dimmable ballast

A dimmable ballast is fundamentally the same as a quick start weight, yet as a rule, it has a capacitor consolidated to give a calculated power closer to solidarity than a standard quick start weight. A QUADRAC sort light dimmer can be utilized with a darkening counterweight, which maintains the warming current while permitting light current to be controlled. A resistor of around 10 kΩ is required to be connected in parallel with the fluorescent tube to permit reliable termination of the QUADRAC at low light levels.

8.9.1.4 Customized start

A customized start counterbalance is a more popular variant of the quick start weight. This counterbalance applies energy to the fibres to start with; it permits the cathodes to preheat and afterward applies voltage to the lights to strike a bend. Light time regularly works up to 100,000 on/off cycles when utilizing customized start counterbalances. Once began, fibre voltage is decreased to increase working efficiency. This counterweight gives the best life and start up. It is favoured for applications with exceptional power cycling, for example, vision examination rooms and restrooms with a movement finder switch.

8.9.1.5 Hybrid

A half-breed balance has a transformer and an electronic switch for the cathode warming circuit. A hybrid unit works at a line power frequency of about 50 Hz in Europe, for instance. These sorts of weights, which are also called cathode-isolated stabilizers, detach the anode warming circuit after they start the lights.

8.9.1.6 Merits

- Greater efficiency
- Better control and design flexibility
- Ability to drive more lamps
- Lighter in weight
- Reduced lamp flicker
- Reduced cooling load

8.10 Brushless DC (BLDC) Motors

Brushless engines satisfy numerous capacities earlier performed by brushed DC engines; however, cost and control of the many-sided engine quality keeps brushless engines from supplanting brushed engines totally. Brushless engines have come to be used in numerous applications, for example, PC hard drives and CD/DVD players. Cooling fans in electronic gear are fuelled only by brushless engines. They can be found in cordless power devices where the expanded productivity of the engine prompts longer times of utilization before the battery should be charged. Low-speed, low-power brushless engines are utilized as part of direct-drive turn tables for records.

8.11 Thermal Management and Heat Sinks

Every electronic device and circuit creates heat and therefore require cooling to enhance reliability and avert untimely issues. The measure of heat yield is equivalent to the power input, if there are no other energy interactions. There are a few procedures for cooling including different styles of heat sinks, thermoelectric coolers, constrained air systems and fans, heat channels, and others. In instances of great low ecological temperatures, it might really be important to warm the electronic segments to accomplish proper operation.

Heat sinks are generally utilized as part of hardware and are turning out to be vital to present-day focal preparing units. It is a metal item carried into contact with an electronic part's hot surface – however as a rule, a slender warm interface material intercedes between the two surfaces. Microchips and power handling semiconductors are examples of hardware that need a heat sink to decrease their temperature through expanded warm mass and heat scattering by conduction and convection and to a lesser degree by radiation. Heat sinks have turned out to be practically fundamental to advanced coordinated circuits such as microchips, digital signal processings, graphical processing units, and that is only the tip of the iceberg.

A heat sink as a rule comprises a metal structure with one or more level surfaces to guarantee more warm contact with the segments to be cooled, and a variety of brush or blade-like projections to increase surface contact with air, and subsequently, the rate of heat dispersal.

> **Summary**
>
> At the end of this chapter, the students will be familiar with the trends in industrial needs for power electronics. Future needs can be predicted and innovative designs can be created with simulation models.

Multiple Choice Questions

1. The induction heating depth of penetration is proportional to
   - (a) Frequency × 2
   - (b) Frequency/2
   - (c) Frequency^2
   - (d) $1/(\text{Frequency})^{1/2}$

2. Which of the following types of heating process is used for surface heating of steel?
   - (a) Dielectric heating
   - (b) Infrared heating
   - (c) Induction heating
   - (d) Resistance heating

3. Uninterruptible supply is used in
   - (a) Computers
   - (b) Communication links
   - (c) Essential instrumentation
   - (d) All of the above
4. EHV-AC is generally preferred for HVDC transmission because
   - (a) HVDC terminal equipment is inexpensive
   - (b) VAR compensators are not needed
   - (c) Of its improved system stability
   - (d) there is less harmonics
5. An SMPS operating at 20 kHz–100 kHz range uses as the main switching elements
   - (a) SCR
   - (b) MOSFET
   - (c) Transistor
   - (d) SIT
6. High-speed circuit breakers are used to
   - (a) Minimize short-circuit current
   - (b) Improve system stability
   - (c) Reduce system stability
   - (d) Maximize short-circuit current
7. HVDC transmissions are preferred over bulk power transmission because of their
   - (a) Less cost
   - (b) Less harmonic issues
   - (c) Minimal power loss
   - (d) Reliable protection
8. In case of DC transmission
   - (a) Sending and receiving end should be synched with each other
   - (b) The impacts of reactance are greater than in AC transmission
   - (c) The effect of inductive and capacitive reactance is greater than in an AC transmission line of the same rating
   - (d) Reactance causes no effects
   - (e) Stability considerations limit power transfer capability
9. Static VAR compensators are used to control
   - (a) Only magnitude of the AC line current from the utility
   - (b) Only phase of the AC line current from the utility
   - (c) Both magnitude and phase of the AC line current from the utility
   - (d) None of the above
10. A metal bar is heated electronically by
    - (a) Emission heating
    - (b) Dielectric heating
    - (c) Induction heating
    - (d) Conductive heating
11. A rod of mild steel kept inside a coil carrying high-frequency current gets heated due to
    - (a) Dielectric heating
    - (b) Induction heating
    - (c) Both (a) and (b)
    - (d) None of these
12. A freshly painted layer may be dried electronically by
    - (a) Conduction heating
    - (b) Induction heating

(c) Dielectric heating (d) None of these

13. High-frequency induction heating is used for
    (a) Ferrous metals only
    (b) Nonferrous metals
    (c) Both ferrous and nonferrous metals
    (d) None of these

14. In dielectric heating, nonuniform heating
    (a) Occurs for higher frequencies
    (b) Occurs for lower frequencies
    (c) Is independent of frequency
    (d) Occurs for higher power factors

15. In dielectric heating, the rate of heating cannot be increased by increasing the potential gradient because
    (a) Coupling problems become highly pronounced
    (b) Very high voltages are not easily available
    (c) Heating becomes nonuniform
    (d) Corona takes place

Review Questions

1. Explain the concept of UPS in detail.
2. Explain in brief the concepts of SMPS.
3. Highlight the applications of various power electronic circuits that meet recent industrial trends.
4. Explain how power electronic circuits are applied in HVDC transmission.

Practice Questions

1. Compare the various features of power electronics converters in detail.
2. Elaborate in detail the various applications of power electronic converters and differentiate them.

9

Introduction to Electrical Drives

Learning Objectives

- ✓ To learn the basic of electrical drives
- ✓ To learn the different types of DC drives and their applications
- ✓ To learn the different types of AC drives and their applications

---

9.1 Introduction

In general, systems being utilized for control of movement are called drives. Drives might utilize any of the prime movers such as diesel motors, steam turbines and electric engines for supplying mechanical energy for movement control. Drives utilizing electric engines are called electrical drives. A drive can also be considered a combination of different systems consolidated together with the end goal of movement control. Electric drives for engines are utilized to draw electrical energy from the mains and supply the electrical energy to the engine at whatever voltage, current and frequency needed to accomplish the desired mechanical output. The basic block diagram of an electrical drive system is shown in Figure 9.1.

The power source gives energy to the electric drive system. The converter connects the engine with the power source and furnishes the engine with flexible voltage, current and frequency. The controller screens the operation of the whole system and guarantees the general system execution and dependability. The mechanical loads are preceded by the operation and the power source is controlled by what is accessible at the place. In any case, we can choose alternate parts such as the electric engine, converter and controller. Converters can change the electric waveform of the power source to a waveform that the engine can utilize. For instance, if the accessible power source is AC and the engine is a DC arrangement engine, then the converter changes the AC into DC. As it is a rectifier, the circuit is set in the system. The engine for the specific application is chosen by considering different parameters such as cost, meeting the power level, and execution required by the load for steady state and element operations. The classifications of drive systems include group drive, individual, and multi motor drive. Drives systems are further classified based on the following parameters.

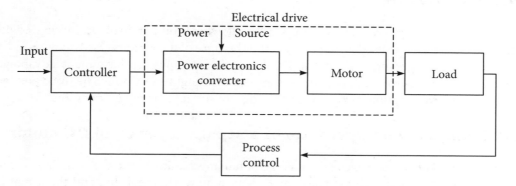

Figure 9.1 Basic Block Diagram of an Electrical Drive System.

Based on supply:
- DC drives
- AC drives

Based on speed:
- Constant speed drives
- Variable speed drives

Based on the number of motors:
- Single motor drive
- Multi motor drive

Based on control:
- Constant torque drives
- Constant power drives

9.1.1 Merits and demerits of electrical drive systems

In contrast to other prime movers, there is no reason to refuel or warm-up the motor. They are accessible in a variety of torque, velocity and power. Electric braking can be utilized. They also have adaptable control qualities. Earlier, induction and synchronous engines were utilized for the most part in steady speed drives. Variable velocity drives utilize DC motor. Nowadays, AC engines also use variable rate drives because of advancement of semiconductor converters. The system requires extensive support. They are big and costly; not suitable for rapid operation because of commutator and brushes and not suitable for a clean environment.

9.2 DC Drives

The drive is moderately basic and less efficient when compared to prompting engine drives. The DC engine itself is more costly; thus because of the various disadvantages of DC engines, they are getting less popular. DC-based drives, however, have many advantages such as ease of control, high beginning torque and straight execution. The disadvantages include high preservation, large and costly, and not suitable for rapid operation.

9.2.1 Steady-state operation of a separately excited DC motor

Field windings are used to excite the flux. To obtain mechanical work, the armature current is applied to the rotor unit via brush and commutator. The field flux and the armature current interact with each other in the rotor unit, thereby resulting in a torque. For instance, when a separately excited motor (Figure 9.2) is triggered by the field current called I_f, the armature current I_a flows through the circuit, where the motor develops a back electro magnetic field with a torque to balance the torque at the load at a particular speed. Each winding is supplied separately, where I_f is independent of I_a. Modification in the value of armature current has no impact on the field current where I_f is usually less than I_a.

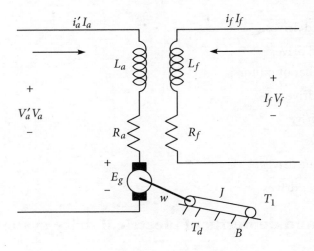

Figure 9.2 Separately Excited DC Motor.

The instantaneous field current is given by

$$V_f = R_f i_f + L_f \frac{di_f}{dt} \qquad (9.1)$$

where R_f and L_f are the field resistor and inductor, respectively.

DC motor equations

$$E_a = \frac{Z\phi N}{60}\frac{P}{A} \qquad N \rightarrow \text{rpm}$$

$$= Z\phi n \frac{P}{A} \qquad n \rightarrow \text{rps}$$

If $\omega_m = 2\pi n$

$$\omega_m = Z - \phi \frac{\omega_m}{2\pi}\frac{P}{A}$$

$$= \left(\frac{Z}{2\pi}\frac{P}{A}\right)\phi\omega_m$$

→ $E_a = K_a \phi \omega_m$, K_a = motor constant = $\left(\left(Z/2\pi\right)\left(P/A\right)\right)$ V/Wb rad/s
Torque developed by the motor:

→ Torque $T = \frac{1}{2\pi} Z \phi I_a \frac{P}{A}$

$$= \left(\frac{Z}{2\pi}\frac{P}{A}\right)\phi I_a$$

$$T = k_a \phi I_a \qquad K_a = \text{N·m / Wb·A}$$

For a DC separately excited motor:

→ Flux, ϕ is constant

→ $E_a = K_m \omega_m K_m$ → motor constant volts/rad/s

→ Torque = $K_a \phi I_a$

→ $T_e = K_m I_a$ K_m → Newton meter/Ampere

For a DC series motor:

$$\emptyset \propto I_a, \phi = cI_a$$

$$E_a = K_a c I_a \omega_m$$

$$E_a = K_1 I_a \omega_m, \quad K_1 \quad \text{motor constant} = \frac{\text{volt s}}{\text{rad.amp}}$$

$$T_e = K_a \phi I_a^2$$

$$= K_a c I_a^2$$

$$T_e = K_1 I_a^2 \quad \rightarrow \quad K_1 \quad \text{N·m/A}^2$$

For a normal operation, the developed torque must be equal to the load torque including friction and inertia. Therefore, $T = J d\omega/dt + B\omega + T_L$, where J is the inertia of the motor, B is the viscous friction constant in N·m/rad/s, and T_L is load torque in N·m. Under steady state operations, time derivatives are zero. Assume that the motor is not saturated.

For field circuit,

$$V_f = I_f R_f \tag{9.2}$$

The back emf is given by

$$Eg = K_v \omega I_f \tag{9.3}$$

The armature circuit,

$$V_a = I_a R_a + Eg = I_a R_a + K_v \omega I_f \tag{9.4}$$

The speed of the motor can be derived by

$$\omega = V_a - I_a R_a / K_v I_f \tag{9.5}$$

When the motor is lightly loaded, I_a will be small. Therefore,

$$\omega = V_a / K_v I_f \tag{9.6}$$

When the field current is kept constant, the motor current depends on the supply voltage. Refer to Figure 9.3.

Hence, the developed torque is given by

$$T_d = K_t I_f I_a = B\omega + T_L \qquad (9.7)$$

Also the power required is,

$$P_d = Td\omega \qquad (9.8)$$

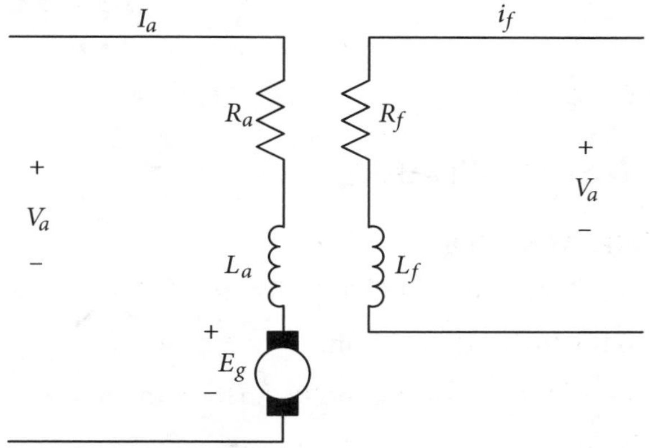

Figure 9.3 Steady State Operation of a DC Motor.

9.2.1.1 Torque and speed control

From the aforementioned equations, important factors can be deduced for the motor to be operated at steady-state operation. Considering a fixed field current I_f, or flux, the need of torque can be satisfied by varying the armature current I_a. The speed of the motor can be varied by voltage control method and field control method. These types of observations lead to the usage of variable DC voltage to control the speed–torque characteristics of the motor.

9.2.1.2 Variable speed operation

Figure 9.4 represents the speed–torque curve at steady-state conditions for a wide range of armature voltage. By applying correct voltage, the speed of the DC motor can be applied. Depending on the armature resistance, speed variation occurs even when the value of no-load is small.

The base speed ω_{base} is the speed which corresponds to the rated V_a, I_a and I_f. The constant torque region is nothing but the $\omega > \omega_{base}$. The armature current and the field current can be maintained at a constant rate to meet the torque demand. When the armature voltage is varied to control the speed value, the power rate increases with respect to speed. At constant power, the armature voltage is maintained in range and is increased to reduce the speed level.

As a result, the power developed by the motor is given by

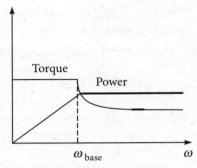

Figure 9.4 Speed–Torque Curve.

Power = torque × speed

9.2.2 Four quadrant operation

First Quadrant Forward Motoring

During the operation in Quadrant 1, the following specifications are observed:

- The motor runs in the forward direction.
- The operating point of the motor is at point A (refer to Figure 9.5).
- The applied voltage is greater than the back emf.
- The current flowing through the armature is positive.
- The power drawn from the supply is always positive.

Second Quadrant Forward Braking

During the operation in Quadrant 2, the following specifications are observed:

- When the motor is running in position A, the supply voltage reduces to a value below the emf value, where the current will start flowing in the reverse direction.
- The operating point of the motor is at point B (refer to Figure 9.5).
- Owing to the current flow in the reverse direction, the power drawn is negative.
- The motor behaves like a generator because the power flow is from the machine to the supply.
- The speed will be reduced due to the combinations of load torque and machine torque.
- Again when the emf falls below the applied voltage, the current value becomes positive again. Hence, the motor settles back to Quadrant 1 but at low speed.

Third Quadrant Reverse Motoring

During the operation in Quadrant 3, the following specifications are observed:

- The operating point of the motor is at point C (refer to Figure 9.5).
- The torque and the speed are negative.
- There is a need to change the polarity of the armature supply.

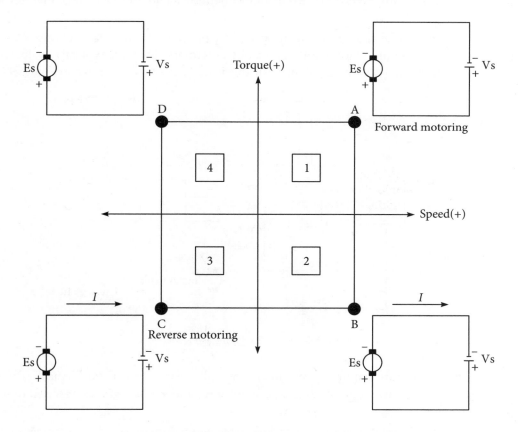

Figure 9.5 Four Quadrant Operation of a DC Motor.

Fourth Quadrant Reverse Braking

During the operation in Quadrant 4, the following specifications are observed:

- The operating point of the motor is at point D (refer to Figure 9.5).
- The operation is same as Quadrant 2, where the torque is positive and the speed is negative.

9.2.3 Single-phase and three-phase DC drive

For minimal effort low-power applications up to around 10 kW, single-phase rectifiers can be used. Low-power conservative drives can likewise be developed utilizing single-phase half-wave rectifiers with free wheeling diodes (FWDs). For higher power drives, up to MW range, three-phase supply with three-phase rectifiers are typically employed. For low to medium-power DC supplied drives, for example, battery, a chopper DC–DC converter is utilized.

It is likewise normal that in a few applications, particularly in pulses, choppers are utilized as a part of conjunction with uncontrolled rectifiers. They are typically evaluated at medium power 100s of kW. One of the basic thyristor-based drive is shown in Figure 9.6.

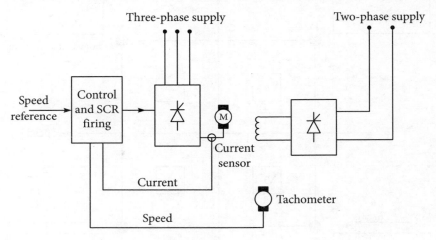

Figure 9.6 Basic Thyristor-Based Drive.

Variable DC voltages are acquired from silicon controlled rectifiers (SCRs). Typically, field rectifiers have much lower appraisals than armature rectifiers. It is just they are used to set up the flux.

The key explanation behind effective DC drive operation is the extensive armature inductance L_a. Substantial L_a takes into consideration nearly constant armature current with little effect due to current shifting impact of L. Normal estimation of the current is zero. No critical impact on the torque will take place. In the event that L_a is not sufficiently substantial, or when the engine is softly stacked, or if supply is single-phase half-wave, alternating current may occur. The effect of alternating current is rise of output voltage of rectifier, where the engine speed goes higher. In open-loop operation, the speed is ineffectively directed. It is beneficial to include additional inductance in arrangement with the armature inductance.

If the armature circuit of a DC engine is connected to the output of a single phase-controlled rectifier, the armature voltage can be changed by fluctuating the delay angle of the converter. The fundamental circuit for a single-phase converter-fed independently energized engine is shown in Figure 9.7. At a low edge, the armature current might be irregular, and this would cause some issues in the engine. A smoothing inductor L is ordinarily associated in arrangement with the armature circuit to decrease the current to an adequate extent. A converter is additionally connected in the field circuit to control the field current by differing the postponement point.

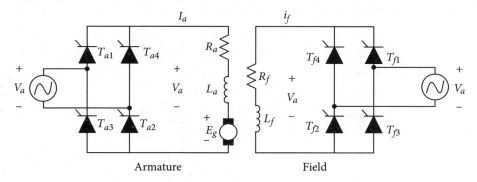

Figure 9.7 Single-Phase DC Drive.

Single-phase DC drive can be classified based on the type of single-phase converters into the following types.

- Half-wave converter drive
- Semiconverter drive
- Full-converter drive
- Dual-converter drive

For continuous conduction mode, the armature and field voltage is given by

$$V_a = 2V_m \cos\alpha/\pi \tag{9.9}$$

The circuit of the armature is purely dependent on the output of the three-phase-controlled rectifier. They are mainly applied for high-control applications up to the range of some megawatts. The operating frequency is higher than the single-phase drives, which require less inductance in the armature circuit to reduce the armature circuit. The armature current is generally maintained constant, whereas the engine excitation is contrasted with single-phase drives.

Similar to the single-phase drives, three-phase drives (refer to Figure 9.8) may also be subdivided into the following groups.

- Three-phase half-wave-converter drives
- Three-phase semiconverter drives
- Three-phase full-converter drives

- Three-phase dual-converter drives.

Three-phase full-wave-converter drive is a two-quadrant drive with no field inversion. It is restricted to applications up to 1500 kW around the heading of power. However, the back emf of the engine is switched by turning around the field excitation. The converter in the field circuit is a solitary or three-phase full converter. The MATLAB model of single phase and three phase DC drives are shown in Figures 9.9 and 9.12. Their speed–torque curves and output responses are shown in Figures 9.10, 9.11, 9.13 and 9.14.

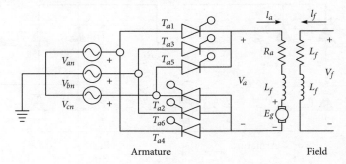

Figure 9.8 Three-Phase DC Drive.

For continuous conduction mode, the armature and field voltage is given by

$$V_a = 3V_m \cos\alpha / \pi \tag{9.10}$$

MATLAB Example 1

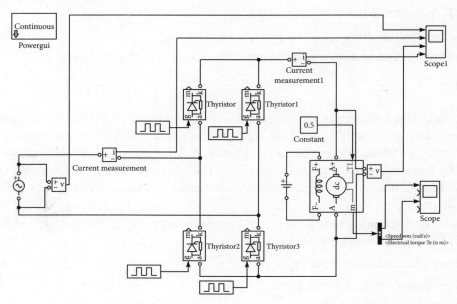

Figure 9.9 Simulation Model of a Single-Phase DC Drive.

Introduction to Electrical Drives 399

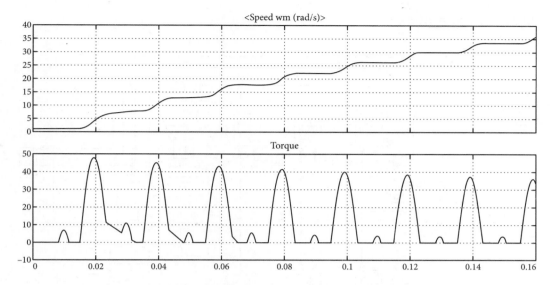

Figure 9.10   Speed–Torque Curves of a Single-Phase DC Drive.

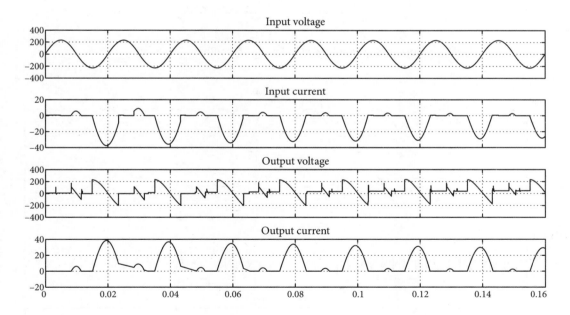

Figure 9.11   Output Waveform of a Single-Phase DC Drive.

MATLAB Example 2

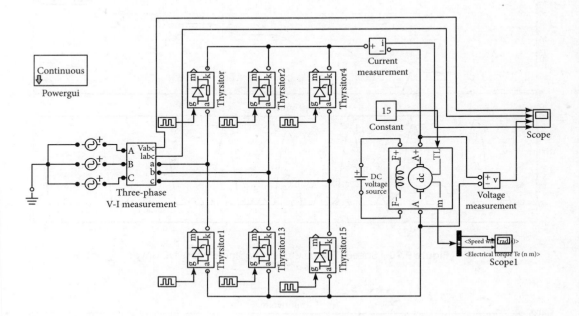

Figure 9.12 Simulation Model of a Three-Phase DC Drive.

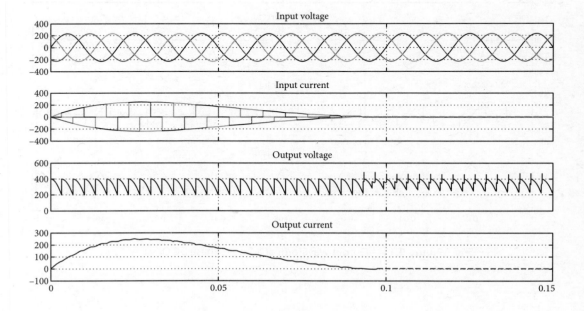

Figure 9.13 Output Waveform of a Three-Phase AC Drive.

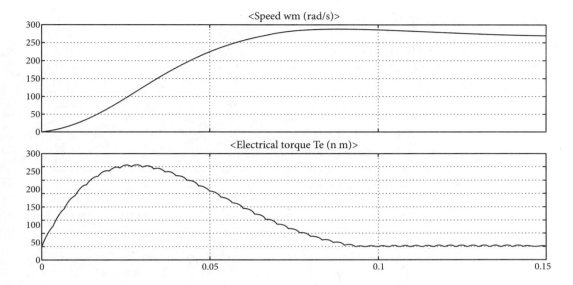

Figure 9.14 Speed–Torque Curves of a Three-Phase DC Drive.

9.2.4 Reversal of DC motor

DC drives are intrinsically bidirectional. Thus, it is possible to invert the heading. It can be an engine or a generator. However, the rectifier is unidirectional as thyristors are unidirectional devices. However, if the rectifier is completely controlled, it can be worked to provide negative DC voltage, by making the terminating edge more prominent than 90°, as in Figure 9.15.

The reversal operation can be fulfilled by armature and field reversal or by the double converter under two-quadrant or four-quadrant operations.

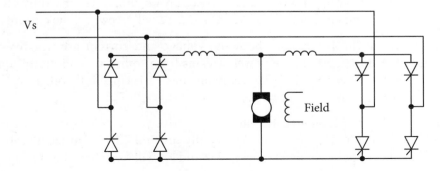

Figure 9.15 Concept of Reversal.

9.2.5 DC chopper drives

A chopper (shown in Figure 9.16) is a static device that converts constant DC input to a variable DC output voltage straightforwardly. A chopper might be considered as an AC transformer since it carries on in an indistinguishable way. It is otherwise called a DC-to-DC converter. It is generally utilized for engine control. It is also utilized as a part of regenerative braking. Basically, a chopper is an electronic switch that is utilized to interfere with one sign under the control of another. The DC/DC converter can be a 1/4, 2/4 and 4/4 quadrant chopper. Refer to Chapter 4 for a detailed explanation of the concepts.

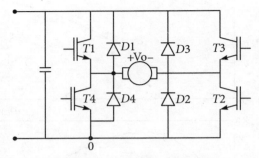

Figure 9.16 DC–DC Chopper Drive.

The switches in the four-quadrant chopper can be exchanged in two distinct modes. The output voltage swings in both bearings, that is, from $+V_{DC}$ to $-V_{DC}$. This method of changing is alluded to as pulse width modulation (PWM) with bipolar voltage exchanging. The output voltage swings either from zero to $+V_{DC}$ or zero to $-V_{DC}$. This method of changing is called PWM with unipolar voltage exchanging. The four-quadrant chopper works in the four quadrants in the following ways. The MATLAB Simulink model of chopper-fed drives and its speed–torque characteristics are shown in Figures 9.17 and 9.18.

Quadrant 1: In the principal quadrant, the voltage and current ensure that the power is certain. For this situation, the power streams from source to load. In this operation, T1 is on, T4 is off, T2 is ceaselessly on, and T3 is persistently off. T1 and T2 are leading in this mode.

Quadrant 2: In the second quadrant, the voltage is still positive yet the current is negative. Along these lines, the power is negative. For this situation, the power streams from load to source and this can happen if the load is an inductive or back emf source, for example, a DC engine. Here T1 is off, T4 is on, T2 is consistently on and T3 is constantly off. As the inductor current cannot be changed quickly, D4 and T2 will freewheel the current.

Quadrant 3: In the third quadrant, both the voltage and current are negative yet the power is sure. For this situation, the power streams from source to load. In this quadrant operation, T3 is on, T2 is off, T4 is ceaselessly on, and T1 is constantly off. T3 and T4 are leading in this mode.

Quadrant 4: In the fourth quadrant, voltage is negative yet current is sure. The power is negative. Here T3 is off, T2 is on, T4 is constantly on, and T1 is consistently off. As the inductor current cannot be changed quickly, D2 and T4 will freewheel the current.

MATLAB Example 3

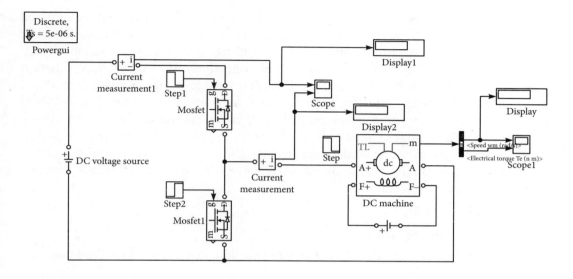

Figure 9.17 Chopper-Fed Drive.

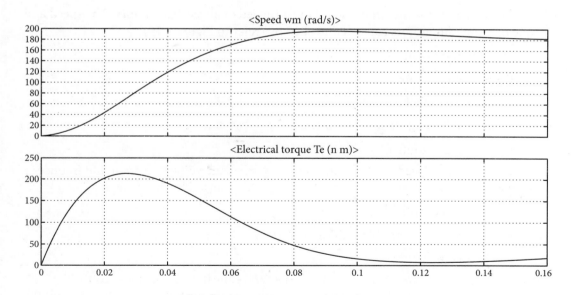

Figure 9.18 Speed–Torque Characteristics.

9.3 AC Drives

AC drives are extremely coupled, nonlinear and multivariable structures as opposed to the less intricate, decoupled structures of autonomously stimulated DC motors. The control of AC drives requires complex control figurings that can be performed by chip or microcomputers near brisk trading control converters. AC motors have different inclinations: they are lightweight and sensitive, and have low upkeep differentiated and DC motors. They need frequency control with voltage and current for variable speed operations. These parameters can be controlled by various power converters to meet the requirements. These power controllers, which are for the most part ideal, require pushed control systems, for instance, model reference, flexible control, sliding mode control and field-arranged control. In any case, the advantages of AC drives surpass the disadvantages. There are two sorts of AC drives:

- Induction motor drives
- Synchronous motor drives

AC drives are replacing DC drives nowadays and are used in many industrial and domestic applications.

9.3.1 Induction motor drive

The induction motor is the simplest and most rugged of all electric motors. The speed and torque of induction motors can be controlled by (refer to Figure 9.19)

- Stator voltage control
- Rotor voltage control
- Frequency control
- Stator voltage and frequency control
- Stator current control

To meet the torque–speed requirement cycle of a drive, the voltage, current and frequency control are ordinarily used. The stator voltage can be changed by three-stage

- AC voltage controllers
- voltage-bolstered variable DC-join inverters
- PWM inverters

However, because of the restricted pace range prerequisites, AC voltage controllers are typically used to provide voltage control. In order to obtain the torque–speed of a drive, the voltage, current and frequency control are used.

The AC voltage controllers are extremely basic. The harmonics are high and the data power factor (PF) of the controllers is low. They are utilized for the most part as a part of low-power applications, for example, fans, blowers and radiating pumps, where the starting torque is low. They are also used for starting high-control incitement engines to constrain the in-surge current.

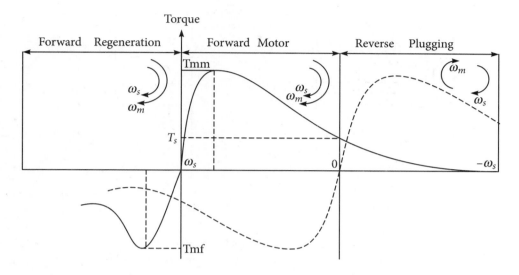

Figure 9.19 Speed–Torque Characteristics of an Induction Motor.

9.3.1.1 Rotor voltage control

For an induction motor drive (shown in Figure 9.20), the created torque might be shifted by fluctuating the resistance R. In the event that R is connected to the stator winding, R_r might be added to decide the created torque. The ordinary torque–speed qualities for varieties in rotor resistance are arrived at. This strategy builds the beginning torque while restricting the beginning current. It is an inefficient strategy and if the resistances in the rotor circuit are not approached, there would be irregular characteristics in voltages and streams. A wound rotor instigation engine is meant to have a low-rotor resistance so that the running effectiveness is high and the full-stack slip is low. The expansion in the rotor resistance does not influence the estimation of most extreme torque but rather builds the slip at greatest torque. The wound rotor engines are broadly utilized as part of applications requiring regular beginning and braking with extensive engine torques. Because of the accessibility of rotor windings for changing the rotor resistance, the wound rotor offers more noteworthy adaptability for control. However, it is expensive and needs support because of slip rings and brushes.

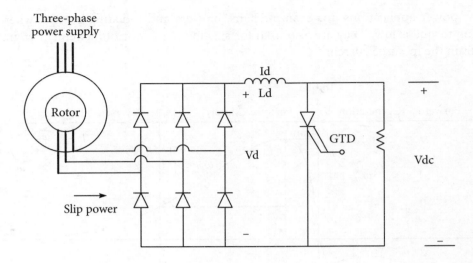

Figure 9.20 Induction Motor Drive.

The wound rotor engine is less generally less used compared to the squirrel-case engine. The three-phase resistor might be supplanted by a three-phase diode rectifier and a DC converter, where the gate turn-off thyristor (GTO) or insulated-gate bipolar transistor (IGBT) works as a DC converter switch. The inductor L_d works as a current source I_d and the DC converter shifts the powerful resistance, which can be found from

$$R_e = R(1-k)$$

where k is the duty cycle of the DC converter and the engine pace can be controlled by differing the duty cycle. The part of the power that is not changed over into mechanical force, is called slip power. The slip force is dispersed in R.

9.3.1.2 Frequency control

The torque and speed of driving engines can be controlled by changing the supply frequency. If voltage is kept constant at its appraised esteem while the frequency is decreased underneath its evaluated esteem, the flux increases. This would bring about immersion of the air flux, and the engine parameters would not be reasonable in deciding the torque–speed attributes. At low frequency, the reactance decreases and the engine current might be too high. This kind of frequency control is not regularly used. It makes the squirrel cage type engine exceptionally suitable for most constant speed applications, where, at times, 3 per cent speed regulation may be worthy. In the event that better speed regulation is required, the squirrel cage engine might work from a closed loop controller, for example, a variable frequency drive.

9.3.1.3 VSI induction-fed drive

DC engines have been utilized in recent years as a part of businesses for variable rate applications, since its flux and torque can be controlled easily by the method for changing the field and armature streams individually. Besides, operation in the four quadrants of the torque–speed plane including provisional break was accomplished. Earlier, drive engines were used in industry because of their strength, ease, high proficiency and less maintenance. The actuation engines were for the most part utilized for basically steady speed applications due to the inaccessibility of the variable frequency voltage supply. The advances in power hardware has made it possible to change the frequency of the voltage supplies, accordingly broadening the utilization of the prompting engine in variable rate drive applications. Yet, because of the natural coupling of flux and torque segments in the driving engine, it cannot make the torque execution on par with the DC engine. In AC matrix-connected engine drives, a rectifier, for the most part a typical diode span giving a pulse DC voltage from the mains, is required. In spite of the fact that the fundamental circuit for an inverter may appear to be straightforward, precisely exchanging these devices causes various difficulties. The most widely recognized exchanging procedure is PWM.

PWM is an effective system for controlling simple circuits with a processor's advanced outputs. It is utilized in a wide assortment of applications such as uninterruptible power supply (UPS), electric drives and high-voltage, direct current (HVDC) receptive power compensators in power systems, extending from estimation and interchanges to power control and change. In AC engine drives, PWM inverters make it possible to control both frequency and magnitude of the voltage and current connected to an engine. Thus, PWM inverter-controlled engine drives are more variable and offer better effectiveness and higher execution when compared to altered frequency engine drives. The energy, which is conveyed by the PWM inverter to the AC engine, is controlled by PWM signals connected to the gates of the power switches at various times to create the output waveform. To enhance the nature of the waveform, variable rate is required; for this, less speed control is required. Contingent upon the kind of load and the sort of velocity, there are diverse techniques for pace control of engines. For step less speed control lower or more than the appraised speed with high torque and harmonics-free output, the PWM inverter prompting engine control is the best suitable one. The square outline of a voltage source inverter (VSI)-fed induction motor is shown in Figure 9.21.

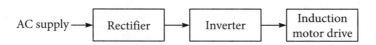

Figure 9.21 VSI-Fed Induction Motor.

9.3.1.4 Three-phase induction-fed drive

The PWM inverter needs to produce a sinusoidal current that can control the voltage and current with 120° distinction in every stage. The controlling signs of three-phase PWM inverters have numerous examples. The operation of the three-phase inverter can be characterized in eight modes, which demonstrates the status of every switch in every operation mode. In inverter operation, the fundamental stage leg-short is actually acknowledged through reverse parallel to parallel diodes in the three-phase span. Similarly, the same gate pulse as in the traditional VSI can be connected. Then again, the switch on the DC join should effectively work. The late headway in power hardware starts to enhance the level of the inverter as opposed to the expanding span of the channel. Refer to Figure 9.22 for a three-phase induction-fed drive.

In a multilevel inverter, the circuit includes parallel association with the inverter. For multiple exchanging, a space vector regulation is required, which depends on vector choice in stationary reference circuits. For a multilevel system, either space vector modulation or sinusoidal triangle regulation might be taken. However, space vector modulation is more favourable because of low harmonics creation. The execution of the multilevel inverter is superior to the traditional inverter. The aggregate harmonic contortion of the traditional inverter is high. The diode-fixed inverter gives numerous voltage levels from an arrangement capacitor bank. The voltage over the switches is just 50 per cent of the DC transport voltage. The power switches used here are bidirectional in nature. The aggregate consonant twisting is broken down between multilevel inverter and other traditional inverters.

The strategy, which uses this drive, has changed the execution of the driving engine like that of the DC engine. The execution of this system however is complex and moreover, in especially complicated circuits, the system is not profoundly sensitive to parameter varieties because of its control system. In the direct torque control (DTC) drive, flux linkage and electromagnetic torque are controlled specifically and autonomously by the determination of ideal inverter exchanging modes.

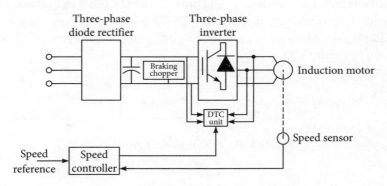

Figure 9.22 Induction Motor Drive.

The circuit of a three-phase inverter system is shown in Figure 9.21. In a three-phase inverter fed drive system, AC is changed over into DC utilizing uncontrolled rectifiers. DC is changed over into variable voltage, variable frequency AC utilizing three-phase PWM inverter.

9.4 Synchronous Motor Drive

Synchronous motors or synchronous engines provide synchronism with the line frequency and keep up a consistent speed, paying little attention to stack and without complex electronic control. The synchronous engine ordinarily gives up to at most 140 per cent of evaluated torque. These plans begin like a stimulation engine yet rapidly quicken from around 90 per cent adjusted speed to synchronous speed. When worked from an AC drive, they require help voltage to create the obliged torque to synchronize rapidly after power application. Also accessible in high-pull engines is the independently energized synchronous engine; this engine requires a load-commutated inverter (LCI). Some extensive engines may have a wound rotor, permitting the engine qualities to be adjusted by including resistors in arrangement with the rotor. More resistance means higher slip and higher beginning torque over the line, while utilizing a low estimation of arrangement resistance brings about lower slip and more prominent proficiency. Frequently, resistors will be available for start-up and after that, turned off while running. For a situation where an injury rotor engine is encouraged by an AC drive, the injury rotor associations is permanent, where no resistors are added in series. The MATLAB Simulink model of a synchronous motor-fed drive, its output waveform and speed-torque curves are shown in Figures 9.23, 9.24 and 9.25

MATLAB Example 4

Since the voltage, in actuality, is not changing above base speed, it is more fitting to characterize torque according to frequency change rather than voltage change. It can be expressed then that torque above base speed drops as the square of the frequency; that is, multiplying the frequency quarters the accessible torque. Frequency and synchronous speed are proportionate; thus, torque might be characterized according to speed.

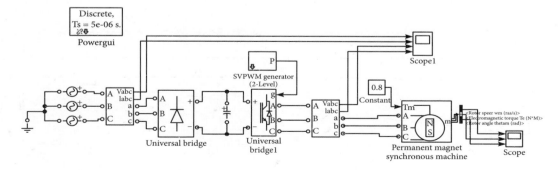

Figure 9.23 Simulink Model of a Synchronous Motor-Fed Drive.

Hence, in the consistent voltage run, engine torque drops off as the reverse of synchronous speed squared, or $1/N^2$. Numerous machine applications are steady strength in their load attributes. The speed–torque characteristics of a synchronous motor-fed drive are shown in Figures 9.23 and 9.24.

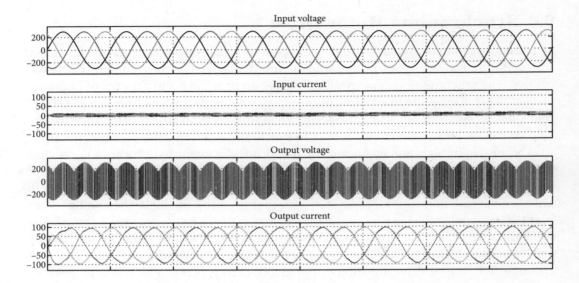

Figure 9.24 Waveform of a Synchronous Motor-Fed Drive.

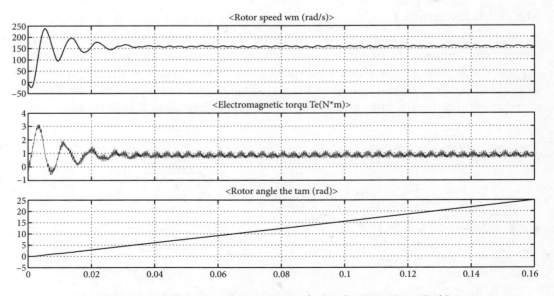

Figure 9.25 Speed–Torque Characteristics of a Synchronous Motor-Fed Drive.

9.5 Phase-Locked Loop (PLL)

Phase-locked loop (PLL) is a control system that produces an output flag whose stage is identified with the period of a data flag. The electronic circuit generally comprises a variable frequency oscillator and a stage locator. The oscillator creates an intermittent flag. The stage finder contrasts the period of that flag and the period of the data intermittent flag and modifies the oscillator to keep the stages coordinated. Bringing the output motion back toward the input motion for investigation is known as the PLL operation.

Keeping the input and output stage in step also infers keeping the input and output frequencies the same. Therefore, notwithstanding synchronizing flags, a PLL can track an input frequency, or it can produce a frequency that is a different from the input frequency. These properties are utilized for PC clock synchronization, demodulation and frequency modulation.

PLLs are utilized in radio, information technologies, PCs, and other electronic applications. They can also be utilized to demodulate a flag, recoup a flag from a difficult correspondence channel, and create a steady frequency at products of an input, for example, microchips. Since a solitary incorporated circuit can give a complete outcome, the system is broadly utilized as a part of present-day electronic devices, with output frequencies from a small amount of hertzs up to numerous gigahertz.

Summary

At the end of this chapter, the users will be able to understand the concepts of solid state drives. Students will be familiar with various types of DC and AC drives where power electronic concepts are applied. The MATLAB examples will help the students design voltage and current-fed drives.

Solved Problems

1. A motor is used to drive a hoist. The motor characteristics are as follows:

    Quadrant I, II, and IV: $T = 200 - 0.2N$ (N·m)

    Quadrant II, III, and IV: $T = -200 - 0.2N$ (N·m)

    The speed is measured in rpm. When the hoist is loaded, the net load torque $T_L = -80$ N·m. Obtain the equilibrium speed for operation in all four quadrants.

    Solution

    The load torque $T_L = T$

    Assuming this condition,

    For Quadrant I,

    $$T_L = 200 - 0.2N = 100 \text{ N·m}$$
    $$0.2N = 100$$

$N = 500$ rpm

For Quadrant II,

$$T_L = 200 - 0.2N = -80 \text{ N·m}$$
$$200 - 0.2N = -80$$
$$200 + 80 = 0.2N$$
$$280 = 0.2N$$
$$N = 1400 \text{ rpm}$$

For Quadrant III,

$$T_L = -200 - 0.2N = -80 \text{ N·m}$$
$$-120 = 0.2N$$
$$N = -600 \text{ rpm}$$

For Quadrant IV,

$$T = -200 - 0.2N = 100$$
$$-200 - 0.2N = 100$$
$$-300 = 0.2N$$
$$N = -1500 \text{ rpm}$$

2. In a separately excited DC motor drive, the polarity of the counter emf is reversed by reversing the field excitation to its excitation value. Calculate delay angle of the field converter, delay angle of the armature converter at 1200 rpm to maintain the armature current at 50 A, the power feedback to the supply during the regenerative braking of motor with 400 V source.

Solution

The maximum negative voltage is obtained at $\alpha = 180°$

Therefore,

$$V_f = (2V_m/\pi) \times \cos\alpha = -2\sqrt{2}V_s/\pi = -360 \text{ V}$$

$$\alpha = 180°$$

When the field current is reversed, the motor back emf is also reversed.

$$V_a = -E_b + I_a R_a$$

$$(2V_m/\pi) \times \cos\alpha = -180.96 + (50 \times 0.2)$$

$$(2 \times \sqrt{2} \times 400/\pi) \times \cos\alpha = -170.96$$

$$\alpha = 118.34$$

During regenerative braking, power feedback to supply $P = V_a I_a$

$$= 70.96 \times 50$$

$$P = 8.54 \text{ kW}$$

3. A 250 V separately excited DC motor has an armature resistance of 2.5 Ω. When driving a load at 600 rpm with constant torque, the armature takes 20 A. The motor is controlled by a chopper circuit with a frequency of 400 Hz and an input voltage of 250 V. What should be the value of the duty ratio if one desires to reduce the speed from 600 to 400 rpm, with the load torque maintained constant?

Solution

The back emf at rated equation is

$$V_s = E_{b1} + I_a R_a$$

$$E_{b1} = V_s - I_a R_a$$

$$= 250 - 20 \times 2.5$$

$$= 200 \text{ V}$$

Back emf at 400 rpm is given by

$$E_{b2}/E_{b1} = N_2/N_1$$

$$E_{b2}/200 = 400/600 = 133.33 \text{ V}$$

Therefore, average output voltage $V_a = E_{b2} + I_a R_a$

$$= 133.33 + 20 \times 2.5$$

$$= 183.33 \text{ V}$$

$$V_a = \delta V_s$$

$$\delta = V_a/V_s$$

$$= 183.33/250$$

$$\delta = 0.733$$

4. A three-phase converter is used to control the speed of 100 HP, 600 V, 1800 rpm with separately excited DC motor. The converter is operated from a three-phase 480 V at 60 Hz supply. The motor parameters are $R_a = 0.2$ Ω, $L_a = 6$ mH, $K_a \Phi = 0.3$ V/rpm. The rated armature current is 120 A. When the motor is operated under rectifier condition, the operating speed is 2500 rpm. Determine the firing angle and the supply PF when the motor current is ripple free.

Solution

The motor terminal equation is given by

$$V_a = E_b + I_a R_a$$

$$E_b = K_a \Phi N$$

$$= 0.3 \times 1500 = 450 \text{ V}$$

$$V_a = 450 + 120 \times 0.2 = 474 \text{ V}$$

For a three-phase full converter,

$$V_a = \frac{3\sqrt{2} \times 480}{2\pi} \cos\alpha$$

$$\alpha = 43°$$

Input PF is given by

$$PF = V_a I_a / \sqrt{3} V_s I_s$$

Supply current

$$I_s = I_a \sqrt{(2/3)}$$

$$= 120\sqrt{(2/3)}$$

$$= 97.97 \text{ A}$$

Input PF = $474 \times 120/(\sqrt{3} \times 480 \times 97.97) = 0.698$ (lag).

Objective Type Questions

1. To receive a smooth voltage waveform, the motor armature is supplied through phase-controlled SCR with
   - (a) High motor speed
   - (b) Low motor speed
   - (c) Rated motor speed
   - (d) Very low motor speed

2. An SCR is used to control the speed of DC motor. At full speed, the motor takes 1 A at 75 V. The maximum forward surge current rating and maximum forward break-over voltage rating, respectively, are of the order of
   - (a) 3 A, 225 V
   - (b) 1 A, 300 V
   - (c) 6 A, 150 V
   - (d) 5 A, 200 V

3. What will happen to the speed of the DC motor when the field opens?
   - (a) Decrease
   - (b) Come to a stop
   - (c) Increase
   - (d) None of these

4. In DC motors, the chopper control provides variation in
   - (a) Input voltage
   - (b) Frequency
   - (c) Both (a) and (b) above
   - (d) None of the above

5. To drive a DC shunt motor in both forward and reverse direction, which one of the following would you use?
   - (a) A half-controlled thyristor bridge
   - (b) A full-controlled thyristor bridge
   - (c) A dual converter
   - (d) A diode bridge

6. A three-stage semiconverter feeds the armature of an independently energized DC motor, providing a nonzero torque. For constant state operation, the motor armature current is found to drop to zero at certain occasions of time. At such occasions, a voltage expects a worth that is
   - (a) Equal to the instantaneous value of the AC phase voltage
   - (b) Equal to the instantaneous value of the motor back emf

(c) Arbitrary

(d) Zero

7. A thyristorized three-stage, completely controlled converter sustains a DC load that draws a steady state current. At that point, the input AC line current to the converter has

   (a) A root mean square (rms) value equal to the DC load current
   (b) An average value equal to the DC load current
   (c) A peak value equal to the DC load current
   (d) A fundamental frequency component, whose rms value is equal to the DC load current

8. If there should be an occurrence of armature controlled independently energized DC motor drive with close-loop control, an inward current circle is valuable since it

   (a) Limits the speed of the motor to a safe value
   (b) Helps in improving the drive energy efficiency
   (c) Limits the peak current of the motor to a permissible value
   (d) Reduces the steady-state speed error

9. The advantage of the tachometer speed control method for DC motors is that it senses

   (a) Back emf
   (b) Armature current
   (c) Armature voltage
   (d) Speed

10. A step-down chopper works from a DC voltage source V_s and sustains a DC motor armature with a back emf E_b. From oscilloscope results, it is found that the current increments for time t_r decreases to zero after some time t_1, and stays zero for time t_o, in each duty cycle. At that point, the normal DC voltage over the FWD is

    (a) $\dfrac{V_s t_r}{(t_r + t_f + t_o)}$

    (b) $\dfrac{(V_s t_r + E_b * t_f)}{(t_r + t_f + t_o)}$

    (c) $\dfrac{(V_s t_r + E_b * t_o)}{(t_r + t_f + t_o)}$

    (d) $\dfrac{(V_s t_r + E_b [t_f + t_o])}{(t_r + t_f + t_o)}$

11. To control the armature voltage of a DC motor, which converter can be used?

    (a) Cycloconverters
    (b) Inverters
    (c) AC–DC converters
    (d) Bridge rectifier circuit with fixed input

12. The speed of a DC shunt motor above normal speed can be controlled by

    (a) Armature voltage control method
    (b) Flux control method
    (c) Both the methods
    (d) None of these

13. For controlling the speed of a DC motor of 150 HP rating, the following types of converters are normally used

    (a) Single-phase full converters
    (b) Single-phase dual converters
    (c) Three-phase full converters
    (d) Three-phase dual converters

14. A motor armature supplied through phase-controlled SCRs receives a smoother voltage shape at
    (a) High motor speed
    (b) Low motor speeds
    (c) Rated normal motor speeds
    (d) None of these

15. A DC chopper circuit controls the average voltage across the DC motor by
    (a) Controlling the input voltage
    (b) Controlling the field current
    (c) Controlling the line current
    (d) Continuously switching on and off the motor for fixed durations of t_{ON} and t_{OFF}, respectively

16. In an AC motor control, the ratio of voltage to frequency is maintained at constant value
    (a) To make maximum use of magnetic circuit
    (b) To make minimum use of magnetic circuit
    (c) To maximise the current from the supply to provide torque
    (d) To minimize the current drawn from the supply to provide torque

17. A single-phase voltage controller feeds an induction motor and a heater.
    (a) In both the loads, only fundamental and harmonics are useful.
    (b) In induction motor, only fundamental and in heater, only harmonics are useful.
    (c) In induction motor, only fundamental and in heater, harmonics as well as fundamental are useful.
    (d) In induction motor, only harmonics and in heater, only fundamental are useful.

18. Which parameter can be varied to control the speed of an induction motor?
    (a) Flux
    (b) Voltage input to stator
    (c) Keeping rotor coil open
    (d) None of these

19. Voltage induced in the rotor of the induction motor when it runs at synchronous speed is
    (a) Very near input voltage to stator
    (b) Slip time the input voltage
    (c) Zero
    (d) None of these

20. The speed and torque of induction motors can be varied by which of the following means?
    (a) Stator voltage control
    (b) Rotor voltage control
    (c) Frequency control
    (d) All of these

21. An inverter is fit for supplying an adjusted three-stage variable voltage/variable frequency, its output sustains a three-stage induction engine evaluated for 50 Hz and 440 V. The stator twisting resistances of the engine are insignificantly small. At starting, the current inrush can be prevented without losing the starting torque by appropriately applying
    (a) Low voltage at rated frequency
    (b) Low voltage keeping V/F ratio constant
    (c) Rated voltage at low frequency
    (d) Rated voltage at rated frequency

22. When the firing angle of SCR is delayed in a controlled rectifier, the speed of the motor will
    (a) Become high
    (b) Become low
    (c) Remain same
    (d) Depend on firing of inverter
23. Thyristor switching circuits are used
    (a) To reduce the stator voltage
    (b) To increase the stator voltage
    (c) To keep the stator voltage control
    (d) None of these
24. Variable speed drives using stator voltage control are normally
    (a) Open-loop system
    (b) Closed-loop system
    (c) Both are correct
    (d) None of these
25. For controlling the speed of three-phase induction motor, the method generally used is
    (a) Fixed voltage fixed frequency method
    (b) Variable voltage variable frequency method
    (c) Fixed voltage variable frequency method
    (d) None of these
26. In variable voltage variable frequency control, to achieve constant torque operation below base speed
    (a) (V/F) has to be kept constant
    (b) Flux has to be increased
    (c) Flux has to be decreased
    (d) None of these
27. PF of synchronous motor can be made leading by adjusting its
    (a) Speed
    (b) Supply voltage
    (c) Excitation
    (d) Supply frequency

Review Questions

1. What are the elements of electrical drive systems?
2. What are the advantages of solid state drives?
3. Give the classification of electrical drives.
4. Explain the four quadrant operations in motor.
5. What are the different modes of operation of an electric drive?
6. What are the uses of phase-controlled rectifiers in DC drives?
7. Write the expression for the average output voltage of a single-phase and three-phase full converter-fed DC drive.
8. Name the methods of speed control applicable on the rotor side of a three-phase induction motor.
9. Compare VSI-fed induction motor drive with CSI-fed drive.
10. Give any two applications of synchronous motor drives.
11. When can a synchronous motor be load commutated?

12. What are the advantages of load commutation over forced commutation?
13. What is meant by PLL?
14. Compare AC and DC drives.

Practice Questions

1. Derive the equations governing motor load dynamics from the basic principles.
2. Develop a criterion for evaluating the steady state stability of an electrical drive.
3. Explain the multiquadrant operation of an electric motor driving a hoist load.
4. Explain the operation of a single-phase fully controlled converter-fed separately excited DC motor with neat waveforms and derive the speed–torque characteristics.
5. Explain in detail the advantages of three-phase drives over single-phase drives?
6. With necessary diagrams, explain the theoretical principles of stator voltage control.
7. Bring out the limitations of stator voltage control scheme.
8. Bring out the advantages of current source inverter (CSI)- over VSI-fed induction motor drives.
9. Explain the concept of self-control mechanism of synchronous motor, which is fed from VSI.
10. Explain the concept of separately excited synchronous feed from VSI.

Appendix 1

Block Parameter Settings

1. Voltage Source

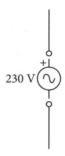

This block represents the AC Voltage Source. Double click on that block. Block parameter settings open like shown here:

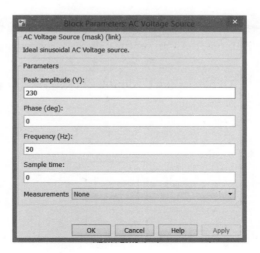

Based on the user's requirement, one can set the values. After setting the values, click **Apply** then **OK**.

2. Pulse Generator

This block represents the Pulse Generator. Double click on that block. Block parameter settings open like shown here:

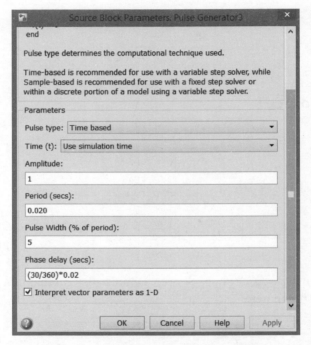

Based on the user's requirement, one can set the values. After setting the values, click **Apply** then **OK**.

Similarly, all the MATLAB blocks can be double clicked and the parameters set based on the user's requirement.

3. Scope

This block represents the Scope. Double click on that block. Block parameter settings open like shown here:

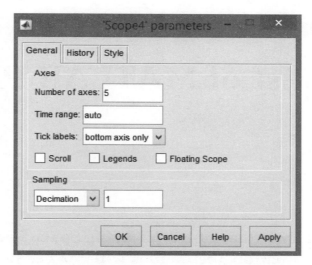

Number of Axes: The Number of Output Signals Being Measured (Variable).

The rest of the settings are kept default (optional). After setting the values, click **Apply** then **OK**.

Appendix 2

List of MATLAB Projects

I. Design of Two-Phase Inverter Using Space Vector Pulse Width Modulation Techniques with MATLAB–Simulink

This work accomplishes space vector pulse width modulation (SVPWM) for a two-stage inverter using MATLAB/Simulink. Modern efforts mainly deal with reducing the losses and improving the performance of electrical devices. One of the numerous possible ways of managing this issue is to utilize an inverter driving strategy called SVPWM. A two-stage SVPWM for three-inverter topologies is proposed in this work. Standards of the SVPWM for three-leg inverters regarding decrease in exchanging can be set. A decrease in exchanging increases productivity and this could be the principle purpose behind the utilization of the SPWM: sinusoidal pulse width modulation.

II. Simulation and Optimization of Diode and Insulated Gate Bipolar Transistor (IGBT) Interaction in a Chopper Cell Using MATLAB and Simulink

Recently, a reproduction technique for force electronic gadgets has been developed, which has high exactness and short running times because of the use of Fourier models of the gadget material science. This work depicts the utilization of Fourier models for diodes and protected entryway bipolar transistors (IGBTs) and usage in MATLAB and Simulink in a formal enhancement method. Specifically, this work researches coupled circuit, diode and IGBT conduct. Conclusions are drawn concerning gadget stacking and circuit plan, especially the part of stray inductance. The rapid recreation inside MATLAB/Simulink is a key to the enhancement process, as the time overhead of the pursuit routine is unimportant. Specifically, the Fourier-based IGBT and diode models are preferably suited for this utilization. Having IGBT and diode models in a nonstop arrangement of the semiconductor mathematical statements also permits temperature impacts to be incorporated easily. It is clear that the joined circuit and gadget advancement are equipped for prompting genuine advantages in both gadgets and circuit utilization. It is fascinating to note that the disadvantages are decreased by expanding the electric field and current thickness stacking, which suggests that numerous current traditionalist plans are not working at full proficiency. The turn-on snubber impact of stray inductance has been demonstrated to be critical, which shows that minimizing stray inductance is not fundamental in the great circuit plan. Circuit creators may utilize such gadget models to help them in settling on educated decisions in their circuit format. It is also clear that the entryway drive configuration must endeavour to decrease the exchanging postponements while, in the same time, holding control of the moves. It is likely that numerous future entryway drives will do this, although higher crest door streams will be vital. The important disadvantages of the model, and the way that the Tabu hunt does not the genuine worldwide least, implies that this streamlining strategy have to be utilized only as an evidence of the bearing in which possible gadget and circuit enhancements may be found. The fundamental point of making outlines less progressive may include more care somewhere else in the configuration, especially to evade voltages. Watchful testing of upgraded configurations will likewise be vital. Considering all things, the methodology here takes into consideration the first run through an exhaustive survey of issues in chopper cell exchanging.

III. Modelling and Simulation of Two-Level SVPWM Inverter Using Photovoltaic Cells as DC Source

An SVPWM system for a two-level inverter is proposed in this work. A two-level inverter utilizing space vector tweak methodology has been displayed and reenacted with a detached *RL* load. Photovoltaic cells are utilized as DC hotspots for data of two-level inverters. Results are exhibited for different operation conditions to check the framework model. In this work, MATLAB/Simulink bundle project has been utilized for demonstrating and reenacting PV (photo voltaic) cells and two-level space vector beat width balance (SVPWM) inverters. In this work, a two-level inverter has been displayed and recreated utilizing Simulink/MATLAB bundle program. Recreation results have been given for different inputs utilizing 1 kHz exchanging recurrence and an inactive burden. Photovoltaic cells are utilized for DC voltage supply of two-level inverters. The proposed control can be effortlessly connected in the two-level inverter. It has been demonstrated that astounding waveforms can be got even with 1 kHz of low exchanging recurrence. The photovoltaic cell is exceptional among the known renewable sources. It has exceptional width application zone. In this work, it has been demonstrated that photovoltaic cells can be utilized as DC hotspots for inverters and that the cells have effective working regions for force electronic applications.

IV. Modelling and Simulation of Three-Phase Voltage Source Inverter using a Model Predictive Current Control

This work concentrates on a combination of three-stage voltage source inverter (VSI) with a prescient current control to give an improved framework to the three-stage inverter that controls the heap current. The present project introduces a finite set-model predictive control (FS-MPC) technique for a two-level three-stage VSI with resistive–inductive burden (*RL* load). Keeping in mind the end goal of decreasing the computational exertion, numerous possible outcomes has been resolved. With diverse cases, the assessment of the framework is finished. First, the framework execution with a long forecast skyline is done. Then, the dynamic reaction of the framework with step change in the adequacy of the reference is explored. Recreations are completed utilizing MATLAB/Simulink to test the viability of an FS-MPC for the two-level VSI with *RL* load. In this work, the FS-MPC for two-level VSIs has been introduced. The proposed control does not have to utilize modulators. The control has been assessed for three distinct cases through reenactment results. Above all else, the vigour of the proposed control system with four forecast steps has been surveyed; the evaluation has been finished by checking the heap current and the lapse between the reference and genuine current for four expectation steps. It has been seen that the control gives great current following conduct. Furthermore, with step change in the reference circuit, the reproduction results demonstrate that the prescient control system has quick element reaction with inborn decoupling in the middle of iL and iT. Recreation results demonstrate that an FS-MPC system executes well under these conditions.

V. Design and Simulation of Super Capacitor Energy Storage System

STATCOM (STATic Synchronous COMpensator) are broadly used to improve power framework strength. They can trade responsive force with the force framework; however, they have restricted capacity to trade genuine force because they exclude vitality stockpiling gadgets. STATCOMs combined with vitality stockpiling gadgets, for example, batteries, have been known for their capacity to trade genuine force. Nonetheless, batteries have an impediment in their greatest deliverable force on account of the moderate substance procedure needed to discharge their vitality. The pattern now is to utilize super capacitor vitality stockpiling frameworks SCESS (super capacitor energy storage system) as vitality stockpiling for STATCOMS. Super capacitors have lower vitality stockpiling but higher force trading ability compared with batteries. This work shows the investigation, plan and control of a super capacitor vitality stockpiling framework (SCESS) for a STATCOM. A top current-mode controller is utilized to control the SCESS framework. Reenactment results of the SCESS framework are presented, which demonstrate the brilliant execution of the proposed SCESS framework. STATCOM–SCESS is a promising innovation for enhancing force framework solidness and quality. A brief introduction about the utilization of

STATCOM–SCESS in different force framework applications is included. The parameters and control plans for the STATCOM–SCESS were discussed. A reproduction model was manufactured and tried for the SCESS framework on MATLAB/Simulink. The test demonstrates that the SCESS framework can keep up the DC join voltage by trading genuine force, which gives the STATCOM–SCESS the capacity to trade genuine force with the framework. This examination can be extended by setting up a reenactment show that incorporates both the SCESS and the STATCOM. It can also be extended by an equipment examination of the proposed SCESS framework.

VI. Simulation of SVPWM for VSI Using MATLAB/Simulink

Nowadays, VSI is used to get variable voltage and recurrence for AC drives. The most well-known PWM systems are bearer beat PWMs; space vector beat width balance are utilized to shift the voltage and recurrence. The space vector beat width regulation is favoured due to its simple digitalization and DC transport voltage use. In this work, a space vector beat width tweak given and actualized by utilizing MATLAB/Simulink.

A basic MATLAB/Simulink model is introduced to actualize SVPWM for three-stage VSI. A brief introduction on space vector representation is also included. A MATLAB/Simulink-based model for usage of SVPWM is displayed. The regulated model is an improvement on the original. The introduced model explains the SVPWM. By modifying the size of the data reference, diverse balance records can be established.

VII. Analysis and Simulations of Z-Source Inverter Control Methods

In this work, two diverse control strategies for Z-source inverters are inspected. The simple boost and the maximum boost control routines for the inverter are analyzed and compared using reenactment and MATLAB/Simulink. The simple boost control with autonomous connection between the adjustment list and the shoot-through requirement proportion is also reproduced and examined. The determination of a high balance file and shoot-through requirement proportion can diminish the inverter's DC join voltage overshoot and expand the force conveyance limit of the inverter. Two control systems with two ST (shoot through) states insertion of Z-source inverter has been analyzed in this work. The support variable, voltage pick up, obligation proportion and voltage stress over the switches for every system have been dissected in point of interest. Reenactment of the Z-source converter under basic control strategy using straight lines in relation to the top estimation of the sinusoidal reference has also been displayed, demonstrating that better execution can be got if adjustment record (M) and shoot-through requirement proportion (D_0) are set high.

VIII. Cosimulation of Generic Power Converter Using MATLAB/Simulink and ModelSim

In this work, we examine the co-recreation of a nonspecific converter utilizing MATLAB and ModelSim. A nonspecific converter is utilized to transform steady DC to variable DC and to bring about DC/AC change as indicated by the example. This reenactment is performed by utilizing two projects: MATLAB/Simulink and ModelSim. Recently, MATLAB has been used both by the framework engineer and the specialist. As the framework designer and analyst work at the same stage, this has decreased the required time taken to decipher work or determine particular specializations. ModelSim utilized hardware description language (HDL) test system. The interface gives a direct bidirectional connection between equipment description language (HDL) simulator, ModelSim, and programming, MATLAB/Simulink, for direct equipment outlay check and co-reproduction. The most widely recognized system for making equipment configuration is HDL, for example, VHDL (VHSIC hardware description language or Verilog). Recreation is a compulsory step in the configuration procedure of a power converter. When the controller is digitally actualized by means of a field programmable gate array (FPGA) or ASIC (application specific integrated circuits), equipment portrayal dialects are typically utilized, for example, VHDL or Verilog. This work shows the reproduction of a framework, which is utilized as a DC/DC converter and also as a DC/AC converter and is called a nonspecific converter. The MATLAB permits performing the flow of framework, whereas the ModelSim runs a project written in HDL dialect VHDL. The sort of recreation done in this work reduces the chances of huge changes of code for future execution in equipment and blunders.

IX. Nonlinear Modelling and Analysis of DC–DC Buck Converter and Comparing with Other Converters

The configuration of power electronic converter circuits with the utilization of a shut circle plan needs demonstration; the converter has to be reproduced using the displayed comparisons. This should effectively be possible with the assistance of state mathematical statements and MATLAB/Simulink as an instrument for reproduction of those state comparisons. An endeavour has been made in this work to reproduce all essential nondisconnected force converters, so that these models can be promptly utilized for any nearby circle outlay (say utilizing PI, fuzzy, or sliding-mode control, and so on). This work examines nonlinear, exchanged, state space models for buck, support, buck-help and Cuk converters. The reproduction environment of MATLAB/Simulink is very suitable for planning the demonstrating circuit, and to understand the dynamic conduct of distinctive converter structures in open circle. The reenactment model in MATLAB/Simulink for the support converter is manufactured for close circle. The reenactment results gotten demonstrate that the yield voltage and inductor current can come back to the steady state even if it is influenced by info voltage and burden variety, with a little overshoot and settling time.

X. MATLAB and Simulink Simulation with FPGA-Based Implementation of Sliding-Mode Control of a Permanent Magnet Synchronous Machine Drive

This work begins with a review of the progress in FPGA innovation followed by a presentation of the approaches, improvement devices and pertinent CAD (computer aided dispatch) situations, including the utilization of convenient equipment portrayal dialects and framework-level programming/configuration instruments. In this work, FPGA-based usage of electrical controls is highlighted. This methodology is used for its idea-measured quality and reusability. In this work, a definite depiction of the structure using the indirect sliding mode of the permanent magnet synchronous machine (PMSM) is seen. Test results obtained from a prototyping stage are given to show the proficiency and advantages of the proposed methodology and the different phases of execution of this structure in FPGA. On account of a sliding mode controlled by the current stator of the perpetual magnet synchronous machine, it is required that the PWM strategy is used. This work shows the execution of sliding-mode control structural engineering on FPGA for PMSM. The relating outlay has thoroughly been improved after a procedure that offers significant points of interest and permits the formation of a library for enhanced reusable modules. Among the benefits of these control structures is that the exchanging recurrence is altered and there is agreeability among the eight vector voltages that can give the voltage inverter. But it also has disadvantages; the general structure of the control calculation is intricate to execute and parameters of the control calculation rely on the inspecting period.

XI. Developing FPGA-Based Embedded Controllers Using MATLAB/Simulink

FPGAs are developing as suitable devices for executing installed control frameworks. They offer focal points, for example, superior and simultaneous processing, which make them appealing in numerous implanted applications. As reconfigurable gadgets, they can be utilized to construct the equipment and programming parts of an installed framework on a solitary chip. Customary FPGA that outlay streams and devices, requiring the use of HDLs, are in an alternate area than standard control framework configuration apparatuses, for example, MATLAB/Simulink. This work outlines the advances in FPGA-based controllers using prominent instruments, for example, MATLAB/Simulink accessible for the configuration and improvement of control frameworks. The configurations of power DSP (digital signal processing) builder is reached by adding to a custom library of control framework building that encourages fast advancement of FPGA-based controllers in the natural MATLAB/Simulink environment. As a contextual analysis, this work exhibits how the instruments can be used to add to an FPGA-based controller for a research centre scale air levitation framework. The utilization of the framework-level device DSP builder for abnormal state advancement of FPGA-based controllers was emphasized. The capacities of the DSP builder device were further extended by adding to the custom control library. The custom library included generally utilized segments, for example, discretized integrators, PID (proportional integral differential) controller, PWM generator and A/D (analog/digital) controller. DSP builder and the custom control library together can be utilized to quickly create controllers in the natural and standard Simulink plan environment for FPGA usage. An execution contextual analysis exhibited the use of DSP builder and the custom control library to add to an FPGA-based controller for an air levitation framework in the MATLAB/Simulink environment.

XII. Electric Vehicle Drive Simulation with MATLAB/Simulink

The work reproduces an essential electric vehicle engine drive framework that is utilized to examine force stream amid both motoring and recovery. The recreation expects a DC perpetual magnet engine, a perfect engine controller consolidated with a relative fundamental controller, and the electric vehicle battery. The model can be utilized to assess the electric commute's vitality stream and proficiency for particular velocity and torque burden conditions. A percentage of the key framework parameters was indicated and others were demonstrated as perfect. A stable MATLAB/Simulink model was created and accepted. It was then used to focus the framework execution and vitality stream over a given arrangement of motoring and recovery speed/torque conditions. The model could be utilized to expand direction in vitality transformation or vehicle frameworks courses. Reenactment is a manifestly obvious and vital piece of electric vehicle advancement and should be coordinated into learning encounters inside the building training. Reenactment-based testing as hardware-in-the-loop testing is also an extremely vital piece of current building improvement particularly in cutting-edge frameworks, for example, half and half and electric vehicle drive frameworks that depend on complex inserted framework subsystems. Understudy learning encounters that incorporate recreation-based plan and testing are important to incorporate undergrad designing training keeping in mind the end goal of getting understudies ready for current industry.

XIII. Design of FPGA-Controlled Power Electronics and Drives Using MATLAB/Simulink

This work shows a straightforward and quick prototyping procedure for FPGAs-based advanced controllers for force gadgets and engine drives utilizing MATLAB's Simulink and HDL coder configuration programming. The MATLAB/Simulink models are upgraded and changed to obtain autonomous, particular, and traceable high speed coordinated circuit hardware description language (VHDL) code for FPGA programming. A sample execution of the space vector beat width balance (SVPWM) system is displayed, representing the outlay of a bland three-stage VSI. Reenactment and co-recreation, the framework-level outlay, and quick prototyping of FPGA-based computerized controllers will help power hardware specialists and analysts to create models in a moderately brief time by making dreary and lengthy manual coding redundant. This increases efficiency and encourages the advancement of force electronic controllers with more mind-boggling control calculations. This work portrays a technique to encourage the advancement and execution of FPGA-based computerized controllers in force electronic converters and drives. The system is faster and gives a more noteworthy level of certainty than conventional manual HDL coding. To delineate the technique, a lab model of a 1 kW FPGA-controlled VSI was portrayed in which the altered point control calculation was naturally produced from Simulink models utilizing the MATLAB HDL coder, and checked before usage on a Xilinx Spartan-6 XC6SLX45 board. Exploratory portrayal of the subsequent VSI converter brought about 3.8 per cent aggregate consonant contortion and a line-to-line peak component of 1.52, well inside the suitable scope of IEEE guidelines. The close ascension in the investigation and recreation demonstrates the adequacy of the technique. The strategy is especially helpful for model advancement of other force electronic converters and electric drives with more complex interfacing and control calculations.

XIV. Modelling and Simulation of Renewable Hybrid Power System Using MATLAB/Simulink Environment

The work shows a sun-powered wind-hydroelectric cross-breed framework in MATLAB/Simulink environment. The application is valuable for examination and recreation of a genuine crossover sun-based wind-hydroelectric framework associated with an open matrix. Application is based on secluded structural planning to encourage simple investigation of the effect every segment module. Squares like wind model, sunlight-based model, hydroelectric model, vitality transformation and burden are executed and the consequences of recreation are also displayed. As an example, a standout among the most vital studies is the conduct of half-breed framework, which permits utilizing renewable and variable in time vitality sources while giving a constant supply. The work displays a recreation model valuable for the investigation of little power frameworks in view of renewable vitality. For the reenactment model improvement, an accumulation of items sorted out in another MATLAB/Simulink library named RegenSim was manufactured. The model reproduced using the new RegenSim library is perfect and can be interconnected with segments of the devoted SimPowerSystems library that is utilized for force frameworks working, recreation, demonstrating and examination. The proposed reenactment model is valuable in the current circumstances permitting studies considering the distinguishing proof, the kind of opportunity, and the execution for force frameworks taking into account renewable energies that are accessible in a certain range. The Mureş County region was used to perform the study for the suitability of utilizing sun-powered, wind and hydro assets accessible here. Other critical studies permitted the created recreation model to be used for steady or transient administrations, with the likelihood of dynamic and responsive force streaming development. The reproduction also shows that it is a valuable instrument in the vitality administration framework area. Further research can be done on current and voltage waveforms and force quality studies.

XV. Modelling and Simulation PMSG-Based on Wind Energy Conversion System in MATLAB/Simulink

This work shows the dynamic model of a permanent magnet synchronous generator (PMSG) in view of wind energy conversion system (WECS). The models of WECS comprise a wind turbine, pitch edge control, drive prepare, PMSG and force converter. A wind turbine model with controller for generator insurance in high wind pace is also displayed in this work. The PMSG and converter model are set up in the d-q model. The displayed model, dynamic reenactment and reproduction results are tried in MATLAB/Simulink. This work exhibited the PMSG taking into account WECS. It also comprises an air conditioning/DC/air conditioning converter model. PMSG and converters are set up in d-q model. The result analysis demonstrated that the pitch point control dynamic changes at high twist speed from control rotor pace to power coefficient (pc) for secure generator.

XVI. Teaching Nonlinear Modelling, Simulation, and Control of Electronic Power Converters Using MATLAB/Simulink

This work depicts an effective strategy to investigate and recreate force electronic converters to college understudies, utilizing framework-level nonlinear state-space models. Framework-level displaying of force electronic converters replicates the ideal exchanging conduct of the semiconductors and is a valuable idea for the numerical reproduction of force converters, since recreations introduce no merging issues and require minimal computational time. Exchanged state-space models, modified in the MATLAB/Simulink programming bundle, can be beneficially used to reenact power converters at the framework level to plan and study their controllers. Exchanged state-space nonlinear models can be obtained by utilizing a hypothetical structure suitable for the upgraded control of variable structure power frameworks. Since the strategy is intrinsically nonlinear, no approximated direct models are required; and since state-space models are utilized, present-day control procedures (sliding mode, neural systems, and fuzzy rationale) for force converters can, without much of a stretch, be utilized. This work outlines the proposed system and gives a few examples. It has laid out and represented a technique to get nonlinear, exchanged state-space models of force converters, suited for recreation and control outlay. As the technique uses state-space models, electronic force converters, relationship of force converters and electromechanical gadgets or drives, with intricate control frameworks, can be successfully mimicked. Relationship of electromechanical frameworks and/or electronic force converters, which are frequently difficult to understand and hard to examine and to join, can likewise, be analyzed. Recreation takes a couple of seconds, and no meeting issues were found. To some degree, the state-space model of the converter, which is a hypothetical edge essential for controller outlay, can be favourably used to perform the reproduction. Along these lines, this approach is a viable instrument to show college understudies to reenact electronic force converters and show their control plan. Utilizing the MATLAB/Simulink programming bundle, understudies or force converter control designers can make an effective reproduction and control instrument for force converters. A few air conditioner/DC and DC/air conditioning converters have been effectively reenacted. The cases given delineate the impressive capability of the introduced methods as educating guides.

XVII. A MATLAB-Based Modelling and Simulation Package for Electric and Hybrid Electric Vehicle Design

This work shows a reproduction and displaying bundle created at Texas A&M University, V-Elph 2.01. VElph encourages top to bottom investigations of electric vehicle (EV) and cross-breed EV (HEV) designs or vitality administration methods through visual creation so as to program parts as progressive subsystems that can be utilized reciprocally as implanted frameworks. V-Elph is made out of small models of four noteworthy sorts of parts: electric engines, inner ignition motors, batteries, and bolster segments that can be incorporated to model and train having every single electric, series crossover, and parallel half and half designs. V-Elph was composed in the MATLAB/Simulink graphical reproduction dialect and is versatile to most PC stages. This work also discusses the approach for planning vehicle drive trains utilizing the V-Elph bundle. An EV, an arrangement HEV, a parallel HEV and a customary inside ignition motor ICE (internal combustion engine) driven commute train have been composed utilizing the reenactment bundle. Reenactment results, for example, fuel utilization, vehicle emanations, and multifaceted nature are discussed for every vehicle. This work also analyses another train created at Texas A&M University utilizing MATLAB/Simulink to study issues identified with EV and HEV plan, for example, vitality effectiveness, mileage, and vehicle emanations. The bundle utilizes visual programming procedures, permitting the client to rapidly change architectures, parameters, and to view yield information graphically. It likewise incorporates point by point models of electric engines, inside ignition motors and batteries. The outlays for four vehicle drive prepares–an EV, parallel HEV, arrangement HEV, and customary ICE vehicle–are displayed. The consequences of applying basic, worker, government urban, and elected expressway drive cycles are looked at. These outcomes delineate the adaptability of the bundle for contemplating different issues identified with electric and half-breed EV outlay. The recreation bundle can run on a PC or a Unix-based workstation.

Appendix 3

MATLAB Functions

Listed are the various sources and elements that are used in electrical power systems for various experiments.

Electrical Sources and Elements

AC Current Source	Implement sinusoidal current source
AC Voltage Source	Implement sinusoidal voltage source
Battery	Implement generic battery model
Controlled Current Source	Implement controlled current source
Controlled Voltage Source	Implement controlled voltage source
DC Voltage Source	Implement DC voltage source
Three-Phase Programmable Voltage Source	Implement three-phase voltage source with programmable time variation of amplitude, phase, frequency and harmonics
Three-Phase Source	Implement three-phase source with internal RL impedance
Linear Transformer	Implement two- or three-winding linear transformer
Grounding Transformer	Implement three-phase grounding transformer providing a neutral in a three-wire system
Saturable Transformer	Implement two- or three-winding saturable transformer
Multi-winding Transformer	Implement multi-winding transformer with taps

Three-Phase Transformer (Three Windings)	Implement three-phase transformer with configurable winding connections
Three-Phase Transformer (Two Windings)	Implement three-phase transformer with configurable winding connections
Three-Phase Transformer (12 Terminals)	Implement three single-phase, two-winding transformers where all terminals are accessible
Three-Phase Transformer Inductance Matrix Type (Three Windings)	Implement three-phase three-winding transformer with configurable winding connections and core geometry
Three-Phase Transformer Inductance Matrix Type (Two Windings)	Implement three-phase two-winding transformer with configurable winding connections and core geometry
Zigzag Phase-Shifting Transformer	Implement zigzag phase-shifting transformer with configurable secondary winding connection
Three-Phase OLTC Regulating Transformer (Phasor Type)	Implement phasor model of three-phase OLTC regulating transformer
Three-Phase OLTC Phase Shifting Transformer Delta-Hexagonal (Phasor Type)	Implement phasor model of three-phase OLTC phase-shifting transformer using delta hexagonal connection
PI Section Line	Implement single-phase transmission line with lumped parameters
Distributed Parameter Line	Implement N-phase distributed parameter transmission line model with lumped losses
Three-Phase PI Section Line	Implement three-phase transmission line section with lumped parameters
Series *RLC* Branch	Implement series *RLC* branch
Series *RLC* Load	Implement linear series *RLC* load
Parallel *RLC* Branch	Implement parallel *RLC* branch
Parallel *RLC* Load	Implement linear parallel *RLC* load
Mutual Inductance	Implement inductances with mutual coupling
Three-Phase Harmonic Filter	Implement four types of three-phase harmonic filters using *RLC* components
Three-Phase Dynamic Load	Implement three-phase dynamic load with active power and reactive power as a function of voltage or controlled from external input
Three-Phase Series *RLC* Branch	Implement three-phase series *RLC* branch

Three-Phase Series *RLC* Load	Implement three-phase series *RLC* load with selectable connection
Three-Phase Parallel *RLC* Branch	Implement three-phase parallel *RLC* branch
Three-Phase Parallel *RLC* Load	Implement three-phase parallel *RLC* load with selectable connection
Three-Phase Mutual Inductance Z1-Z0	Implement three-phase impedance with mutual coupling among phases
Breaker	Implement circuit breaker opening at current zero crossing
Three-Phase Breaker	Implement three-phase circuit breaker opening at current zero crossing
Three-Phase Fault	Implement programmable phase-to-phase and phase-to-ground fault breaker system
Ground	Provide connection to ground
Neutral	Implement common node in circuit
Surge Arrester	Implement metal-oxide surge arrester
Connection Port	Create physical modeling connector port for subsystem

Motors and Generators

AC1A Excitation System	Implement IEEE type AC1A excitation system model
AC4A Excitation System	Implement IEEE type AC4A excitation system model
AC5A Excitation System	Implement IEEE type AC5A excitation system model
DC1A Excitation System	Implement IEEE type DC1A excitation system model
DC2A Excitation System	Implement IEEE type DC2A excitation system model
Excitation System	Provide excitation system for synchronous machine and regulate its terminal voltage in generating mode
ST1A Excitation System	Implement IEEE type ST1A excitation system model
ST2A Excitation System	Implement IEEE type ST2A excitation system model
Asynchronous Machine	Model the dynamics of three-phase asynchronous machine, also known as induction machine
DC Machine	Implement wound-field or permanent magnet DC machine

Generic Power System Stabilizer	Implement generic power system stabilizer for synchronous machine
Hydraulic Turbine and Governor	Model hydraulic turbine and proportional-integral-derivative (PID) governor system
Multiband Power System Stabilizer	Implement multiband power system stabilizer
Permanent Magnet Synchronous Machine	Model the dynamics of a three-phase permanent magnet synchronous machine with sinusoidal or trapezoidal back electromotive force, or the dynamics of a five-phase permanent magnet synchronous machine with sinusoidal back electromotive force
Simplified Synchronous Machine	Model the dynamics of a simplified three-phase synchronous machine
Single-Phase Asynchronous Machine	Model the dynamics of a single-phase asynchronous machine with squirrel cage rotor
Steam Turbine and Governor	Model the dynamics of a speed governing system, steam turbine and multimass shaft
Stepper Motor	Implement stepper motor model
Switched Reluctance Motor	Model the dynamics of a switched reluctance motor
Synchronous Machine	Model the dynamics of a three-phase round-rotor or salient-pole synchronous machine

Power Electronics

Diode	Implement diode model
GTO	Implement gate turn off (GTO) thyristor model
Ideal Switch	Implement ideal switch device
IGBT	Implement insulated gate bipolar transistor (IGBT)
IGBT/Diode	Implements ideal IGBT, GTO or MOSFET and antiparallel diode
MOSFET	Implement MOSFET model
Three-Level Bridge	Implement three-level neutral point clamped (NPC) power converter with selectable topologies and power switching devices
Thyristor	Implement thyristor model

| Universal Bridge | Implement universal power converter with selectable topologies and power electronic devices |

Sensors and Measurements

Voltage Measurement	Measure voltage in circuit
Current Measurement	Measure current in circuit
Three-Phase V–I Measurement	Measure three-phase currents and voltages in circuit
Multimeter	Measure voltages and currents specified in dialog boxes of SimPowerSystems blocks
Impedance Measurement	Measure impedance of circuit as a function of frequency
Fourier	Perform Fourier analysis of signal
Fundamental (PLL-Driven)	Compute fundamental value of signal
Mean	Compute mean value of signal
Mean (Phasor)	Compute mean value of input phasor over a running window of one cycle of specified frequency
Mean (Variable Frequency)	Compute mean value of signal
Positive-Sequence (PLL-Driven)	Compute positive-sequence component of three-phase signal at fundamental frequency
Power	Compute active and reactive powers of voltage–current pair at fundamental frequency
Power (3ph)	Compute three-phase instantaneous active and reactive powers
Power (3ph, Phasor)	Compute three-phase active and reactive powers using three-phase voltage and current phasors
Power (PLL-Driven, Positive-Sequence)	Compute positive-sequence active and reactive powers
Power (Phasor)	Compute active and reactive powers using voltage and current phasors
Power (Positive-Sequence)	Compute positive-sequence active and reactive powers
Power (dq0, Instantaneous)	Compute three-phase instantaneous active and reactive powers
RMS	Compute true root mean square (rms) value of signal
Sequence Analyzer	Compute positive-, negative- and zero-sequence components of three-phase signal

Sequence Analyzer (Phasor)	Compute sequence components (positive, negative and zero) of three-phase phasor signal
THD	Compute total harmonic distortion (THD) of signal

Control and Signal Generation

Overmodulation	Add third harmonic or triple harmonic zero-sequence signal to three-phase signal
PWM Generator (2-Level)	Generate pulses for PWM-controlled 2-level converter
PWM Generator (3-Level)	Generate pulses for PWM-controlled three-level converter
PWM Generator (DC–DC)	Generate pulse for PWM-controlled DC–DC converter
Pulse Generator (Thyristor)	Generate pulses for twelve-pulse and six-pulse thyristor converters
Sawtooth Generator	Generate sawtooth wave format regular intervals
Stair Generator	Generate signal changing at specified transition times
SVPWM Generator (2-Level)	Generate pulses for SVPWM-controlled two-level converter
Three-Phase Programmable Generator	Generate three-phase signal with programmable time variation of amplitude, phase, frequency and harmonics
Three-Phase Sine Generator	Generate three-phase balanced signal, amplitude, phase and frequency controlled by block inputs
Triangle Generator	Generate symmetrical triangle wave format regular intervals
Alpha-Beta-Zero to dq0, dq0 to Alpha-Beta-Zero	Perform transformation from dq0 stationary reference frame to dq0 rotating reference frame or the inverse
abc to Alpha-Beta-Zero, Alpha-Beta-Zero to abc	Perform transformation from three-phase (abc) signal stationary reference frame or the inverse
abc to dq0, dq0 to abc	Perform transformation from three-phase (abc) signal to dq0 rotating reference frame or the inverse
First-Order Filter	Implement first-order filter
Lead-Lag Filter	Implement first-order lead-lag filter
Second-Order Filter	Implement second-order filter
Second-Order Filter (Variable-Tuned)	Implement second-order variable-tuned filter

PLL	Determine frequency and fundamental component of signal-phase angle
PLL (3ph)	Determine frequency and fundamental component of three-phase signal-phase angle
Bistable	Implement prioritized S-R flip-flop (bistable multivibrator)
Edge Detector	Detect change in logical signal state
Monostable	Implement monostable flip-flop (one-shot multivibrator)
On/Off Delay	Implement switch-on or switch-off delay
Sample and Hold	Sample first input and hold its value based on value of second input
Discrete Shift Register	Implement serial-in, parallel-out shift register
Discrete Variable Time Delay	Delay signal by variable time value

Electric Drives

Four-Quadrant Chopper DC Drive	Implement four-quadrant chopper DC drive
Four-Quadrant Single-Phase Rectifier DC Drive	Implement single-phase dual-converter DC drive with circulating current
Four-Quadrant Three-Phase Rectifier DC Drive	Implement three-phase dual-converter DC drive with circulating current
One-Quadrant Chopper DC Drive	Implement one-quadrant chopper (buck converter topology) DC drive
Two-Quadrant Chopper DC Drive	Implement two-quadrant chopper (buck–boost converter topology) DC drive
Two-Quadrant Single-Phase Rectifier DC Drive	Implement two-quadrant single-phase rectifier DC drive
Two-Quadrant Three-Phase Rectifier DC Drive	Implement two-quadrant three-phase rectifier DC drive
Brushless DC Motor Drive	Implement brushless DC motor drive using permanent magnet synchronous motor (PMSM) with trapezoidal back electromotive force (BEMF)
DTC Induction Motor Drive	Implement direct torque and flux control (DTC) induction motor drive model
Field-Oriented Control Induction Motor Drive	Implement field-oriented control (FOC) induction motor drive model

Five-Phase PM Synchronous Motor Drive	Implement five-phase permanent magnet synchronous motor vector control drive
PM Synchronous Motor Drive	Implement permanent magnet synchronous motor (PMSM) vector control drive
Self-Controlled Synchronous Motor Drive	Implement self-controlled synchronous motor drive
Six-Step VSI Induction Motor Drive	Implement six-step inverter-fed induction motor drive
Space Vector PWM VSI Induction Motor Drive	Implement space vector PWM VSI induction motor drive
Mechanical Shaft	Implement mechanical shaft
Speed Reducer	Implement speed reducer
Battery	Implement generic battery model
Fuel Cell Stack	Implement generic hydrogen fuel cell stack model
Super capacitor	Implement generic super capacitor model

Flexible AC Transmission Systems and Renewable Energy Systems Power Electronics (FACTS)

Static Synchronous Compensator (Phasor Type)	Implement phasor model of three-phase static synchronous compensator
Static Synchronous Series Compensator (Phasor Type)	Implement phasor model of three-phase static synchronous series compensator
Static Var Compensator (Phasor Type)	Implement phasor model of three-phase static var compensator
Unified Power Flow Controller (Phasor Type)	Implement phasor model of three-phase unified power flow controller

Renewable Energy Systems

Wind Turbine	Implement model of variable pitch wind turbine
Wind Turbine Doubly-Fed Induction Generator (Phasor Type)	Implement phasor model of variable speed doubly-fed induction generator driven by wind turbine

| Wind Turbine Induction Generator (Phasor) | Implement phasor model of squirrel cage induction generator driven by variable pitch wind turbine |

Interface to Simscape

Current–Voltage Simscape Interface	Ideal coupling between SimPowerSystems and Simscape electrical circuits
Current–Voltage Simscape Interface (gnd: ground)	Ideal coupling between SimPowerSystems and Simscape electrical circuits
Voltage–Current Simscape Interface	Ideal coupling between SimPowerSystems and Simscape electrical circuits
Voltage–Current Simscape Interface (gnd: ground)	Ideal coupling between SimPowerSystems and Simscape electrical circuits

Simulation and Analysis

| GUI:Graphical User Interface | Environment block for SimPowerSystems models |
| LoadFlowBus | Identify and parameterize loadflowbus |

Appendix 4

Useful Formulae

Mathematics

Matrix

- If $|A| = 0 \rightarrow$ Singular matrix; $|A| \neq 0$ Nonsingular matrix
- **Scalar Matrix** is a diagonal matrix with all diagonal elements being equal
- **Unitary Matrix** is a scalar matrix with diagonal element as '1' ($A^Q = (A^*)^T = A^{-1}$)
- If the product of two matrices are zero matrix, then at least one of the matrices has det zero
- Orthogonal matrix if $AA^T = A^T \cdot A = I \Rightarrow A^T = A^{-1}$
- $A = A^T \rightarrow$ Symmetric
- $A = -A^T \rightarrow$ Skew symmetric

Properties: (If A and B are Symmetrical)

- $A + B$ symmetric
- BA is symmetric
- $AB + BA$ symmetric
- AB is symmetric if $AB = BA$
- For any 'A' $\rightarrow A + A^T$ symmetric; $A - A^T$ skew symmetric
- Diagonal elements of skew symmetric matrix are zero
- If A skew symmetric $A^{2n} \rightarrow$ symmetric matrix; $A^{2n-1} \rightarrow$ skew symmetric
- If 'A' is null matrix, then Rank of $A = 0$.

Consistency of Equations
- $r(A, B) \neq r(A)$ is consistent
- $r(A, B) = r(A)$ consistent and
- if $r(A)$ = number of unknowns, then unique solution
- $r(A)$ < number of unknowns, then ∞ solutions.

Hermitian, Skew Hermitian, Unitary, and Orthogonal Matrices

$A^T = A^* \rightarrow$ then Hermitian

$A^T = -A^* \rightarrow$ then Hermitian

- Diagonal elements of skew Hermitian matrix must be purely imaginary or zero
- Diagonal elements of Hermitian matrix are always real.
- A real Hermitian matrix is a symmetric matrix.

$$|KA| = K^n |A|$$

Eigenvalues and Vectors

Char. Equation $|A - \lambda I| = 0$.

Roots of characteristic equation are called eigenvalues. Each eigenvalue corresponds to nonzero solution X such that $(A - \lambda I)X = 0$. X is called an eigenvector.

- Sum of eigenvalues is sum of diagonal elements (trace).
- Product of eigenvalues is equal to determinant of matrix.
- Eigenvalues of A^T and A are the same.

If λ is eigenvalue of A, then $\dfrac{1}{\lambda} \rightarrow A^{-1}$ and $\dfrac{|A|}{\lambda}$ is eigenvalue of adj A.

If $\lambda_1, \lambda_2, ..., \lambda_n$ are eigenvalues of A, then

$$KA \rightarrow K\lambda_1, K\lambda_2, ..., K\lambda_n$$

$$A^m \rightarrow \lambda_1^m, \lambda_2^m, ..., \lambda_n^m.$$

$$A + KI \rightarrow \lambda_1 + K, \lambda_2 + K, ..., \lambda_n + K$$

$$(A - KI)^2 \rightarrow (\lambda_1 - K)^2, ..., (\lambda_n - K)^2$$

- Eigenvalues of orthogonal matrix have absolute value of 1.
- Eigenvalues of symmetric matrix are also purely real.
- Eigenvalues of a skew symmetric matrix are purely imaginary or zero.

If $\lambda_1, \lambda_2, ..., \lambda_n$ are distinct eigenvalues of A, then the corresponding eigenvectors are $X_1, X_2, ..., X_n$ for a linearly independent set.

$$\text{adj}(\text{adj } A) = |A|^{n-2} \; ; \; |\text{adj}(\text{adj } A)| = |A|^{(n-1)^2};$$

Complex Algebra

- Cauchy–Riemann equations

$$\left.\begin{aligned} \frac{\partial u}{\partial x} &= \frac{\partial v}{\partial y}; \frac{\partial u}{\partial y} = -\frac{\partial v}{\partial x} \\ \frac{\partial u}{\partial r} &= \frac{1}{r}\frac{\partial v}{\partial \theta} \\ \frac{\partial v}{\partial r} &= -\frac{1}{r}\frac{\partial u}{\partial \theta} \end{aligned}\right\} \text{Neccessary and sufficient conditions for } f(z) \text{ to be analytic}$$

$$\int_c \frac{f(z)}{(Z-a)^{n+1}} dz = \frac{2\pi i}{n!}\left[f^n(a)\right] \text{ if } f(z) \text{ is analytic in region c and } Z = a \text{ is single point}$$

$$f(z) = f(z_0) + f'(z_0)\frac{(z-z_0)}{1!} + f''(z_0)\frac{(z-z_0)^2}{2!} + \cdots + f^n(z_0)\frac{(z-z_0)^n}{n!} + \cdots$$

is the Taylor Series

If $Z_0 = 0$, then the Maclaurin Series $f(z) = \sum_0^\infty a_n(z-z_0)^n$; when $a_n = \frac{f_n(z_0)}{n!}$

- If $f(z)$ is analytic in closed curve 'C' except at finite number of poles, then

$$\int_c f(z)dz = 2\pi i \text{ (sum of residues at singular points within 'C')}$$

$$\text{Res } f(a) = \lim_{z \to a}(Z - af(z))$$

$$= \frac{\Phi(a)}{\Phi'(a)}$$

$$= \lim_{z \to a} \frac{1}{(n-1)!} \frac{d^{n-1}}{dz^{n-1}}\left((Z-a)^n f(z)\right)$$

Calculus

Rolle's Theorem

If $f(x)$ is

(a) Continuous in $[a, b]$

(b) Differentiable in (a, b)

(c) $f(a) = f(b)$, then there exists at least one value $C \in (a, b)$ such that $f'(c) = 0$.

Lagrange's Mean Value Theorem

If $f(x)$ is continuous in $[a, b]$ and differentiable in (a, b), then there exists at least one value 'C' in (a, b) such that $f'(c) = \dfrac{f(b) - f(a)}{b - a}$.

Cauchy's Mean Value Theorem

If $f(x)$ and $g(x)$ are two function such that

(a) $f(x)$ and $g(x)$ is continuous in $[a, b]$

(b) $f(x)$ and $g(x)$ is differentiable in (a, b)

$$g'(x) \neq 0 \,\forall\, x \text{ in } (a,b),$$

then there exists at least one value C in (a, b) such that

$$\frac{f'(c)}{f'(c)} = \frac{f(b) - f(a)}{g(b) - g(a)}$$

Properties of Definite Integrals

$$a < c < b \quad \int_a^b f(x)\cdot dx = \int_a^c f(x)\cdot dx + \int_c^b f(x)\cdot dx$$

$$\int_0^a f(x)\,dx = \int_0^a f(a - x)\,dx$$

$$\int_{-a}^a f(x)\cdot dx = 2\int_0^a f(x)\,dx \quad f(x) \text{ is even}$$

$$= 0 \quad f(x) \text{ is odd}$$

$$\int_0^a f(x) \cdot dx = 2\int_0^a f(x)dx \quad \text{if} \quad f(x) = f(2a-x)$$

$$= 0 \text{ if } f(x) = -f(2a-x)$$

$$\int_0^{na} f(x) \cdot dx = n\int_0^a f(x)dx \quad \text{if} \quad f(x) = f(x+a)$$

$$\int_a^b f(x) \cdot dx = \int_a^b f(a+b-x)dx$$

$$\int_0^a xf(x) \cdot dx = \frac{a}{2}\int_0^a f(x)dx \quad \text{if} \quad f(a-x) = f(x)$$

$$\int_0^{\pi/2} \sin^n x = \int_0^{\pi/2} \cos^n x = \frac{(n-1)(n-3)(n-5)\ldots 2}{n(n-2)(n-4)\ldots 3} \quad \text{if} \quad n \text{ is odd}$$

$$= \frac{(n-1)(n-3)\ldots 1}{n(n-2)(n-4)\ldots 2}\left(\frac{\pi}{2}\right) \quad \text{if} \quad n \text{ is even}$$

$$\int_0^{\pi/2} \sin^m x \cdot \cos^n x \cdot dx = \frac{\{(m-1)(m-3)\ldots(m-5)\ldots 2\text{ or }1\}\{(n-1)(n-3)\ldots 2\text{ or }1\} \cdot K}{(m+n)(m+n-2)(m+n-4)\ldots 2\text{ or }1}$$

where $K = \frac{\pi}{2}$ when both m and n are even, otherwise $k = 1$.

Maxima and Minima

A function $f(x)$ has maximum at $x = a$ if $f'(a) = 0$ and $f''(a) < 0$
A function $f(x)$ has minimum at $x = a$ if $f'(a) = 0$ and $f''(a) > 0$

Constrained Maximum or Minimum

To find maximum or minimum of $u = f(x, y, z)$ where x, y, z are connected by $\Phi(x, y, z) = 0$

Working Rule

Write $F(x, y, z) = f(x, y, z) + \lambda \Phi(x, y, z)$

Obtain $F_x = 0, F_y = 0, F_z = 0$

(i) Solve the aforementioned equations along with $\phi = 0$ to get stationary point.

Laplace Transform

$$L\left\{\frac{d^n}{dt^n}f(s)\right\} = s^n f(s) - s^{n-1} f(0) - s^{n-2} f'(0) \ldots f^{n-1}(0)$$

$$L\{t^n f(t)\} = (-1)^n \frac{d^n}{ds^n} f(s)$$

$$\frac{f(t)}{t} \Leftrightarrow \int_s^\infty f(s)\,ds$$

$$\int_0^t f(u)\,du \Leftrightarrow \frac{f(s)}{s}$$

Inverse Transforms

$$\frac{s}{(s^2 + a^2)^2} = \frac{1}{2a} t \sin at$$

$$\frac{s^2}{(s^2 + a^2)^2} = \frac{1}{2a} [\sin at + at \cos at]$$

$$\frac{1}{(s^2 + a^2)^2} = \frac{1}{2a^3} [\sin at - at \cos at]$$

$$\frac{s}{s^2 - a^2} = \cosh at$$

$$\frac{a}{s^2 - a^2} = \sinh at$$

Laplace transform of periodic function: $L\{f(t)\} = \dfrac{\int_0^T e^{-st} f(t)\,dt}{1 - e^{-sT}}$

Numerical Methods

Bisection Method

Take two values of x_1 and x_2 such that $f(x_1)$ is +ve and $f(x_2)$ is −ve, then $x_3 = \dfrac{x_1 + x_2}{2}$. Find $f(x_3)$. If $f(x_3)$ is +ve then root lies between x_3 and x_2, otherwise it lies between x_1 and x_3.

Regular Falsi Method

Same as bisection except $x_2 = x_0 - \dfrac{x_1 - x_0}{f(x_1) - f(x_0)} f(x_0)$

Newton Raphson Method

$$x_{n+1} = x_n - \dfrac{f(x_n)}{f'(x_n)}$$

Pi Cards Method

$$y_{n+1} = x_n + \int_{x_0}^{x} f(x, y_n) \qquad \leftarrow \dfrac{dy}{dx} = f(x, y)$$

Taylor Series Method

$$\dfrac{dy}{dx} = f(x, y) \quad y = y_0 + (x - x_0)(y')_0 + \dfrac{(x - x_0)^2}{2!}(y'')_0 + \cdots + \dfrac{(x - x_0)^n}{n!}(y)_0^n$$

Euler's Method

$$y_1 = y_0 + h f(x_0, y_0) \leftarrow \dfrac{dy}{dx} = f(x, y)$$

$$y_1^{(1)} = y_0 + \dfrac{h}{2}\left[f(x_0, y_0) + f(x_0 + h, y_1)\right]$$

$$y_1^{(2)} = y_0 + \dfrac{h}{2}\left[f(x_0, y_0) + f(x_0 + h, y_1^{(1)})\right]$$

Calculate till two consecutive value of y agree

$$y_2 = y_0 + h f(x_0 + y_1)$$

$$y_2^{(1)} = y_0 + \dfrac{h}{2}\left[f(x_0 + h, y_1) + f(x_0 + 2h, y_2)\right]$$

Runge's Method

$$k_1 = hf(x_0, y_0)$$
$$k_2 = hf\left(x_0 + \frac{h}{2}, y_0 + \frac{k_1}{2}\right)$$
$$k' = hf(x_0 + h, y_0 + k_1)$$
$$k_3 = h(f(x_0 + h, y_0 + k'))$$

finally compute $k = \frac{1}{6}(k_1 + 4k_2 + k_3)$

Trapezoidal Rule

$$\int_{x_0}^{x_0+nh} f(x) \cdot dx = \frac{h}{2}\left[(y_0 + y_n) + 2(y_1 + y_2 + \cdots + y_{n-1})\right]$$

$f(x)$ takes values $y_0, y_1, \ldots$ at $x_0, x_1, x_2, \ldots$.

Simpson's One-Third Rule

$$\int_{x_0}^{x_0+nh} f(x) \cdot dx = \frac{h}{3}\left[(y_0 + y_n) + 4(y_1 + y_3 + \cdots + y_{n-1}) + 2(y_2 + y_4 + \cdots + y_{n-2})\right]$$

Simpson's Three-Eighth Rule

$$\int_{x_0}^{x_0+nh} f(x) \cdot dx = \frac{3h}{8}\left[(y_0 + y_n) + 3(y_1 + y_2 + y_4 + y_5 + \cdots + y_{n-1}) + 2(y_3 + y_6 + \cdots + y_{n-3})\right]$$

Differential Equations

Variable and Separable

General form is $f(y)dy = \Phi(x)dx$

Solution: $\int f(y)dy = \int \Phi(x)dx + c$

Homogenous Equations

General form is $\dfrac{dy}{dx} = \dfrac{f(x,y)}{\Phi(x,y)}$ $f(x,y)$ and $\Phi(x,y)$ Homogenous of the same degree

Solution: Put $y = Vx \Rightarrow \dfrac{dy}{dx} = V + x\dfrac{dv}{dx}$ and above

Reducible to Homogeneous

General form is $\dfrac{dy}{dx} = \dfrac{ax+by+c}{a'x+b'y+c'}$

$\dfrac{a}{a'} \neq \dfrac{b}{b'}$

Solution: Put $x = X + h \quad y = Y + k$

$\Rightarrow \dfrac{dy}{dx} = \dfrac{ax+by+(ah+bk+c)}{a'x+b'y+(a'h+b'k+c')}$. Choose h, k such that $\dfrac{dy}{dx}$ becomes homogenous, then solve by $Y = VX$

$\dfrac{a}{a'} = \dfrac{b}{b'}$

Solution: Let $\dfrac{a}{a'} = \dfrac{b}{b'} = \dfrac{1}{m}$

$\dfrac{dy}{dx} = \dfrac{ax+by+c}{m(ax+by)+c}$

Put $ax + by = t \Rightarrow \dfrac{dy}{dx} = \dfrac{(\dfrac{dt}{dx} - a)}{b}$

Then by variable and separable method, solve the equation.

Leibnitz Linear Equation

General form is $\dfrac{dy}{dx} + py = Q$, where p and Q are functions of x

$\text{I.F.} = e^{\int p \cdot dx}$

Solution: $y(\text{I.F.}) = \int Q \cdot (\text{I.F.}) dx + C.$

Exact Differential Equations

General form $M dx + N dy = 0 \; M \to f(x, y)$

$N \to f(x, y)$

If $\dfrac{\partial M}{\partial y} = \dfrac{\partial n}{\partial x}$, then

Solution: $\int M \cdot dx + \int (\text{terms of } N \text{ containing } x) dy = C(y \text{ constant})$

Rules for Finding Particular Integral

$$\frac{1}{f(D)}e^{ax} = \frac{1}{f(a)}e^{ax}$$

$$= x\frac{1}{f'(a)}e^{ax} \quad \text{if } f(a) = 0$$

$$= x^2\frac{1}{f''(a)}e^{ax} \quad \text{if } f'(a) = 0$$

$$\left.\begin{array}{l}\dfrac{1}{f(b^2)}\sin(ax+b) = \dfrac{1}{f(-a^2)}\sin(ax+b) \quad f(-a^2) \neq 0 \\[6pt] \qquad\qquad = x\dfrac{1}{f'(-a^2)}\sin(ax+b) \quad f(-a^2) = 0 \\[6pt] \qquad\qquad = x^2\dfrac{1}{f''(-a^2)}\sin(ax+b)\end{array}\right\} \text{Also applicable for } \cos(ax+B)$$

$$\frac{1}{f(D)}x^m = [f(D)]^y x^m$$

$$\frac{1}{f(D)}e^{ax}f(x) = e^{ax}\frac{1}{f(D+a)}f(x)$$

Green's Theorem

$$\int_c (\Phi\, dx + \varphi\, dy) = \left(\frac{\partial \psi}{\partial x} - \frac{\partial \Phi}{\partial x}\right) dxdy$$

This theorem converts a line integral around a closed curve into a double integral, which is a special case of Stokes theorem.

Series Expansion

Taylor Series

$$f(x) = f(a) + \frac{f'(a)}{1!}(x-a) + \frac{f''(a)}{2!}(x-a)^2 + \cdots + \frac{f^n(a)}{n!}(x-a)^n$$

$$f(x) = f(0) + \frac{f'(0)}{1!}(x) + \frac{f''(0)}{2!}x^2 + \cdots + \frac{f^n(0)}{n!}(x)^n + \cdots (mc \text{ lower series})$$

$$(1+x)^n = 1 + nx + \frac{n(n-1)}{2}x^2 + \cdots \quad |nx| < 1$$

$$e^x = 1 + x + \frac{x^2}{2!} + \cdots$$

$$\sin x = x - \frac{x^3}{3!} + \frac{x^5}{5!} + \cdots$$

$$\cos x = 1 - \frac{x^2}{2!} + \frac{x^4}{4!} + \cdots$$

Microprocessors

- Clock frequency = ½ crystal frequency
- Hardware interrupts

  TRAP (RST 4.5)   0024H   both edge level

  RST 7.5 → Edge triggered   003CH
  RST 6.5   0034 H
  RST 5.5   level triggered 002C
  INTR   Nonvectored

- Software interrupts RST 0 0000 H

  RST 1 0008 H

  ` 2 0010 H   Vectored

  : 0018 H

  : 7 0038 H

-

S1	S0	
0	0	Halt
0	1	Write
1	0	Read
1	1	Fetch

- HOLD and HLDA used for direct memory access, which has the highest priority over all interrupts.

Flag Registers

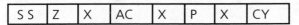

- **Sign flag:** After arithmetic operation, MSB is resolved for sign flag. $S = 1 \rightarrow -ve$ result
- If $Z = 1 \Rightarrow$ Result = 0
- AC: Carry from one stage to an other stage; then AC = 1
- P: $P = 1 \Rightarrow$ even number of ones in result.
- CY: if arithmetic operation results in carry, then CY = 1
- For INX and DCX, no flags effected
- In memory, mapped I/O; I/O devices are treated as memory locations. You can connect max of 65536 devices in this technique.
- In I/O mapped I/O, I/O devices are identified by separate 8-bit address. Same address can be used to identify i/p and o/p device.
- Maximum of 256 i/p and 256 o/p devices can be connected.

Programmable Interfacing Devices

- 8155 → Programmable peripheral interface with 256 bytes RAM and 16-bit counter
- 8255 → Programmable interface adaptor
- 8253 → Programmable interval timer
- 8251 → Programmable communication interfacing device (USART)
- 8257 → Programmable DMA controller (four channels)
- 8259 → Programmable interrupt controller
- 8272 → Programmable floppy disk controller
- CRT controller
- Key board and display interfacing device

RLC: Each bit shifted to adjacent left position. D_7 becomes D_0.

CY flag modified according to D_7

RAL: Each bit shifted to adjacent left position. D_7 becomes CY and CY becomes D_0.

ROC: CY flag modified according to D_0

RAR: D_0 becomes CY and CY becomes D_7

CALL & RET vs. PUSH & POP

CALL & RET
- When CALL executes, μp automatically stores 16-bit address of instruction next to CALL on the Stack
- CALL executed, SP decremented by 2
- RET transfers contents of top 2 of SP to PC
- RET executes "SP" incremented by 2

PUSH & POP
* Programmer uses PUSH to save the contents rp on stack
* PUSH executes "SP" decremented by "2."
* Same here but to specific "rp."
* Same here.

Some Instruction Set Information

CALL Instruction

CALL → 18T states   SRRWW

CC → Call on carry   9 – 18 states

CM → Call on minus   9–18

CNC → Call on no carry

CZ → Call on zero; CNZ call on nonzero

CP → Call on +ve

CPE → Call on even parity

CPO → Call on odd parity

RET: - 10 T

RC: - 6/ 12 "T" states

Jump Instructions

JMP → 10 T

JC → Jump on carry 7/10 T states

JNC → Jump on no carry

JZ → Jump on zero

JNZ → Jump on nonzero

JP → Jump on positive

JM → Jump on minus

JPE → Jump on even parity

JPO → Jump on odd parity.
- **PCHL**: Move HL to PC 6T
- **PUSH**: 12 T; POP : 10 T
- **SHLD**: Address : store HL directly to address 16 T
- **SPHL**: Move HL to SP 6T
- **STAX**: R_p store A in memory 7T
- **STC**: Set carry 4T
- **XCHG**: Exchange DE with HL "4T"
- **XTHL**: Exchange stack with HL 16 T
- For "AND" operation, "AY" flag will be set and "CY" reset
- For "CMP", if A <Reg/mem : CY → 1 and Z → 0 (nothing but A-B)

A > Reg/mem: CY → 0 and Z → 0
A = Reg/mem: Z → 1 and CY → 0

- "DAD" Add HL + RP (10T) → fetching, busidle, busidle
- DCX, INX will not affect any flags (6T)
- DCR, INR affects all flags except carry flag. "Cy" will not be modified
- "LHLD" load "HL" pair directly
- "RST" → 12T states
- SPHL, RZ, RNZ ..., PUSH, PCHL, INX, DCX, CALL → Fetching has 6T states
- PUSH – 12 T; POP – 10T

Power Electronics

1. Turn on time of SCR = $t_d + t_r + t_s$

   where t_d = delay time

   t_r = rise time

   t_s = settling time

   Device turn off time, $t_q = t_{rr} + t_{gr}$

   where t_{rr} = Reverse recovery time and

   t_{gr} = gate recovery time

2. SERIES OPERATION:

   → SCRs are connected in series to increase the voltage rating.

String efficiency

$$\eta_S = \frac{\text{Voltage rating of string}}{\text{No. of SCRs in the string * voltage rating of each SCR}} = \frac{V_1+V_2}{2V_1}$$

Derating factor = 1 − string efficiency

→ Static equalizing resistance $Rs = \dfrac{nV_{bm} - V_s}{(n-1)\Delta I_b}$

→ Dynamic equalizing capacitance $C = \dfrac{(n-1)\Delta Q}{nV_{bm} - V_s}$

3. Parallel operation is applied for SCRs, with higher current ratings

String efficiency

$$\eta S = \frac{\text{Current rating of string}}{\text{No. of SCRs in the string} \times \text{voltage rating of each SCR}} = \frac{I_1+I_2}{2I_1}$$

1. Single-Phase Half-Wave Rectifier R Load

Circuit turn off time $t_c = \dfrac{\pi}{\omega}$

Average output voltage $V_0 = \dfrac{V_m}{2\pi[1+\cos\alpha]}$

RMS output voltage $V_{or} = \dfrac{V_m}{(2\sqrt{\pi})\left[\pi - \alpha + \dfrac{(\sin 2\alpha)}{2}\right]^{\frac{1}{2}}}$

Average output current, $I_0 = \dfrac{V_0}{R}$

→ RMS output current, I_{or}/R

Power factor of input supply = $\dfrac{\text{Power supplied to load}}{\text{Source volt ampere}}$

$PF = \dfrac{\dfrac{V_{or}^2}{R}}{V_S I_S}$ $\qquad PF = \dfrac{V_{or}}{V_S}$

2. Single-Phase Half-Wave Rectifier R–L Load

Circuit turn off time $= t_c = \dfrac{(2\pi - \beta)}{\omega}$

Average output voltage $V_0 = \left(\dfrac{V_m}{2\pi}\right)[\cos\alpha - \cos\beta]$

RMS output voltage $V_{or} = \dfrac{V_m}{(2\sqrt{\pi})\left[\beta - \alpha + \dfrac{1}{2}(\sin 2\alpha - \sin 2\beta)\right]^{\frac{1}{2}}}$

Average output current, $I_0 = \dfrac{V_0}{R}$

RMS value of output current, $I_{or} = \dfrac{V_{or}}{R}$

Power factor of supply, $PF = \dfrac{V_{or}}{V_S}$

3. Single-Phase Half-Wave Rectifier *RL* Load and Free Wheeling Diode

Circuit turn off time $= t_c = \dfrac{\pi}{\omega}$

Average output voltage $V_0 = \left(\dfrac{V_m}{2\pi}\right)[1 + \cos\alpha]$

RMS output voltage, $V_{or} = \dfrac{V_m}{2\pi\left[\pi - \alpha + \dfrac{1}{2(\sin 2\alpha)}\right]^{1/2}}$

Average output current, $I_0 = \dfrac{V_0}{R}$

RMS value of output current, $I_{or} = \dfrac{V_{or}}{R}$

Power factor of supply, $PF = \dfrac{V_{or}}{V_S}$

4. Single-Phase Half-Wave Rectifier *RLE* Load

Circuit turn off time, $t_c = 2\pi + \theta_1 - \beta$, $\theta_1 = \sin^{-1}\left(\dfrac{E}{V_m}\right)$ and $\theta_2 = 180° - \theta_1$

→ Average output voltage, $V_0 = E + I_0 R$

Average output current, $I_0 = \left(\dfrac{1}{2\pi R}\right)[V_m(\cos\alpha - \cos\beta) - E(\beta - \alpha)]$

Supply power factor, $PF = \dfrac{I_{or}^2 R + I_0 E}{V_s I_{or}}$

→ RMS value of output current

$I_{or} = 2\sqrt{\dfrac{1}{2\pi R^2}\left\{\left(V_S^2 + E^2\right)(\beta - \alpha) - \dfrac{V_S^2}{2}(\sin 2\beta - 2\sin 2\alpha) - 2V_m E(\cos\alpha - \cos\beta)\right\}}$

5. Single-Phase Full-Wave Rectifier: Mid Point Converter Type

Circuit turn off time, $t_c = \dfrac{\pi}{\omega}$

Average output voltage, $V_0 = \dfrac{2V_m \cos\alpha}{\pi}$

6. Single-Phase Full-Wave Bridge-Type Rectifier

Circuit turn off time, $t_c = \dfrac{\pi}{\omega}$

Average output voltage, $V_0 = \dfrac{2V_m \cos\alpha}{\pi}$

RMS output voltage, $V_{or} = V_s$

The inductance of source results in an lesser value of voltage, $V_0 = \dfrac{2V_m}{\pi}\cos\alpha - \dfrac{\omega L_s}{\pi} I_0$

7. Single-Phase Semiconverter

Circuit turn off time, $t_c = \dfrac{\pi}{\omega}$

Average output voltage $V_0 = \dfrac{V_m}{\pi}[1 + \cos\alpha]$

RMS output voltage, $V_{or} = \dfrac{V_s}{(\sqrt{\pi})\left[\pi - \alpha + \dfrac{1}{2}(\sin 2\alpha)\right]^2}$

8. Three-Phase-Controlled Half-Wave Rectifier with R Load

Average output voltage, $V_0 = \dfrac{3}{2\pi}V_{ml}\cos\alpha = \dfrac{3\sqrt{3}}{2\pi}V_{mp}\cos\alpha$

Average output current, $I_0 = \dfrac{V_0}{R}$

RMS output voltage, $V_{or} = \dfrac{V_{ML}}{R}\left[\dfrac{1}{6}+\dfrac{\sqrt{3}}{8\pi}(\cos 2\alpha)\right]^{\frac{1}{2}}$

RMS value of output current, $I_{or} = \dfrac{V_{or}}{R}$

9. Three-Phase Full Converter

Voltage V_0 average output $= \dfrac{3}{\pi}V_{ml}\cos\alpha$

RMS output voltage, $V_{or} = V_{ML}\sqrt{\dfrac{3}{2\pi}\left[\dfrac{\pi}{3}+\dfrac{\sqrt{3}}{2}(\cos 2\alpha)\right]^{\frac{1}{2}}}$

$i_s = i_0\sqrt{\dfrac{3}{2}}$

10. Three-Phase Full-Converter

Voltage V_0 average output $= \dfrac{3}{2\pi}V_{ml}(1+\cos\alpha)$

RMS output voltage, $V_{or} = \dfrac{V_{ML}}{2}\sqrt{\dfrac{3}{\pi}\left[\dfrac{2\pi}{3}+\dfrac{\sqrt{3}}{2}(1+\cos 2\alpha)\right]^{\frac{1}{2}}}$

11. For a 3-ϕ Converter, the inductance of source results in an lesser value of voltage

$V_0 = \dfrac{3\sqrt{6}}{\pi}V_{ph}\cos\alpha - \dfrac{3\omega L_s}{\pi}I_0$

Choppers

a. **Step-Up Chopper**

Duty cycle $= \left(\dfrac{T_{on}}{T_{on}+T_{off}}\right) = \left(\dfrac{T_{on}}{T}\right)$

Average output voltage across load, $V_0 = \dfrac{V_s}{1-\alpha}$

Energy supplied by inductor $= W_{out} = (V_o - V_s)\dfrac{(I_{min}+I_{max})}{2} \times T_{off}$

b. **Step-Down Chopper**

Duty cycle $= \left(\dfrac{T_{on}}{T_{on}+T_{off}}\right) = \left(\dfrac{T_{on}}{T}\right)$

Average output voltage across load, $V_o = V_s\left(\dfrac{T_{on}}{T}\right) = fT_{on}V_s = \alpha V_s$

where V_s = supply voltage $\dfrac{1}{T} = f$

Average output current through load, $I_o = \alpha \dfrac{V_s}{R}$

RMS value of output voltage $= \sqrt{\alpha}\, V_s$

RMS value of thyristor current $= \dfrac{\sqrt{\alpha}\, V_s}{R}$

Effective input resistance of chopper $= \dfrac{R}{\alpha}$

The minimum and maximum values of load current is given by

$$I_{max} = \dfrac{\dfrac{V_s}{R}\left[1-e^{-\tfrac{T_{on}}{T_a}}\right]}{\left[1-e^{-\tfrac{T}{T_a}}\right]} - \left(\dfrac{E}{R}\right)$$

$$I_{min} = \dfrac{\dfrac{V_s}{R}\left[e^{\tfrac{T_{on}}{T_a}}-1\right]}{\left[e^{\tfrac{T}{T_a}}-1\right]} - \left(\dfrac{E}{R}\right)$$

$$\text{Ripple } \Delta I = \frac{V_s}{R}\left[\frac{\left(1-e^{-\frac{T_{on}}{T_a}}\right)\left(1-e^{-\frac{T_{off}}{T_a}}\right)}{\left[1-e^{-\frac{T}{T_a}}\right]}\right]$$

$$\text{Per unit ripple} = \frac{\left(1-e^{-\frac{\alpha T}{T_a}}\right)\left(1-e^{-\frac{(1-\alpha)T}{T_a}}\right)}{\left[1-e^{-\frac{T}{T_a}}\right]}$$

Maximum value of ripple current is given by $(\Delta I)_{max} = \dfrac{V_s}{4fL}$

$$\text{Ripple factor} = \frac{\text{AC ripple voltage}}{\text{DC voltage}} = \sqrt{\left(\frac{1}{\alpha}\right)-1}$$

Voltage-Commutated Chopper

Minimum turn on time of chopper $\to \pi\sqrt{LC}$ seconds

Minimum duty cycle of voltage-commutated chopper $\alpha_{min} = \pi f \sqrt{LC}$

The output current, $I_o = \dfrac{CV_s - (-V_s)}{2t_c}$

where $C = \dfrac{I_o t_c}{V_s}, \ L \geq \left(\dfrac{V_s}{I_o}\right)^2 c$

The current through the circuit is $i_c = \dfrac{V_s}{\omega_o L}\sin\omega_o t$

Peak capacitor current $I_{cp} = \dfrac{V_s}{\omega_o L} = V_s\sqrt{\dfrac{C}{L}}$

The minimum time is required for changing the polarity of capacitor from V_s to $-V_s$, that is, $t_1 = \dfrac{\pi}{\omega} t_1 = \pi\sqrt{LC}$

In a voltage-commutated chopper, average value of output voltage is given by

$$V_o = \frac{V_s}{T}(T_{on} + 2t_c)$$

$$= \frac{V_s}{T}\left(2\frac{cv_s}{I_o}\right)$$

Current-Commutated Chopper

The values of L and C are given by

$$L = \frac{V_s t_c}{XI_o\left[\pi - 2\sin^{-1}\left(\frac{1}{x}\right)\right]}$$

$$C = \frac{XI_o t_c}{V_s\left[\pi - 2\sin^{-1}\left(\frac{1}{x}\right)\right]}$$

where $X = \dfrac{I_{cp}}{I_o}$

$$t_c = \left[\pi - 2\sin^{-1}\left(\frac{1}{x}\right)\right]\sqrt{LC}$$

$$t_c = I_{cp}\sin\omega t$$

At $\omega t = \theta_1, i_c = I_o = I_{Cp}\sin\theta_1, \theta_1 = \sin^{-1}\left[\dfrac{I_o}{I_{Cp}}\right] = \sin^{-1}\left[\dfrac{1}{x}\right]$

Peak capacitor voltage $= V_S + I_o\sqrt{\dfrac{L}{C}}$

Inverters

Fourier Analysis of Single-Phase Inverter Output Voltage

1-Phase Half-Bridge Inverter

$$V_o = \sum_{n=1,3}^{\infty} \frac{2V_s}{n\pi}\sin\omega t$$

$$i_o = \sum_{n=1,3}^{\infty} \frac{2V_s}{n\pi Z_n} \sin(n\omega t - \phi_n)$$

1-Phase Full Bridge Inverter

$$V_o = \sum_{n=1,3}^{\infty} \frac{4V_s}{n\pi Z_n} \sin n\omega t$$

$$i_o = \sum_{n=1,3}^{\infty} \frac{4V_s}{n\pi Z_n} \sin(n\omega t - \phi_n)$$

Z_n is the impedance offered to nth harmonic

$$Z_n = \sqrt{R^2 + \left(n\omega L - \frac{1}{n\omega c}\right)^2}$$

$$\phi_n = \tan^{-1}\left(\frac{n\omega L - 1/n\omega c}{R}\right)$$

$V_{or} = V_s$ – full-bridge inverter

$V_{or} = \frac{V_s}{2}$ – half-bridge inverter

Harmonic factor of nth harmonic, $HF_n = \frac{V_n}{V_1}$

→ where V_n = rms value of nth harmonic component

V_1 = fundamental component rms value.

→ Total harmonic distortion (THD) → it is a measure of closeness in shape between a waveform and its fundamental component.

$$THD = \frac{1}{V_1}\left(\sum_{n=2,3}^{\infty} V_n^2\right)^{\frac{1}{2}}$$

$$= \frac{\sqrt{V_{or}^2 - V_1^2}}{V_1}$$

Distortion factor of nth harmonic is defined as $\frac{V_n}{V_1 \cdot n^2}$

Three-Phase Bridge Inverter

$$V_{ab} = \sum_{n=1,3}^{\infty} \frac{4V_s}{n\pi} \cos\frac{n\pi}{6} \sin n\left(\omega t + \frac{\pi}{6}\right)$$

$$V_{bc} = \sum_{n=1,3}^{\infty} \frac{4V_s}{n\pi} \cos\frac{n\pi}{6} \sin n\left(\omega t - \frac{\pi}{2}\right)$$

$$V_{ca} = \sum_{n=1,3}^{\infty} \frac{4V_s}{n\pi} \cos\frac{n\pi}{6} \sin n\left(\omega t + \frac{5\pi}{6}\right)$$

→ All triple n harmonics are absent from the line voltages.

$$V_1 = \sqrt{\frac{1}{\pi}\int_0^{2\pi/3} V_s^2 d(\omega t)} = V_s\sqrt{\frac{2}{3}} = 0.8165 V_s$$

$$V_{ph} = \frac{0.8165 V_s}{\sqrt{3}} = 0.4714 V_s$$

$$V_{a0} = \sum_{n=1,3}^{\infty} \frac{2V_s}{n\pi} \cos\frac{n\pi}{6} \sin n\left(\omega t + \frac{\pi}{6}\right)$$

$$V_{b0} = \sum_{n=1,3}^{\infty} \frac{2V_s}{n\pi} \cos\frac{n\pi}{6} \sin n\left(\omega t - \frac{\pi}{2}\right)$$

$$V_{c0} = \sum_{n=1,3}^{\infty} \frac{2V_s}{n\pi} \cos\frac{n\pi}{6} \sin n\left(\omega t + \frac{5\pi}{6}\right)$$

$$\sqrt{V_{ph}} = \sqrt{\frac{1}{\pi}\int_0^{2\pi/3}\left(\frac{V_s}{2}\right)^2 d(\omega t)} = V_s\sqrt{\frac{1}{6}}$$

$$V_{ph} = 0.4082 V_s, \quad V_1 = 0.4082 V_s \times \sqrt{3} = 0.7071 V_s$$

Voltage Control in 1-ϕ Inverter

1. Single-Pulse Modulation

$$V_{or} = V_s \sqrt{2d/\pi}$$

$$V_0 = \sum_{n=1,3}^{\infty} \left(\frac{4V_s}{n\pi} \sin \frac{n\pi}{2} \cdot \sin nd \right) \sin(n\omega t)$$

$\downarrow$

Maximum value of nth harmonic
→ To eliminate nth harmonic, $nd = \pi$
→ i.e., width of the pulse = $2d = 2\pi/n$

2. Multiple-Pulse Modulation

$$V_{or} = \sqrt{2}\, d/\pi$$

$$V_o = \sum_{n=1,3}^{\infty} \frac{8V_s}{n\pi} \sin n\gamma \sin \frac{nd}{2} \sin(n\omega t)$$

Number of pulses per half cycle = $\dfrac{\text{Length of half cycle of reference waveform}}{\text{Width of one cycle of triangle carrier wave}}$

$$N = \frac{1/2f_r}{1/f_c} = \frac{f_c}{2f_r} = \frac{\omega_c}{2\omega_r}$$

$$N = \omega_c/2\omega_r \text{ per half cycle}$$

→ where f_c is carrier-wave frequency and f_r is reference-wave frequency

AC Voltage Controllers
Single-Phase Voltage Controller with R Load

Circuit turn off time = π/ω s

$$V_C = \sum_{n=1,3,5}^{\omega} A_n \sin n\omega t + \sum_{n=1,3,5}^{\omega} B_n \cos n\omega t \, d(\omega t)$$

where $A_n = \dfrac{V_m}{\pi}\left[\dfrac{\sin(n+1)\alpha}{(n+1)} - \dfrac{\sin(n-1)\alpha}{(n-1)}\right]$

$$B_n = \dfrac{V_m}{\pi}\left[\dfrac{\cos(n+1)\alpha - 1}{(n+1)} - \dfrac{\sin(n-1)\alpha - 1}{(n-1)}\right]$$

$$V_{nm} = \sqrt{A_n^2 + B_n^2}$$

$$\phi_n = \tan^{-1}\dfrac{B_n}{A_n}$$

For $n = 1$, V_{or} of output voltage is given by

$$V_{or} = \dfrac{V_m}{\sqrt{2\pi}}\left[(\pi - \alpha) + \dfrac{1}{2}\sin 2\alpha\right]^{1/2}$$

$$I_{or} = \dfrac{V_{or}}{R}$$

$$P = I_{or}^2 R = \dfrac{V_{or}^2}{R} = \dfrac{V_m^2}{2\pi R}(\pi - \alpha) + \dfrac{1}{2}\sin 2\alpha$$

$$= \dfrac{V_s^2}{\pi R}\left[(\pi - \alpha) + \dfrac{1}{2}\sin 2\alpha\right]$$

$$\text{Power factor} = \dfrac{\text{Real power}}{\text{Apparent power}} = \dfrac{V_{or}^2/R}{V_s V_{or}/R} = \dfrac{V_{or}}{V_s}$$

$$= \left\{\dfrac{1}{\pi}\left((\pi - \alpha) + \dfrac{1}{2}\sin 2\alpha\right)\right\}^{1/2}$$

Phase Voltage Controller with *RL* Load

$$i_0 = \frac{V_m}{2}\sin(\omega t - \phi) - \frac{V_m}{2}\sin(\alpha - \phi)\cdot\exp\left[R\alpha/\omega L\right]\cdot e^{-RT/L}$$

$$i_0 = \frac{V_m}{2}\sin(\omega t - \phi) - \frac{V_m}{2}\sin(\alpha - \phi)\cdot\exp\left[\frac{-R}{\omega L}(\omega t - \alpha)\right]$$

DC and AC Drives

DC Motor Equations

$$E_a = \frac{Z\phi N}{60}\frac{P}{A} \quad N \rightarrow \text{rpm}$$

$$= Z\phi n\frac{P}{A} \quad n \rightarrow \text{rps}$$

If $\omega_m = 2\pi n$

$$\omega_m = Z - \phi\frac{\omega_m}{2\pi}\frac{P}{A}$$

$$= \left(\frac{Z}{2\pi}\frac{P}{A}\right)\phi\omega_m$$

$E_a = K_a\phi\omega_m$, K_a = motor constant = $\left((Z/2\pi)(P/A)\right)$ V/Wb·rad/s

$$\text{Torque } T = \frac{1}{2\pi}Z\phi I_a\frac{P}{A}$$

$$= \left(\frac{Z}{2\pi}\frac{P}{A}\right)\phi I_a$$

$T = K_a\phi I_a$ K_a = Nm/Wb·A

For a DC Separately Excited Motor

→ Flux, ϕ is constant

$E_a = K_m\omega_m K_m \quad \longrightarrow$ motor constant V/rad/s

$$\text{Torque} = K_a \phi I_a$$

$$\rightarrow T_e = K_m I_a \quad K_m \rightarrow \text{Nm/A}$$

For a DC Series Motor

$$\phi \propto I_a, \quad \phi = c I_a$$

$$E_a = K_a c I_a \omega_m$$

$$E_a = K_1 I_a \omega_m \quad K_1 \rightarrow \text{motor constant} \frac{\text{volt s}}{\text{rad.amp}}$$

$$T_e = K_a \phi I_a^2$$

$$= K_a c I_a^2$$

$$T_e = K_1 I_a^2 \quad K_1 \rightarrow \text{Nm/A}^2$$

1-ϕ Half-Wave Converter Drive

I_a is asumed to be constant for speed control of the drive.

$$V_t = V_0 = \frac{V_m}{2\pi}(1 + \cos\alpha_1)$$

$$V_f = \frac{V_m}{\pi}(1 + \cos\alpha_2)$$

$$I_{s\ rms} = \sqrt{\frac{1}{2\pi}\int_\alpha^\pi I_a^2 d(\omega t)}$$

$$= I_a \sqrt{\frac{1}{2\pi}(\pi - \alpha)}$$

$$I_{s\ rms} = I_a \left[\frac{(\pi - \alpha)}{2\pi}\right]^{1/2}$$

$$I_{F.D.R} = I_a \left[(\pi + \alpha) / 2\pi \right]^{1/2}$$

$$\text{Supply PF} = \frac{\text{Power delivered to load}}{\text{Source VA}}$$

$$= \frac{E_a I_a + I_a^2 r_a}{V_s I_{sr}}$$

$$= \frac{I_a [E_a + I_a r_a]}{V_s I_{sr}}$$

→ Input supply PF.

1-Ø Half-Controlled Converter Drive

$$V_0 = V_t = \frac{V_m}{\pi}[1 + \cos\alpha_1]$$

$$V_f = \frac{V_m}{\pi}[1 + \cos\alpha_2]$$

$$I_{sr} = I_a \sqrt{(\pi - \alpha)/\pi}$$

$$I_{fdr} = I_a \sqrt{2\alpha/2\pi} = I_a \sqrt{\alpha/\pi}$$

$$I_{Tr} = I_a \sqrt{(\pi - \alpha)/2\pi} = I_{or}$$

→ Input supply PF $= \dfrac{V_t I_a}{V_s I_{sr}}$

Single-Phase Full-Wave Converter Drive

$$V_0 = V_t = \frac{2V_m}{\pi} \cos\alpha_1$$

$$V_f = \frac{2V_m}{\pi} \cos\alpha_2$$

$$I_{sr} = I_a$$

$$I_{Tr} = I_a\sqrt{\pi/2\pi} = I_a\sqrt{2}$$

$$PF = \frac{V_t I_a}{V_s I_{sr}}$$

3-ϕ Half-Wave Converter Drive

$$V_o = V_t = \frac{3\sqrt{6}}{2\pi} V_{ph} \cos\alpha_1$$

$$I_{sr} = I_a\sqrt{120/360} = I_a\sqrt{1/3}$$

$$I_{Tr} = I_a\sqrt{120/360} = I_a\sqrt{1/3}$$

3-ϕ Full-Wave Converter Drive

$$V_o = V_t = \frac{3\sqrt{6}}{\pi} V_{ph} \cos\alpha$$

$$I_{sr} = I_a\sqrt{240/360} = I_a\sqrt{2/3}$$

$$I_{Tr} = I_a\sqrt{120/360} = I_a\sqrt{1/3}$$

3-ϕ Semiconverter Drive

$$V_o = V_t = \frac{3\sqrt{6}}{\pi} V_{ph} (1+\cos\alpha)$$

For $\alpha_1 < 60°$, $I_{sr} = I_a\sqrt{2/3}$, $I_{Tr} = I_a\sqrt{1/3}$

For $\alpha_1 > 60°$, $I_{sr} = I_a\sqrt{(180-\alpha_1)/180}$, $I_{Tr} = I_a\sqrt{(180-\alpha_1)/360}$

Static Rotor Resistance Control

$$R_{eff} = R \times \frac{T_{off}}{T}$$

$$= R \times (T - T_{on})/T$$

$$\rightarrow R_{eff} = R(1-\alpha) \quad \alpha \rightarrow \text{duty cycle of the chopper}$$

Control System

Time Response of Second-Order System

Step i/p :

$$C(t) = 1 - \frac{e^{-\zeta\omega_n t}}{\sqrt{1-\zeta^2}}\left(\sin\omega_n\sqrt{1-\zeta^2}\, t \pm \tan^{-1}\left(\frac{\sqrt{1-\zeta^2}}{\zeta}\right)\right)$$

$$e(t) = \frac{e^{-\zeta\omega_n t}}{\sqrt{1-\zeta^2}}\left(\sin\omega_d\, t \pm \tan^{-1}\left(\frac{\sqrt{1-\zeta^2}}{\zeta}\right)\right)$$

$$e_{ss} = \lim_{t\to\infty} \frac{e^{-\zeta\omega_n t}}{\sqrt{1-\zeta^2}}\left(\sin\omega_d\, t \pm \tan^{-1}\left(\frac{\sqrt{1-\zeta^2}}{\zeta}\right)\right)$$

$\rightarrow \zeta \rightarrow$ Damping ratio ; $\zeta\omega_n \rightarrow$ Damping factor

$\zeta < 1$ (Underdamped):

$$C(t) = 1 - \frac{e^{-\zeta\omega_n t}}{\sqrt{1-\zeta^2}} \sin\left(\omega_d t \pm \tan^{-1}\left(\frac{\sqrt{1-\zeta^2}}{\zeta}\right)\right)$$

$\zeta < 1$ (Undamped):

$$C(t) = 1 - \cos\omega_n t$$

$\zeta = 1$ **(Critically damped):**

$$C(t) = 1 - e^{-\omega_n t}(1 + \omega_n t)$$

$\zeta > 1$ **(Overdamped):**

$$C(t) = 1 - \frac{e^{-\left(\zeta - \sqrt{\zeta^2 - 1}\right)\omega_n t}}{2\sqrt{\zeta^2 - 1}\left(\zeta - \sqrt{\zeta^2 - 1}\right)}$$

$$T = \frac{1}{\left(\zeta - \sqrt{\zeta^2 - 1}\right)\omega_n}$$

$$T_{\text{undamped}} > T_{\text{overdamped}} > T_{\text{underdamped}} > T_{\text{critical damp}}$$

Time Domain Specifications

- Rise time $t_r = \dfrac{\pi - \phi}{\omega_n\sqrt{1-\zeta^2}}$ $\phi = \tan^{-1}\left(\dfrac{\sqrt{1-\zeta^2}}{\zeta}\right)$

- Peak time $t_p = \dfrac{n\pi}{\omega_d}$

- Max over shoot % $M_p = e^{\frac{-\zeta\omega_n}{\sqrt{1-\zeta^2}}} \times 100$

- Settling time $t_s = 3T$ 5% tolerance $= 4T$ 2% tolerance

- Delay time $t_d = \dfrac{1 + 0.7\zeta}{\omega_n}$

- Damping factor² $\zeta^2 = \dfrac{(\ln M_p)^2}{\pi^2 + (\ln M_p)^2}$

- Time period of oscillations $T = \dfrac{2\pi}{\omega_d}$

- Number of oscillations $= \dfrac{t_s}{2\pi/\omega_d} = \dfrac{t_s \times \omega_d}{2\pi}$]

$t_r \approx 1.5 t_d$ $t_r = 2.2T$

- Resonant peak $M_r = \dfrac{1}{2\zeta\sqrt{1-\zeta^2}}; \omega_r = \omega_n\sqrt{1-2\zeta^2}\quad \begin{matrix}\omega_n > \omega_r \\ \omega_b > \omega_n\end{matrix}\quad \omega_r < \omega_n < \omega_b$

- Bandwidth $\omega_b = \omega_n\left(1-2\zeta^2 + \sqrt{4\zeta^4 - 4\zeta^2 + 2}\right)^{1/2}$

Static Error Coefficients

- Step i/p : $e_{ss} = \lim\limits_{t\to\infty} e(t) = \lim\limits_{s\to 0} sE(s) = \lim\limits_{s\to 0}\dfrac{SR(s)}{1+GH}$

  $e_{ss} = \dfrac{1}{1+K_p}$ (positional error) $K_p = \lim\limits_{s\to 0} G(s)H(s)$

- Ramp i/p (t): $e_{ss} = \dfrac{1}{K_v}\quad K_v = \lim\limits_{s\to 0} sG(s)H(s)$

- Parabolic i/p $\left(t^2/2\right)$: $e_{ss} = \dfrac{1}{K_a}\quad K_a = \lim\limits_{s\to 0} s^2 G(s)H(s)$

  Type $< i/p \to e_{ss} = \infty \to e_{ss} = \infty$

  Type $= i/p \to e_{ss} =$ finite $\to e_{ss} =$ finite

  Type $> i/p \to e_{ss} = 0 \to e_{ss} = 0$

- Sensitivity $S = \dfrac{\partial A/A}{\partial K/K}$ sensitivity of A with respect to K.

- Sensitivity of over all T/F with respect to forward path T/F $G(s)$:

  Open loop: $S = 1$

  Closed loop: $S = \dfrac{1}{1+G(s)H(s)}$

- Minimum S value preferable

- Sensitivity of over all T/F with respect to feedback T/F $H(s)$: $S = \dfrac{G(s)H(s)}{1+G(s)H(s)}$

Stability

RH Criterion

- Take characteristic equation $1+ G(s) H(s) = 0$
- All coefficients should have same sign
- There should not be missing 's' term. Term missed means presence of at least one +ve real part root.
- If char. equation contains either only odd/even terms, indicates roots have no real part and possess only imaginary parts; therefore sustained oscillations in response.
- Row of all zeroes occur if
  (a) Equation has at least one pair of real roots with equal image but opposite sign
  (b) Has one or more pair of imaginary roots
  (c) Has pair of complex conjugate roots forming symmetry about origin.

Appendix 5

Table of Laplace and Z Transforms

Table of Laplace and Z Transforms

	$X(s)$	$x(t)$	$x(kT)$ or $x(k)$	$X(z)$
1.	-	-	Kronecker delta $\delta_0(k)$ 1 $k=0$ 0 $k \neq 0$	1
2.	-	-	$\delta_0(n-k)$ 1 $n=k$ 0 $n \neq k$	z^{-k}
3.	$\dfrac{1}{s}$	$1(t)$	$1(k)$	$\dfrac{1}{1-z^{-1}}$
4.	$\dfrac{1}{s+a}$	e^{-at}	e^{-akT}	$\dfrac{1}{1-e^{-aT}z^{-1}}$
5.	$\dfrac{1}{s^2}$	t	kT	$\dfrac{Tz^{-1}}{(1-z^{-1})^2}$
6.	$\dfrac{2}{s^3}$	t^2	$(kT)^2$	$\dfrac{T^2 z^{-1}(1+z^{-1})}{(1-z^{-1})^3}$
7.	$\dfrac{6}{s^4}$	t^3	$(kT)^3$	$\dfrac{T^3 z^{-1}(1+4z^{-1}+z^{-2})}{(1-Z^{-1})^4}$

8.	$\dfrac{a}{s(s+a)}$	$1-e^{-at}$	$1-e^{-akT}$	$\dfrac{(1-e^{-aT})z^{-1}}{(1-z^{-1})(1-e^{-aT}z^{-1})}$
9.	$\dfrac{b-a}{(s+a)(s+b)}$	$e^{-at}-e^{-bt}$	$e^{-akT}-e^{-bkT}$	$\dfrac{(e^{-aT}-e^{-bT})z^{-1}}{(1-e^{-aT}z^{-1})(1-e^{-bT}z^{-1})}$
10.	$\dfrac{1}{(s+a)^2}$	te^{-at}	kTe^{-akT}	$\dfrac{Te^{-aT}z^{-1}}{(1-e^{-aT}z^{-1})^2}$
11.	$\dfrac{s}{(s+a)^2}$	$(1-at)e^{-at}$	$(1-akT)e^{-akT}$	$\dfrac{1-(1+aT)e^{-aT}z^{-1}}{(1-e^{-aT}z^{-1})^2}$
12.	$\dfrac{2}{(s+a)^3}$	$t^2 e^{-at}$	$(kT)^2 e^{-akT}$	$\dfrac{T^2 e^{-aT}(1+e^{-aT}z^{-1})z^{-1}}{(1+e^{-aT}z^{-1})^3}$
13.	$\dfrac{a^2}{s^2(s+a)}$	$at-1+e^{-at}$	$akT-1+e^{-akT}$	$\dfrac{[(aT-1+e^{-aT})+(1-e^{-aT}-aTe^{-aT})z^{-1}]z^{-1}}{(1-z^{-1})(1-e^{-aT}z^{-1})}$
14.	$\dfrac{\omega}{s^2+\omega^2}$	$\sin\omega t$	$\sin\omega kT$	$\dfrac{z^{-1}\sin\omega T}{1-2z^{-1}\cos\omega T+z^{-2}}$
15.	$\dfrac{s}{s^2+\omega^2}$	$\cos\omega t$	$\cos\omega kT$	$\dfrac{z^{-1}\cos\omega T}{1-2z^{-1}\cos\omega T+z^{-2}}$
16.	$\dfrac{\omega}{(s+a)^2+\omega^2}$	$e^{-at}\sin\omega t$	$e^{-akT}\sin\omega kT$	$\dfrac{e^{-aT}z^{-1}\sin\omega T}{1-2e^{-aT}z^{-1}\cos\omega T+e^{-2aT}z^{-2}}$
17.	$\dfrac{s+a}{(s+a)^2+\omega^2}$	$e^{-at}\cos\omega t$	$e^{-akT}\cos\omega kT$	$\dfrac{1-e^{-aT}z^{-1}\cos\omega T}{1-2e^{-aT}z^{-1}\cos\omega T+e^{-2aT}z^{-2}}$
18.	–	–	a^k	$\dfrac{1}{1-az^{-1}}$
19.	–	–	a^{k-1}, $k=1,2,3,\ldots$	$\dfrac{z^{-1}}{1-az^{-1}}$
20.	–	–	ka^{k-1}	$\dfrac{z^{-1}}{(1-az^{-1})^2}$
21.	–	–	$k^2 a^{k-1}$	$\dfrac{z^{-1}(1+az^{-1})}{(1-az^{-1})^3}$

22.	–	–	$k^3 a^{k-1}$	$\dfrac{z^{-1}(1+4az^{-1}+a^2z^{-2})}{(1-az^{-1})^4}$
23.	–	–	$k^4 a^{k-1}$	$\dfrac{z^{-1}(1+11az^{-1}+11a^2z^{-2}+a^3z^{-3})}{(1-az^{-1})^5}$
24.	–	–	$a^k \cos k\pi$	$\dfrac{1}{1+az^{-1}}$

$x(t) = 0$ for $t < 0$

$x(kT) = x(k)$ for $k < 0$

Unless otherwise noted, $k = 1, 2, 3, \ldots$

Definition of Z Transform

$$Z\{x(k)\} = X(z) = \sum_{k=0}^{\infty} x(k) z^{-k}$$

Important Properties and Theorems of the Z Transform

	$x(t)$ or $x(k)$	$Z\{x(t)\}$ or $Z\{x(k)\}$
1.	$ax(t)$	$aX(z)$
2.	$ax_1(t) + bx_2(t)$	$aX_1(z) + bX_2(z)$
3.	$x(t+T)$ or $x(k+1)$	$zX(z) - zx(0)$
4.	$x(t+2T)$	$z^2 X(z) - z^2 x(0) - zx(T)$
5.	$x(k+2)$	$z^2 X(z) - z^2 x(0) - zx(1)$
6.	$x(t+kT)$	$z^k X(z) - z^k x(0) - z^{k-1}x(T) - K - zx(kT-T)$
7.	$x(t-kT)$	$z^{-k} X(z)$
8.	$x(n+k)$	$z^k X(z) - z^k x(0) - z^{k-1}x(1) - K - zx(kT-1)$
9.	$x(n-k)$	$z^{-k} X(z)$
10.	$tx(t)$	$-Tz \dfrac{d}{dz} X(z)$

11.	$kx(k)$	$-z\dfrac{d}{dz}X(z)$
12.	$e^{-at}x(t)$	$X(ze^{aT})$
13.	$e^{-ak}x(k)$	$X(ze^{a})$
14.	$x(0)$	$\lim\limits_{z\to\infty} X(z)$ if the limit exists
15.	$x(\infty)$	$\lim\limits_{z\to 1}[(1-z^{-1})X(z)]$ if $(1-z^{-1})X(z)$ is analytic on and outside the unit circle
16.	$\nabla x(k) = x(k) - x(k-1)$	$(1-z^{-1})X(z)$
17.	$\Delta x(k) = x(k+1) - x(k)$	$(z-1)X(z) - zx(0)$
18.	$\sum\limits_{k=0}^{n} x(k)$	$\dfrac{1}{1-z^{-1}}X(z)$
19.	$\dfrac{\partial}{\partial a}x(t,a)$	$\dfrac{\partial}{\partial a}X(z,a)$
20.	$\sum\limits_{k=0}^{n} x(kT)y(nT-kT)$	$X(z)Y(z)$

Appendix 6

Gate Questions

1. Silicon-based rectifiers are preferred over germanium-based rectifiers because
   (A) Si is available easily compared to Ge
   (B) Only Si has a stable off state
   (C) Ge is very temperature sensitive
   (D) Si only has the characteristics $\alpha_1 + \alpha_2 < 1$ at low collector currents and reaches 1 at high currents

   Which of the above statements are true?
   (a) A, B, D
   (b) B, D
   (c) B only
   (d) D only
   (e) None of the above options

2. In a TRIAC,
   (A) The triggering pulse to main terminal 1 should be of the same polarity as the anode potential between MT1 and MT2
   (B) The triggering pulse should be of opposite polarity to that of the anode potential
   (C) When the triggering pulse is positive and the anode is positive, it is operating in the first quadrant
   (D) When the triggering pulse is negative and the anode is negative, its sensitivity is highest.

   Which of the above statements are true?
   (a) A, D
   (b) A, C

(c) C, D
(d) A, C, D
(e) None of the above options

3. In a single-phase full-wave SCR circuit with *RL* load
   (A) Power is delivered to the source for firing angle of less than 90°
   (B) The SCR changes from inverter to converter at $\alpha = 90°$
   (C) The negative DC voltage is maximum at $\alpha = 180°$
   (D) To turn off the SCR, the maximum delay angle must be less than 180°

   Which of the above statements are true?
   (a) C, D
   (b) C only
   (c) D only
   (d) A, B
   (e) None of the above options

4. While comparing TRIAC and SCR,
   (A) Both are unidirectional devices
   (B) TRIAC requires more current for turn on than SCR at particular voltages
   (C) A TRIAC has less time for turn off than SCR
   (D) Both are available with comparable voltage and current ratings

   Which of the above statements are true?
   (a) A, C
   (b) B, C
   (c) A, B
   (d) D only
   (e) None of the above options

5. Which of the following statements are true?
   (A) If the SCR, even with proper gate excitation and anode–cathode voltage does not conduct for a particular load resistance, then it would be necessary to decrease the load resistance to turn on the SCR
   (B) The SCR would be turned off by voltage reversal of the applied anode–cathode AC supply of frequencies up to 30 kHz
   (C) If the gate current of the SCR is increased, then the forward breakdown voltage will decrease
   (a) A, B, C
   (b) B, C
   (c) A, C

(d) A, B
(e) None of the above options
6. Which of the following statements are true?
  When gate triggering is employed, an SCR can withstand higher values of di/dt, if the
  (A) Gate current is increased
  (B) Rate of rise of gate current is increased
  (C) Gate current is increased
  (D) Rate of rise of gate current is decreased
   (a) C, D
   (b) A, D
   (c) B, C
   (d) A, B
   (e) None of the above options
7. In an SCR-based converter, the freewheeling diode is used to
  (A) Add to the conduction current of thyristors
  (B) Oppose the SCR conduction
  (C) Conduct current during the off period of the SCR
  (D) Protect the SCR by providing a shunt path
8. When an inductance is inserted in the load circuit of SCR
  (A) The turn on time of SCR is increased
  (B) Output voltage is reduced for the same firing angle
  (C) Conduction continues even after reversal of phase of input voltage
  (D) A freewheeling diode is connected in such circuits
  Which of the above statements are true?
   (a) A, D
   (b) B, C, D
   (c) A, B, C, D
   (d) C, D
9. Snubber circuit is used to limit the rate of
  (a) Rise of current
  (b) Conduction period
  (c) Rise of voltage across SCR
  (d) None of the above
10. Provision of a freewheeling diode across an inductive load is
  (a) To restore conduction angle on phase
  (b) To avoid negative reversal voltage drop

(c) To reduce the PRV
(d) None of the above

11. While working in series operation, equalizing circuits are added across each SCR to provide uniform
    (a) Current distribution
    (b) Firing of SCRs
    (c) Voltage distribution
    (d) None of the above

12. When the SCR conducts, the forward voltage drop
    (A) Is 0.7 V
    (B) Is 1–1.5 V
    (C) Increases slightly with load current
    (D) Remains constant with load current

    Which of the above statements are true?
    (a) A only
    (b) B, C
    (c) D only
    (d) A, C

13. The latching current of an SCR is 18 mA. Its holding current will be
    (a) 6 mA
    (b) 18 mA
    (c) 54 mA
    (d) 12 mA

14. The turn-off time is longer than turn-on time because
    (A) The anode and cathode junctions get reverse biased while gate junction is still forward biased
    (B) There is flow of reverse current
    (C) The gate pulse has been removed
    (D) The forward break over voltage is high

    Which of the above statements are true?
    (a) B, C
    (b) C, D
    (c) A, B
    (d) A, B, C, D

15. The thyristor will turn on faster with
    (A) Pulse signal applied to the gate terminal of the SCR
    (B) Continuous signal applied to the gate terminal of the SCR
    (C) Both A & B
    (D) Pulse signal but with minimum duration

    Which of the above statements are true?
    (a) A only
    (b) B only
    (c) A, D
    (d) None of the above

16. In an SCR
    (A) The holding current is less than latching current
    (B) The holding current is greater than latching current
    (C) The two currents are equal
    (D) The latching current is about three times the holding current

    Which of the above statements are true?
    (a) A only
    (b) D only
    (c) A, C, D
    (d) A, D
    (e) B only

17. When a positive voltage is applied to the gate of a reverse-biased SCR
    (A) It injects more electrons into junction J1
    (B) It increases reverse leakage current into anode
    (C) Heating of junction is unaffected
    (D) Failure of junctions occurs due to thermal runaway

    Which of the above statements are true?
    (a) B only
    (b) A, B, D
    (c) B, D
    (d) D only

18. During forward-blocking state, the SCR has
    (a) Low current, medium voltage
    (b) Low current, large voltage

(c) Medium current, large voltage

(d) Large current, low voltage

19. Once SCR starts conducting a forward current, its gate loses control over

   (a) Anode circuit current only

   (b) Anode circuit voltage only

   (c) Anode circuit voltage and current

   (d) Anode circuit voltage, current and time

20. In SCRs,

   (a) Both latching current and holding current are associated with turn-off process

   (b) Latching current is associated with turn-off process and holding current with turn-on process

   (c) Holding current is associated with turn-off process and latching current with turn-on process

   (d) Both latching current and holding current are associated with turn-on process

21. The SCR can be termed as

   (a) DC switch

   (b) AC switch

   (c) Square-wave switch

   (d) Either A or B

22. In a thyristor, the magnitude of anode current will

   (a) Increase if gate current is increased

   (b) Increase if gate current is decreased

   (c) Decrease if gate current is decreased

   (d) Not change with any variation in gate current

23. Turn-on time of an SCR can be reduced by using a

   (a) Rectangular pulse of high amplitude and narrow width

   (b) Rectangular pulse of low amplitude and wide width

   (c) Triangular pulse

   (d) Trapezoidal pulse

24. Turn-off time of an SCR in series with RL circuit can be reduced by

   (A) Increasing circuit resistance R

   (B) Decreasing circuit resistance R

   (C) Increasing circuit inductance L

   (D) Decreasing circuit inductance L

      (a) B, C

      (b) A, D

(c) B, D
(d) D only
25. A forward voltage can be applied to an SCR after its
    (a) Anode current reduces to zero
    (b) Gate recovery time
    (c) Reverse recovery time
    (d) Anode voltage reduces to zero
26. Turn-off time of an SCR is measured from the instant
    (a) Anode current becomes zero
    (b) Anode voltage becomes zero
    (c) Anode voltage and anode current become zero at the same time
    (d) Gate current becomes zero
27. In an SCR, anode current flows over a narrow region near the gate during
    (a) Delay time t_d
    (b) Rise time t_r and spread time t_p
    (c) t_d and t_p
    (d) t_d and t_r
28. Gate characteristic of a thyristor
    (a) Is a straight line passing through origin
    (b) Is of the type $V_g = a + bI_g$
    (c) Is a curve between V_g and I_g
    (d) Has a spread between two curves of V_g–I_g
29. Surge current rating of an SCR specifies the maximum
    (a) Repetitive current with sine wave
    (b) Nonrepetitive current with rectangular wave
    (c) Nonrepetitive current with sine wave
    (d) Repetitive current with rectangular wave
30. The di/dt rating of an SCR is specified for its
    (a) Decaying anode current
    (b) Decaying gate current
    (c) Rising gate current
    (d) Rising anode current
31. For an SCR, dv/dt protection is achieved through the use of
    (a) RL in series with SCR
    (b) RC across SCR

(c) *L* in series with SCR
(d) *RC* in series with SCR

32. For an SCR, *di/dt* protection is achieved through the use of
    (a) *R* in series with SCR
    (b) *RL* in series with SCR
    (c) *L* in series with SCR
    (d) *L* across SCR

33. When a line-commutated converter operates in the inverter mode
    (a) It draws both real and reactive power from the AC supply
    (b) It delivers both real and reactive power to the AC supply
    (c) It delivers real power to the AC supply
    (d) It draws reactive power from the AC supply

34. A chopper operating at a fixed frequency is feeding an *RL* load. As the duty ratio of the chopper is increased from 25% to 75%, the ripple in the load current
    (a) Remains constant
    (b) Decreases, reaches a minimum at 50% duty ratio and then increases
    (c) Increases, reaches a maximum at 50% duty ratio and then decreases
    (d) Keeps on increasing as the duty ratio is increased

35. To turn off an SCR, the reverse bias should be applied for a period .... the turn-off time of the SCR
    (a) Equal to
    (b) Longer than
    (c) Less than
    (d) Irrespective of

36. In class A and class B commutation, the resonating circuit has to be
    (a) Overdamped
    (b) Critically damped
    (c) Underdamped
    (d) Negatively damped

37. In phase-controlled rectification, power factor (PF)
    (a) Remains unaffected
    (b) Improves with increase of firing angle α
    (c) Deteriorates with increase of α
    (d) Is unrelated to α

38. Comparing the four-diode rectifier with the full-wave rectifier using two diodes, the four-diode bridge rectifier has the dominant advantage of
    (a) Higher current carrying
    (b) Lower peak inverse voltage requirement
    (c) Lower ripple factor
    (d) Higher efficiency
39. An SCR is rated at 75 A peak, 20 A average. The greatest possible delay in the trigger angle if the DC is at rated value is
    (a) 47.5°
    (b) 30°–45°
    (c) 74.5°
    (d) 137°
40. The applied sine voltage to an SCR is $V_m = 200$ V and $R = 10\ \Omega$. If the gate trigger lags the AC supply by 120°, the average load current is
    (a) $15/\pi$ A
    (b) $5/\pi$ A
    (c) $-5/\pi$ A
    (d) $-15/\pi$ A
41. A sine voltage of 200 V rms at 50 Hz is applied to an SCR through a 100 Ω resistor. The firing angle is 60°. Consider no voltage drop. The output voltage in rms is
    (a) 89.7 V
    (b) 126.7 V
    (c) 166.7 V
    (d) $200\sqrt{2}$ V
42. A 100 V DC is applied to the inductive load through an SCR. The SCR's specified latching current is 100 mA. The minimum required width of gating pulse to turn on the SCR is
    (a) 100 μS
    (b) 100 S
    (c) 1 mS
    (d) 50 μS
43. A cycloconverter is a
    (a) Frequency changer from higher to lower frequency with one-state conversion
    (b) Frequency changer from higher to lower frequency with two-stage conversion
    (c) Frequency changer from lower to higher frequency with one-stage conversion
    (d) Either a or c

44. The cycloconverter requires natural or forced commutation as
    (a) Natural commutation in both step-up and step down cycloconverter
    (b) Forced commutation in both step-up and step-down cycloconverter
    (c) Forced commutation in step-up cycloconverter
    (d) Forced commutation in step-down cycloconverter
45. In the synchronized UJT triggering of an SCR, voltage VC across capacitor reaches UJT threshold thrice in each half-cycle so that there are three firing pulses during each half-cycle. The firing angle of the SCR can be controlled
    (a) Once in each half-cycle
    (b) Thrice in each half-cycle
    (c) Twice in each half-cycle
    (d) Four times in each half-cycle
    (e) None of the above
46. In a GTO, anode current begins to fall when gate current
    (a) Is negative peak at time $t = 0$
    (b) Is negative peak at $t =$ storage period t_s
    (c) Just begins to become negative at $t = 0$
    (d) Is negative peak at $t = (t_s +$ fall time)
47. The SCR can be turned on by
    (A) Applying anode voltage at a sufficiently fast rate
    (B) Applying sufficiently large anode voltage
    (C) Increasing the temperature of SCR to a sufficiently large value
    (D) Applying sufficiently large gate current
        (a) A, B
        (b) C, D
        (c) B, C
        (d) A, B, C, D
        (e) None of the above options
48. In an SCR,
    (a) Gate current is directly proportional to forward breakover voltage
    (b) As gate current is raised, forward breakover voltage reduces
    (c) Gate current has to be kept on continuously for conduction
    (d) Forward breakover voltage is low in the forward blocking state
49. In an SCR, at switch on, there is a sudden increase of current, then
    (A) The tolerance limit of di/dt varies from 3 to 30 A/μs
    (B) di/dt upper limit is 3 A/μs

(C) *di/dt* rating can be increased by having an inductive load
(D) *di/dt* rating can be increased by a resistive load

Which of the above statements are true?
(a) A, C
(b) C, D
(c) A, B, C, D
(d) A, B

50. The following PNPN device has a terminal for synchronizing purpose
    (a) SUS
    (b) DIAC
    (c) TRIAC
    (d) Schmitt trigger

51. In a single-phase full-wave-controlled rectifier, maximum output voltage is obtained at conduction angle .... and minimum at conduction angle ...
    (a) 0°, 180°
    (b) 180°, 0°
    (c) 0°, 0°
    (d) 180°, 180°

52. A thyristor is triggered by a pulse train of 5 kHz. The duty ratio is 0.4. If the allowable average power is 100 W, the maximum allowable gate drive power is
    (a) $100\sqrt{2}$ W
    (b) 50 W
    (c) 150 W
    (d) 250 W

53. The consequence of introducing inductance in the load circuit of SCR is
    (A) Anode current through SCR rises slowly
    (B) Anode current reached is more than without it
    (C) SCR has to be derated for anode current
    (D) SCR's current rating can be raised

    Which of the above statements are true?
    (a) A, C
    (b) A, B, C, D
    (c) B, D
    (d) A, D

54. When an SCR is in the forward-blocking state
    (a) All the three junctions are reverse biased
    (b) The anode and cathode junctions are forward biased but the gate junction is reverse biased
    (c) The anode junction is forward biased but the cathode and gate junctions are reverse biased
    (d) The anode and gate junctions are forward biased but the cathode junction is reverse biased
55. When we need to drive a DC shunt motor at different speeds in both directions (forward and reverse) and also to brake it in both directions, which one of the following would you use?
    (a) A half-controlled SCR bridge
    (b) A full-controlled thyristor bridge
    (c) A dual converter
    (d) A diode bridge
56. The frequency of the ripple present in the output voltage of the three-phase half-controlled bridge rectifier depends on the
    (a) Firing angle
    (b) Load inductance
    (c) Load resistance
    (d) Supply frequency
57. A fully-controlled thyristor bridge drives a DC motor. The system is capable of
    (a) Motoring and braking in both directions
    (b) Only motoring in both directions, no braking
    (c) Motoring in forward direction and braking in reverse direction
    (d) Only motoring in forward direction, no braking
58. Three-phase to three-phase cycloconverters employing 18 SCRs and 36 SCRs have the same voltage and current ratings for their component thyristors. The ratio of VA rating of 36 SCR device to that of 18 SCR device is
    (a) 1/2
    (b) 1
    (c) 2
    (d) 4
59. Three-phase to three-phase cycloconverters employing 18 SCRs and 36 SCRs have the same voltage and current ratings for their component thyristors. The ratio of power handled by the 36 SCR device to that handled by the 18 SCR device is
    (a) 4
    (b) 2

(c) 1
(d) 1/2

60. The number of SCRs required for single-phase to single-phase cycloconverter of the midpoint type and for three-phase to three-phase type cycloconverter are respectively
    (a) 4, 6
    (b) 8, 18
    (c) 4, 18
    (d) 4, 36

61. For the dynamic equalizing circuit used for series-connected SCRs, the choice of C is based on
    (A) Reverse recovery characteristics
    (B) Turn-on characteristics
    (C) Turn-off characteristics
    (D) Rise-time characteristics
        (a) A
        (b) B, C
        (c) D
        (d) A, B, C, D
        (e) None of the above options

62. The function of the Zener diode in an UJT circuit, used for the triggering of SCRs, is to
    (A) Expedite the generation of triggering pulses
    (B) Delay the generation of triggering pulses
    (C) Provide a constant voltage to UJT to prevent erratic firing
    (D) Provide a variable voltage to UJT as the source voltage changes
        (a) B, C
        (b) C
        (c) A
        (d) D
        (e) None of the above options

63. SCR is the solid state equivalent of
    (a) Transistor
    (b) Thyratron
    (c) Vacuum diode
    (d) Crystal diode

64. With gate open, if the supply voltage exceeds the breakover voltage of SCR, then SCR will conduct
    (a) False
    (b) True
    (c) For DC
    (d) For AC

65. The SCR is turned off when the anode current falls below
    (a) Forward current rating
    (b) Break over voltage
    (c) Holding current
    (d) Latching current

66. In an SCR circuit, the angle of conduction can be changed by changing
    (a) Anode voltage
    (b) Anode current
    (c) Forward current rating
    (d) Gate current

67. The normal way to close an SCR is by appropriate
    (a) Gate current
    (b) Cathode current
    (c) Anode current
    (d) Forward current

68. An SCR has ...... PN junctions
    (a) Two
    (b) Four
    (c) Three
    (d) One

69. If gate current is increased, the anode–cathode voltage at which SCR closes is.....
    (a) Increased
    (b) Decreased
    (c) Maximum
    (d) Least

70. A conducting SCR can be opened by reducing ..... to zero
    (a) Supply voltage
    (b) Grid voltage
    (c) Grid current
    (d) Anode current

71. With gate open, an SCR can be turned on by making supply voltage....
    (a) Minimum
    (b) Reverse
    (c) Equal to cathode voltage
    (d) Equal to breakover voltage

72. If firing angle in an SCR rectifier is increased, output is....
    (a) Increased
    (b) Maximum
    (c) Decreased
    (d) Unaffected

73. An SCR is a ... switch
    (a) Two directional
    (b) Unidirectional
    (c) Three directional
    (d) Four directional

74. The anode of SCR is always maintained at ... potential with respect to cathode
    (a) Positive
    (b) Zero
    (c) Negative
    (d) Equipotential

75. If the chopper switching frequency is 200 Hz and T_{on} time is 2 ms, the duty cycle is
    (a) 0.4
    (b) 0.8
    (c) 0.6
    (d) None of the above

76. Speed control of induction motor can be effected by
    (a) Varying flux
    (b) Varying voltage input to stator
    (c) Keeping rotor coil open
    (d) None of the above

77. In a DC motor, if the field coil gets opened, the speed of the motor will
    (a) Decrease
    (b) Come to stop
    (c) Increase
    (d) None of the above

78. The reverse recovery time of the diode is defined as the time between the instant the diode current becomes zero and the instant reverse recovery current decays to
    (a) Zero
    (b) 10% of the reverse peak current (I_{RM})
    (c) 25% of (I_{RM})
    (d) 15% of (I_{RM})
79. The cut-in voltage and forward-voltage drop of the diode are respectively
    (a) 0.7 V, 0.7 V
    (b) 0.7 V, 1 V
    (c) 0.7 V, 0.6 V
    (d) 1 V, 0.7 V
80. The softness factor for soft-recovery and fast-recovery diodes are respectively
    (a) 1, >1
    (b) <1, 1
    (c) 1, 1
    (d) 1, <1
81. Reverse recovery current in a diode depends on
    (a) Forward field current
    (b) Storage charge
    (c) Temperature
    (d) PIV
82. The three terminals of power MOSFET are
    (a) Collector, emitter, base
    (b) Drain, source, base
    (c) Drain, source, gate
    (d) Collector, emitter, gate
83. The three terminals of IGBT are
    (a) Collector, emitter, base
    (b) Drain, source, base
    (c) Anode, cathode, gate
    (d) Collector, emitter, gate
84. The three terminals of MCT are
    (a) Anode, cathode, gate
    (b) Collector, emitter, gate
    (c) Drain, source, base
    (d) Drain, source, gate

85. Compared to power MOSFET, the power BJT has
    (a) Lower switching losses but higher conduction loss
    (b) Higher switching losses and higher conduction loss
    (c) Higher switching losses but lower conduction loss
    (d) Lower switching losses and lower conduction loss
86. Which one of the following statement is true
    (a) MOSFET has positive temperature coefficient whereas BJT has negative temperature coefficient
    (b) Both MOSFET and BJT have positive temperature coefficient
    (c) Both MOSFET and BJT have negative temperature coefficient
    (d) MOSFET has negative temperature coefficient whereas BJT has positive temperature coefficient
87. Which one of the following statement is true
    (a) Both MOSFET and BJT are voltage-controlled devices
    (b) Both MOSFET and BJT are current-controlled devices
    (c) MOSFET is a voltage-controlled device and BJT is current-controlled device
    (d) MOSFET is a current-controlled device and BJT is voltage-controlled device
88. Practical way of getting static voltage equalization in series-connected SCRs is by the use of
    (A) One resistor across the string
    (B) Resistors of different values across each SCR
    (C) Resistors of the same value across each SCR
    (D) One resistor in series with the string
        (a) A and D
        (b) A
        (c) C
        (d) B
        (e) None of the above options
89. For series-connected SCRs, dynamic equalizing circuit consists of
    (A) Resistor R and capacitor C in series but with a diode D across C
    (B) Series R and D circuit but with C across R
    (C) Series R and C circuit but with D across R
    (D) Series C and D circuit but with R across C
        (a) A
        (b) B
        (c) C

(d) D
   (e) None of the above options
90. During forward blocking of two series-connected SCRs, a thyristor with
   (A) High leakage impedance shares lower voltage
   (B) High leakage impedance shares higher voltage
   (C) Low leakage impedance shares higher voltage
   (D) Low leakage impedance shares lower voltage
   (a) B
   (b) D
   (c) B, D
   (d) A, C
   (e) None of the above options
91. Thyristor A has rated gate current of 1 A and thyristor B, a rated gate current of 100 mA
   (A) A is a GTO and B is a conventional SCR
   (B) B is a GTO and A is a conventional SCR
   (C) A may operate as a transistor
   (D) B may operate as a transistor
   (a) A
   (b) B
   (c) C
   (d) A, C
   (e) B, D
92. A resistor connected across the gate and cathode of an SCR increases its
   (A) dv/dt rating
   (B) Holding current
   (C) Noise immunity
   (D) Turn-off time
   (a) A, C
   (b) A, B
   (c) B, C
   (d) A, B, C
   (e) D
93. An SCR is considered to be a semicontrolled device because
   (a) It can be turned off but not on with a gate pulse
   (b) It conducts only during one half-cycle of an alternating current wave

(c) It can be turned on but not off with a gate pulse
(d) It can be turned on only during one half-cycle of an alternating voltage wave

94. The current source inverter shown in the following figure is operated by alternatively turning on thyristor pairs (T_1, T_2) and (T_3, T_4). If the load is purely resistive, the theoretical maximum output frequency obtainable will be

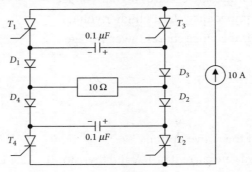

(a) 125 kHz
(b) 250 kHz
(c) 500 kHz
(d) 50 kHz

95. In the chopper circuit shown, the main thyristor (T_M) is operated at a duty ratio of 0.8, which is much larger than the communication interval. If the maximum allowable reapplied dv/dt on T_M is 50 V/μs, what should be the theoretical minimum value of C_1? Assume current ripple through L_0 to be negligible.

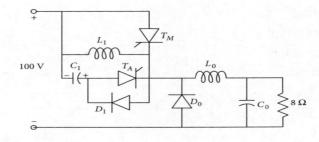

(a) 0.2 μF
(b) 0.02 μF
(c) 2 μF
(d) 20 μF

96. Match the switch arrangements on the top row to the steady-state V–I characteristics on the lower row. The steady state operating points are shown by large black dots.

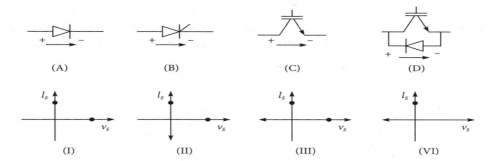

(a) A-I, B-II, C-III, D-IV
(b) A-II, B-IV, C-I, D-III
(c) A-I, B-II, C-III, D-II
(d) A-IV, B-III, C-II, D-I

97. Six MOSFETs connected in a bridge configuration (having no other power device) MUST be operated as a voltage source inverter (VSI). This statement is
    (a) True, because being majority carrier devices, MOSFETs are voltage driven
    (b) True, because MOSFETs have inherently antiparallel diodes
    (c) False, because it can be operated both as current source inverter (CSI) or a VSI
    (d) False, because MOSFETs can be operated as excellent constant current sources in the saturation region

98. A single-phase full-wave half-controlled bridge converter feeds an inductive load. The two SCRs in the converter are connected to a common DC bus. The converter has to have a freewheeling diode
    (a) Because the converter inherently does not provide for freewheeling
    (b) Because the converter does not provide for freewheeling for high values of triggering angles
    (c) Or the freewheeling action of the converter will cause shorting of the AC supply
    (d) Or if a gate pulse to one of the SCRs is missed, it will subsequently cause a high load current in the other SCR

99. A single-phase fully controlled thyristor bridge AC–DC converter is operating at a firing angle of 25°, and an overlap angle 10° with constant DC output current of 20 A. The fundamental power factor (displacement factor) at input AC mains is
    (a) 0.78
    (b) 0.827
    (c) 0.866
    (d) 0.9

100. A three-phase fully controlled thyristor bridge converter is used as a line-commutated inverter to feed 50 kW power at 420 V DC to a three-phase 415 V (line), 50 Hz mains. Consider DC link current to be constant. The rms current of the thyristor is
    (a) 119.05 A
    (b) 79.37 A
    (c) 68.73 A
    (d) 39.68 A

101. $G_c(s)$ is a lead compensator if
    (a) $a = 1, b = 2$
    (b) $a = 3, b = 2$
    (c) $a = -3, b = -1$
    (d) $a = 3, b = 1$

102. The phase of the lead compensator in Problem 101 is maximum at
    (a) $\sqrt{2}$ rad/s
    (b) $\sqrt{3}$ rad/s
    (c) $\sqrt{6}$ rad/s
    (d) $\sqrt{1/3}$ rad/s

103. A three-phase to single-phase conversion device employs a six-pulse bridge cycloconverter. For an input voltage of 200 V/phase, the fundamental rms value of output voltage is
    (a) $600/\pi$ V
    (b) $300\sqrt{3}/\pi$ V
    (c) $300/\pi$ V
    (d) $600\sqrt{3}/\pi$ V

104. A three-phase to single-phase cycloconverter consists of positive and negative group of converters. In this device, one of the two component converters would operate as a
    (A) Rectifier if the output voltage VO and output current IO have the same polarity
    (B) Inverter if VO and IO have the same polarity
    (C) Rectifier if VO and IO are of opposite polarity
    (D) Inverter if VO and IO are of opposite polarity
        (a) A, D
        (b) A
        (c) D
        (d) C, D

105. A three-phase to three-phase cycloconverter requires
    (A) 18 SCRs for a three-pulse device
    (B) 18 SCRs for a six-pulse device

(C) 36 SCRs for a three-pulse a device
(D) 36 SCRs for a six-pulse device
   (a) A
   (b) A, D
   (c) B, D
   (d) D

106. Which of the following statements are correct for a cycloconverter?
   (A) Step-down cycloconverter works on natural commutation
   (B) Step-up cycloconverter requires forced commutation
   (C) Load-commutated cycloconverter works on line commutation
   (D) Load-commutated cycloconverter requires a generated emf in the load circuit
      (a) B, D
      (b) A, B, C
      (c) B, C
      (d) A, B, D

107. The functions of a resistor connected in series with a gate-cathode circuit and a Zener diode across the gate-cathode circuit are, respectively, to protect the gate circuit from
   (a) Over voltages, over currents
   (b) Over currents, over voltages
   (c) Over currents, noise signals
   (d) Noise signals, over voltages

108. Thermal runaway is not possible in FET because as the temperature of FET increases
   (a) The mobility decreases
   (b) The transconductance increases
   (c) The drain current increases
   (d) None of the above

109. The ripple frequency from a full-wave rectifier is
   (a) Twice that from a half-wave circuit
   (b) The same as that from a half-wave circuit
   (c) Half that from a half-wave circuit
   (d) 1/4 that from a half-wave circuit

110. In a full-wave rectifier using two ideal diodes, V_{DC} and Vm are the DC and peak values of the voltage, respectively, across a resistive load. If PIV is the peak inverse voltage of the diode, then the appropriate relationships for the rectifier is
   (a) $V_{DC} = Vm/\pi$, PIV $= 2Vm$
   (b) $V_{DC} = 2Vm/\pi$, PIV $= 2Vm$

(c) $V_{DC} = 2Vm/\pi$, PIV = Vm
(d) $V_{DC} = Vm/\pi$, PIV = Vm

111. The output of a rectifier circuit without filter is
    (a) 50 Hz AC
    (b) Smooth DC
    (c) Pulsating DC
    (d) 60 Hz AC

112. An advantage of full-wave bridge rectification is
    (a) It uses the whole transformer secondary for the entire AC input cycle
    (b) It costs less than other rectifier types
    (c) It cuts off half of the AC wave cycle
    (d) It never needs a filter

113. The best rectifier circuit for the power supply designed to provide high power at low voltage is
    (a) Half-wave arrangement
    (b) Full-wave, centre tap arrangement
    (c) Quarter-wave arrangement
    (d) Voltage doubler arrangement

114. If a half-wave rectifier is used with 165 Vpk AC input, the effective DC output voltage is
    (a) Considerably less than 165 V
    (b) Slightly less than 165 V
    (c) Exactly 165 V
    (d) Slightly more than 165 V

115. If a full-wave bridge circuit is used with a transformer whose secondary provides 50 V rms, the peak voltage that occurs across the diodes in the reverse direction is approximately
    (a) 50 Vpk
    (b) 70 Vpk
    (c) 100 Vpk
    (d) 140 Vpk

116. The main disadvantage of voltage doubler power supply circuit is
    (a) Excessive current
    (b) Excessive voltage
    (c) Insufficient rectification
    (d) Poor regulation under heavy loads

117. A source follower using a FET usually has a voltage gain which is
    (a) Greater than +100
    (b) Slightly less than unity but positive
    (c) Exactly unity but negative
    (d) About −10
118. A metal oxide varistor (MOV) is used for protecting
    (A) Gate circuit against overcurrents
    (B) Gate circuit against overvoltages
    (C) Anode circuit against overcurrents
    (D) Anode circuit against overvoltages
        (a) A
        (b) A, B
        (c) C, D
        (d) D
        (e) None of the above options
119. In a dual converter, the circulating current
    (a) Allows smooth reversal of load current, but increases the response time
    (b) Does not allow smooth reversal of load current, but reduces the response time
    (c) Allows smooth reversal of load current with improved speed of load response
    (d) Follows only if there is no interconnecting inductor
120. The snubber circuit connected across an SCR is to
    (a) Suppress dv/dt
    (b) Increase dv/dt
    (c) Decrease dv/dt
    (d) Keep transient overvoltage at a constant value
121. The object of connecting resistance and capacitance across gate circuit is to protect the SCR gate against
    (a) Overvoltages
    (b) dv/dt
    (c) Noise signals
    (d) Over currents
122. Which of the following statements are correct?
    (A) Thyristor is current-driven device
    (B) GTO is current-driven device
    (C) GTR is current-driven device
    (D) SCR is a pulse-triggered device

(a) A and B
(b) A, B, C
(c) All
(d) D only

123. Which of the following statements are correct?
    (A) GTO is a pulse-triggered device
    (B) MOSFET is unipolar device
    (C) SCR is a bipolar device
    (D) Continuous gate signal is not required to maintain the SCR to be in on state
    (a) A, B, D only
    (b) A, B only
    (c) D only
    (d) All

124. Which of the following is not a fully controlled semiconductor device?
    (a) MOSFET
    (b) IGBT
    (c) IGCT
    (d) SCR

125. Which of the following is not associated with p–n junction
    (a) Junction capacitance
    (b) Charge storage capacitance
    (c) Depletion capacitance
    (d) Channel length modulation

126. In a p–n junction diode under reverse bias, the magnitude of the electric field is maximum at
    (a) The edge of the depletion region on the p side
    (b) The edge of the depletion region on the n side
    (c) The p–n junction
    (d) The centre of the depletion region on the n-side

127. An n-channel JFET has IDSS = 2 mA, and $V_p = -4$ V. Its transconductance gm = 2 (in mA/V) for an applied gate to source voltage $V_{GS} = -2$ V is
    (a) 0.25
    (b) 0.5
    (c) 0.75
    (d) 1

128. The MOSFET switch in its on state may be considered equivalent to
    (a) Resistor
    (b) Inductor
    (c) Capacitor
    (d) Battery
129. The effective channel length of a MOSFET in a saturation decreases with increase in
    (a) Gate voltage
    (b) Drain voltage
    (c) Source voltage
    (d) Body voltage
130. The early effect in a bipolar junction transistor is caused by
    (a) Fast turn on
    (b) Fast turn off
    (c) Large collector–base reverse bias
    (d) Large emitter–base forward bias
131. MOSFET can be used as a
    (a) Current-controlled capacitor
    (b) Voltage-controlled capacitor
    (c) Current-controlled inductor
    (d) Voltage-controlled inductors
132. The number of p–n junctions in an SCR is
    (a) 1
    (b) 2
    (c) 3
    (d) 4
133. In the SCR, when the anode terminal is positive with respect to the cathode terminal, the number of blocked p–n junctions is
    (a) 1
    (b) 2
    (c) 3
    (d) 4
134. In the SCR, when the cathode terminal is positive with respect to the anode terminal, the number of blocked p–n junctions is
    (a) 1
    (b) 2
    (c) 3
    (d) 4

135. In the SCR, the anode current is made up of
    (a) Electrons only
    (b) Electrons or holes
    (c) Electrons and holes
    (d) Holes only

136. When the SCR is triggered, it will change from the forward-blocking state to the conduction state if its anode to cathode voltage is equal to
    (a) Peak repetitive off-state forward voltage
    (b) Peak working off-state forward voltage
    (c) Peak working off-state reverse voltage
    (d) Peak nonrepetitive off-state forward voltage

137. Which of the following statements are true about VI characteristic of SCR?
    (A) Holding current is more than latching current
    (B) SCR will be triggered if the applied voltage exceeds forward break over voltage
    (C) SCR can be triggered without gate current
    (D) When the SCR is in reverse biased, small leakage current will flow
        (a) A, B, and C
        (b) All are true
        (c) B, C, D
        (d) C, D

138. Which of the following statements are true about BJT?
    (A) It has more power handling capability than MOSFET
    (B) Has higher switching speed than IGBT and MOSFET
    (C) Has low on state conduction resistance
    (D) Has second breakdown voltage problem
        (a) All are true
        (b) A, B, C, D
        (c) A, C, D
        (d) B, C, D

139. For a JFET, when VDS is increased beyond the pinch off voltage, the drain current
    (a) Increases
    (b) Decreases
    (c) Remains constant
    (d) First decreases and then increases

140. n-channel FETs are superior to p-channel FETs, because
    (a) They have higher input impedance
    (b) They have high switching time
    (c) They consume less power
    (d). Mobility of electrons is greater than that of holes
141. Which of the following is true about the diodes
    (A) During forward bias, small amount of voltage drop will appear across the anode and the cathode
    (B) If the reverse voltage exceeds V_{RRM}, the diode will be destroyed
    (C) t_{rr} depends on the softness factor
    (D) Schottky diodes have low t_{rr}
        (a) All are true
        (b) A, B, D
        (c) A, B, C
        (d) B, C, D
142. The MOSFET has
    (A) Higher power handling capability than BJT
    (B) Faster switching speed than BJT
    (C) High on state resistance
    (D) Secondary breakdown voltage problem
        Which of the above statements are incorrect?
        (a) A, C, D
        (b) B, C
        (c) All of the above
        (d) B, C, D
143. Which of the following is called an uncontrolled semiconductor device?
    (a) Diode
    (b) Thyristor
    (c) GTO
    (d) MOSFET
144. Which of the following is a half-controlled semiconductor device?
    (a) MOSFET
    (b) GTO
    (c) MCT
    (d) SCR

145. Which of the following abbreviation is not a power semiconductor device?
    (A) SIT
    (B) SITH
    (C) MCT
    (D) IGCT
        (a) A and D
        (b) A only
        (c) A, B, D
        (d) All are power semiconductor device
146. Which of the following statements are correct?
    (A) IGBT is A current-driven device
    (B) IGCT is A voltage-driven device
    (C) MOSFET is A voltage-driven device
    (D) GTO is A minority carrier device
        (a) A, B, C
        (b) B, C, D
        (c) All are correct
        (d) None are correct
147. The breakdown mechanism in a lightly doped p–n junction under reverse biased condition is called
    (a) Avalanche breakdown
    (b) Zener breakdown
    (c) Breakdown by tunnelling
    (d) High-voltage breakdown
148. For large values of |VDS|, a FET behaves as A
    (a) Voltage-controlled resistor
    (b) Current-controlled current source
    (c) Voltage-controlled current source
    (d) Current-controlled resistor
149. In a full-wave rectifier without filter, the ripple factor is
    (a) 0.482
    (b) 1.21
    (c) 1.79
    (d) 2.05
150. Space charge region around a p–n junction
    (a) Does not contain mobile carriers
    (b) Contains both free electrons and holes

(c) Contains one type of mobile carriers depending on the level of doping of the p or n regions

(d) Contains electrons only as free carriers

151. In a JFET, at pinch-off voltage applied on the gate
    (a) The drain current becomes almost zero
    (b) The drain current begins to decrease
    (c) The drain current is almost at saturation value
    (d) The drain to source voltage is close to zero volts

152. The value of the ripple factor of a half-wave rectifier without filter is approximately
    (a) 1.2
    (b) 0.2
    (c) 2.2
    (d) 2.0

153. The transformer utilization factor of a half-wave rectifier is approximately
    (a) 0.6
    (b) 0.3
    (c) 0.9
    (d) 1.1

154. Transistor is a
    (a) Current-controlled current device
    (b) Current-controlled voltage device
    (c) Voltage-controlled current device
    (d) Voltage-controlled voltage device

155. The output current I_c depends on the input current I_b. If the output voltage of a bridge rectifier is 100 V, the PIV of the diode will be
    (a) $100\sqrt{2}$ V
    (b) $200/\pi$ V
    (c) 100π V
    (d) $100/2$ V

156. Peak inverse voltage = maximum secondary voltage = $V_{DC} = 2Vm/\pi = 100\ Vm = 100\pi/2$. The thyristor can be brought to forward-conducting state with gate-circuit open when the applied voltage exceeds
    (a) The forward breakover voltage
    (b) Reverse breakdown voltage
    (c) 1.5 V
    (d) Peak nonrepetitive off-state voltage

157. In an SCR, the holding current is
    (a) More than latching current
    (b) Less than latching current
    (c) Equal to latching current
    (d) Very small
158. When an SCR gets turned on, the gate drive
    (a) Should not be removed as it will turn off the SCR
    (b) May or may not be removed
    (c) Should be removed
    (d) Should be removed to avoid increased losses and higher junction temperature
159. For the normal SCR, the turn on time is
    (a) Less than turn-off time
    (b) More than turn-off time
    (c) Equal to turn-off time
    (d) Half of turn-off time
160. The forward voltage drop during SCR on state is 1.5 V. This voltage drop
    (a) Remains constant and is independent of load current
    (b) Increases slightly with load current
    (c) Decrease slightly with load current
    (d) Varies linearly with load current
161. The average output voltage (V_{DC}) of the full-wave diode bridge rectifier is
    (a) $Vm/2$
    (b) $2Vm/\pi$
    (c) $3Vm/\pi$
    (d) $4Vm/\pi$
162. The typical ratio of latching current to holding current in a 20 A thyristor is
    (a) 5.0
    (b) 2.0
    (c) 1.0
    (d) 0.5
163. A half-controlled single-phase bridge rectifier is supplying an *RL* load. It is operated at a firing angle (alpha) and load current is continuous. The fraction of cycle that the freewheeling diode conducts is
    (a) 1/2
    (b) $[1 - (\alpha/\pi)]$
    (c) $\alpha/2\pi$
    (d) α/π

164. For an SCR with turn-on time of 5 μS, an ideal trigger pulse should have
   (a) Short rise time with pulse width = 3 μS
   (b) Long rise time with pulse width = 6 μS
   (c) Short rise time with pulse width = 6 μS
   (d) Long rise time with pulse width = 3 μS

165. Turn-on time for an SCR is 10 μS. If an inductance is inserted in the anode circuit, then the turn-on time will be
   (a) 10 μS
   (b) Less than 10 μS
   (c) More than 10 μS
   (d) About 10μS

166. On-state voltage drop across a thyristor used in a 250 V supply system is of the order of
   (a) 100–110 V
   (b) 240–250 V
   (c) 1–1.5 V
   (d) 0.5–1 V

167. In an SCR, ratio of latching current to holding current is
   (a) 0.4
   (b) 1.0
   (c) 2.5
   (d) 6.0

168. The fully controlled thyristor converter in the figure is fed from a single-phase source. When the firing angle is 0°, the DC output voltage of the converter is 300 V. What will be the output voltage for a firing angle of 60°, assuming continuous conduction?

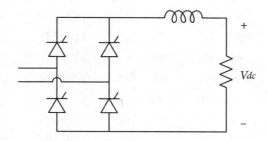

   (a) 150 V
   (b) 210 V
   (c) 300 V
   (d) 100π V

169. Circuit turn-off time of an SCR is defined as the time
    (a) Taken by the SCR to turn off
    (b) Required for SCR current to become zero
    (c) For which the SCR is reverse biased by the commutation circuit
    (d) For which the SCR is reverse biased to reduce its current below the holding current

Answers of Gate Questions

1. (b)	26. (a)	51. (a)	76. (b)	101. (a)	126. (c)	151. (c)
2. (c)	27. (d)	52. (d)	77. (c)	102. (a)	127. (b)	152. (d)
3. (a)	28. (d)	53. (d)	78. (c)	103. (a)	128. (c)	153. (b)
4. (b)	29. (c)	54. (b)	79. (c)	104. (a)	129. (b)	154. (a)
5. (c)	30. (d)	55. (c)	80. (d)	105. (b)	130. (c)	155. (d)
6. (a)	31. (b)	56. (d)	81. (a)	106. (d)	131. (b)	156. (a)
7. (c)	32. (c)	57. (d)	82. (c)	107. (b)	132. (c)	157. (b)
8. (c)	33. (c)	58. (c)	83. (d)	108. (a)	133. (a)	158. (d)
9. (c)	34. (a)	59. (a)	84. (a)	109. (a)	134. (b)	159. (a)
10. (a)	35. (b)	60. (c)	85. (c)	110. (b)	135. (c)	160. (b)
11. (c)	36. (c)	61. (a)	86. (a)	111. (c)	136. (b)	161. (b)
12. (b)	37. (c)	62. (b)	87. (c)	112. (a)	137. (c)	162. (b)
13. (a)	38. (b)	63. (b)	88. (c)	113. (b)	138. (c)	163. (d)
14. (c)	39. (b)	64. (b)	89. (c)	114. (a)	139. (c)	164. (c)
15. (c)	40. (a)	65. (c)	90. (c)	115. (a)	140. (c)	165. (c)
16. (d)	41. (b)	66. (d)	91. (b)	116. (d)	141. (c)	166. (c)
17. (c)	42. (a)	67. (a)	92. (b)	117. (a)	142. (a)	167. (c)
18. (b)	43. (d)	68. (c)	93. (c)	118. (d)	143. (a)	168. (a)
19. (c)	44. (c)	69. (b)	94. (c)	119. (c)	144. (d)	169. (c)
20. (c)	45. (a)	70. (a)	95. (a)	120. (a)	145. (d)	
21. (a)	46. (b)	71. (d)	96. (c)	121. (c)	146. (b)	
22. (d)	47. (d)	72. (c)	97. (d)	122. (c)	147. (a)	
23. (a)	48. (b)	73. (b)	98. (b)	123. (d)	148. (c)	
24. (d)	49. (a)	74. (a)	99. (a)	124. (d)	149. (a)	
25. (b)	50. (a)	75. (a)	100. (d)	125. (d)	150. (a)	

Resources for MATLAB

I. Information Sites

www.mathworks.in

www.colorado.edu

www.masteringmatlab.com

www.users.abo

www.tutorialspoint.com

www.ocw.mit.edu/courses

www.cyclismo.org/tutorial/matlab/

http://ctms.engin.umich.edu/

http://www.math.utah.edu/

http://www.math.mtu.edu/

http://www.mathtools.net/

www.matlabtutorials.com

II. Magazines with MATLAB Articles

1. Recktenwald, Gerald. 2001. *Introduction to Numerical Methods and MATLAB: Implementations and Applications.* Upper Saddle River, NJ: Prentice Hall.
2. Hanselman, Duane and Bruce R. Littlefield. 2000. *Mastering MATLAB 6.* Upper Saddle River, NJ: Prentice Hall.
3. Magrab, Edward B. and Shapour Azarm. 2000. *An Engineer's Guide to Matlab.* Upper Saddle River, NJ: Prentice Hall.

4. Etter, Delores M. 1996. *Introduction to MATLAB for Engineers and Scientists*. Upper Saddle River, NJ: Prentice Hall.
5. Pratap, Rudra. 2002. *Getting Started with MATLAB 5: A Quick Introduction for Scientists and Engineers*. New York: Oxford University Press.
6. King, Joe. 1988. *Matlab for 5.0 Engineers*. Boston: Addison-Wesley Pub Co.
7. Palm, William J. 2000. *Introduction to Matlab 6 for Engineers*. New York: McGraw-Hill.
8. Chapman, Stephen J. 1999. *MATLAB(r) Programming for Engineers*. Pacific Grove, CA: Brooks/Cole Pub Co.
9. Austin, Mark and David Chancogne. 1999. *Introduction to Engineering Programming in C, MATLAB and JAVA*. Hoboken, NJ: John Wiley & Sons.
10. Etter, Dolores, David Kuncicky and Doug Hull. 2002. *Introduction to Matlab 6*. Upper Saddle River, NJ : Prentice Hall.

A bibliography of Matlab-based books from the MathWorks.

1. MATLAB Programming Contest-MATLAB Central.
2. MATLAB Student Version—The Journal.
3. MATLAB: An Introduction | diGIT Magazine SriLanka.
4. www.admin-magazine.com/HPC/Articles/Matlab

III. MATLAB Journals Link

1. http://iosrjournals.org
2. http://asdfjournals.com
3. http://redwood.psych.cornell.edu

Index

120° Conduction Mode, 274
180° Conduction Mode, 273
1st Quadrant Forward Motoring, 394
2nd Quadrant Forward Braking, 374
3rd Quadrant Reverse Motoring, 395
4th Quadrant-Reverse Braking, 395

AC Analysis, 42
AC choppers, 239
AC Drives, 389, 404
AC Voltage Controllers and its Classification, 291
Active Elements, 40
Algorithm for ANN, 357
Angular period, 188
Applications of Choppers, 240
Applications of Controlled Rectifiers, 189
Applications of Inverter, 280
Applications of Power electronics, 63
Applications of Power Semiconductor Devices, 121
Applications of voltage controllers, 319
Arrays, 21
Assigning values to variables, 6
average load current, 157–158, 161–162, 164, 166, 170–171, 179, 181, 186
average load voltage, 222

Background of Neural Networks, 353
Based on number of phases, 153
Based on number of pulses, 153
Based on period of conduction, 152
Basic Arithmetic operations, 5
Basic features of MATLAB, 10
Basic Plotting, 24
Battery charger, 63, 153, 373, 382
Bipolar Junction Transistors, 111, 270
BLDC Motors, 384
Boost Converter, 228, 229, 231
Buck Converter, 226–227, 231, 347, 350–351
Buck-Boost Converter, 230
Building the model, 34

Center tap rectifier, 149–150
Characteristics of Fluorescent Lamps, 383
Characterization of Forced Commutation Methods, 97
Chopper Commutation, 231
Choppers and their Classification, 207
Choppers Configuration, 215
Circuit Descriptions, 33
Circuit Elements, 35, 39
Circulating Current Mode of Operation, 188
Circulating Mode, 313, 314
Class A: Self Commutated by a Resonating Load, 97

514 Index

Class B: Self Commutated by a *LC* Circuit, 97
Class C: C or L-C Switched by another Load Carrying SCR, 98
Class D: L-C or C Switched by an Auxiliary SCR, 98
Class E: External Pulse Source for Commutation, 99
Class F: AC Line Commutation, 97, 99
Classification based on association with other devices, 258
Classification as per the technique for substitution, 258
Classification based on output voltage, 257
classification based on input source, 257
Command History, 3, 4
Commutation angle, 189
Commutation techniques of thyristor, 96
Complex Functions, 11
Concatenating of matrices, 20
Continuous Conduction mode, 181–182
Control characteristics, 147
Control Strategies of chopper, 208
Control Structures and Operators, 28
Converter Control using Microprocessor and Microcontroller, 362
Creating m-files, 26
Creating Plots, 32
Cuk Converter, 231
Current Commutated Chopper, 234–235
Current derating, 92
Current folder, 3–4
Current Limit Control, 208–209
Current Source Inverters, 109, 257, 275
Customized start, 384
Cyclo converters and its types, 302

DC Analysis, 41
DC Chopper Drives, 402
DC Drives, 389–390, 398, 401
Debugging M-Files, 31
Defuzzification, 345–347, 350
Dependent Sources, 40
Derating, 92

Description and Design of FLC, 347
dI/dt protection, 84
Diac, 73, 76, 107
Dimmable ballast, 384
Discontinuous Conduction mode, 181–182
Distortion factor (DF), 265
Dual converter:, 187, 397–398
dV/dt triggering, 93–94

Efficiency, 91–92, 157, 357, 384
Electrical circuit description, 34
Electrical Drive, 125, 388–389
Electronic Ballast, 383
Emitter Turn-off Thyristor, 123
Evaluation, 28, 75, 83, 87, 195, 240
Exponential and Logarithmic Functions, 11

Fast Recovery Diode, 75–76
Features of power electronics, 61
Features of Step down and Step up choppers, 215
Fly-back Mode SMPS, 377–378
Forced Commutation, 96–97, 315
Form factor, 157
Format and Layout, 33
Forward Biasing, 69–70, 102
Forward Mode SMPS, 377–378
Forward Voltage Triggering, 93–94
Four Quadrant Operation, 207, 394, 401
Frequency Control, 208, 404, 406
Full wave Bridge Rectifier, 151, 163–167
Full Wave Rectifier, 147, 149, 152–153, 303–305
Fully controlled Rectifiers, 147, 152
Functions and Scripts, 27
Fuzzification, 348–348, 350
Fuzzy Logic Principles, 341
Fuzzy logic tool box:, 341

Gate characteristics, 82
Gate Triggereing Methods, 116
Gate triggering, 93, 95, 116, 270
Gate Turn-Off Thyristors, 102
GTO based converter, 47

Half controlled Rectifiers, 152, 178
Half Wave Rectifier, 147–148, 152, 396
Harmonic factor of nth Harmonic, 265
High Voltage DC Transmission, 281, 380
Holding current, 77, 82, 85, 96, 226
How to start with MATLAB, 2
Hybrid, 292, 373, 384
Hyperbolic Functions, 11

Implementation, 346, 355
Independent Sources, 40
Indexing of a matrix, 17
Induction motor drive, 404–405
Inference Method, 350
input power factor, 157
Installing and Activation, 3
Insulated-Gate Bipolar Transistor (IGBT), 61, 105, 108, 114, 270, 280, 290, 373, 406
Integrated Gate Commutated Thyristors, 61, 108
Inverse of a matrix, 18
Inverters and their Classification, 257
Investigation of a MATLAB function, 10

Jones Chopper, 237–238

Labeling and annotations in plotting, 25
LabVIEW, 52
Latching current, 82
Light Activated SCR, 95, 123
Light triggering, 93, 95
Line Interactive UPS, 373, 375–376
Load Commutated Chopper, 207, 236
Load Commutated Cycloconverter, 317
Lowest Order Harmonics (LOH), 266

M file functions, 27
Mathematical functions, 10, 12
MATLAB/Simulink model of semiconductor devices in Electronics, 109
MATLAB: A calculator, 4
Matrix converter, 292, 317–318
Matrix operation, 12, 16
maximum output voltage, 157–158, 161, 164, 166, 170–171, 175, 179

Maximum value of ripple current, 215
McMurray Inverter (Auxiliary Commutated Inverter), 266
Merits and demerits of Electrical Drive Systems, 389
Merits and demerits of MATLAB, 54
Minimum duty cycle of voltage commutated chopper, 232
Modified McMurray Full Bridge Inverter, 270
Modified McMurray Half Bridge and Full Bridge Inverter, 267
Modified McMurray Half Bridge Inverter, 269
Moment start, 383
Morgan Chopper, 238
MOS Controlled Thyristor, 108
MOSFET, 46, 65, 100
Multidimensional arrays, 23
Multiphase chopper, 240
Multiple Pulse Width Modulation, 271

Natural Commutation, 96, 317
Neural Network Principles, 352
Noncirculating mode, 313
Noncirculating Mode and Circulating Current Mode, 313
NonCirculating Current Mode of Operation, 188
NPN Transistors, 112
Numeric Functions, 12

Off-Line UPS, 373
On-line UPS, 374
ON-OFF Control, 291
One Pulse Converter, 115
Optocoupler, 119
Other Semiconductor devices, 107
Output ac power, 157
Output dc power, 157, 212

Parallel operation of thyristors, 87, 91
Parallel Resonant Converters, 279
Passive Elements, 40
Performance parameters of inverters, 265

Performance Parameters of Step up and Step down choppers, 213
Phase Control, 96, 153, 155, 291, 303
Phase Locked Loop (PLL), 411
PNP Transistors, 108, 112
Power diode and its chracteristics, 66
Power Electronic Converters, 63, 146
Power Electronic Systems, 64, 125
Power MOSFET, 100, 103, 106, 108, 270
Power Semiconductor Devices in MATLAB/Simulink, 64
Principle of operation of step down chopper configuration, 211
Principle of operation of Step up chopper configuration, 212
Programming with Matlab, 26
Protection of thyristor circuits, 83
PSIM, 52–53
PSpice, 52
Pulse Transformers, 119
Pulse Width Modulation or Constant Frequency System, 208
PWM Inverters, 257, 270, 404, 407–408

Quick start, 383–384

Ramp-Pedestrial Triggering, 120
Random Number Functions, 12
Rectification and its Classification, 147
Regulation of pulse generator using thyristor converter, 50
Regulation of Zener Diode, 49
Resistance Firing Circuit, 116
Resitance-Capacitance Firing Circuit, 116–117
Resonant Converters, 277, 279
Resonant Switch converter using MOSFET, 46
Reversal of DC motor, 401
Reverse biasing, 71–72, 102
Reverse Control Thyristors, 123
Reverse Recovery Characteristic, 67
RF Heating, 383
Ripple factor, 215
RMS load current, 299

RMS load voltage, 157, 159, 161–162, 165, 167, 170–171, 175, 179
Root Mean Square (rms), 149, 214, 261, 290
Rotor Voltage Control, 404–405

Schottky Diode, 109
Scilab, 52–53
Selection of components from simulink library browser, 258
Series and Parallel operation of thyristors, 87
Series operation of thyristors, 87
Series Resonant Converters, 277
Silicon Bilateral Switch, 123
Silicon Carbide Devices, 124
Silicon Controlled Switch, 123
Silicon Unilateral Switch, 123
Simulating the Model, 38
Simulink Features, 33
Simulink Library Browser, 34–35, 65, 68, 103, 105, 209, 258,
Single Phase-Single Phase Step down Cycloconverters, 303, 306
Single Phase AC Voltage Controllers, 292
Single Phase and Three phase DC Drive, 396, 398
Single phase Capacitor Commutated Current Source Inverter with R load, 275
Single Phase Cycloconverters, 303, 308
Single Phase Full Bridge Inverter with R Load, 262–263
Single Phase full-wave AC Voltage Controller with RL Load, 299–300
Single Phase Full-Wave Bridge Rectifier with RL Load, 165–166
Single Phase Full-Wave Bridge Rectifier with RLE Load, 167
Single Phase Half Bridge Inverter with R Load, 260, 263
Single Phase half-wave AC Voltage Controller with R Load, 293, 296–297
Single Phase Half-Wave Controlled Rectifier, 155, 157, 159, 161
Single Phase Half-Wave Controlled Rectifier with R Load, 155

Single Phase Half-Wave Controlled Rectifier with *RL* Load, 157
Single Phase Mid Point Bridge Rectifier, 169, 171–172
Single Phase Mid Point Bridge Rectifier with Freewheeling Diode, 172
Single Phase Mid Point Bridge Rectifier with *R* Load, 169
Single Phase Mid Point Bridge Rectifier with *RL* Load, 171
Single Phase Semi Converter Half Controlled Bridge Rectifier with *R* Load, 173
Single Phase Semi Converter Half Controlled Bridge Rectifier with *RLE* Load, 176
Single Phase Voltage Source Inverters, 260
Single Pulse Width Modulation, 271
Six pulse Converter, 183, 186
Six pulse Converter with *R* Load, 183
Six pulse Converter with *RL* Load, 186
Source Filter, 186
Start button, 3
Static Systems, 373
Statistical Functions, 12
Steady state Analysis of a Linear circuit, 45
Steady State operation of a Separately Excited DC motor, 390
String Efficiency, 91–92
String Functions, 12
Switch Mode Power Supply, 377
Switch mode welding, 383
Switching Mode Regulators, 226
Synchronous Motor drive, 404, 409

Temperature Triggering, 93–94
THD, 33–34, 210, 259, 265, 373
The minimum and maximum values of load current, 214
The Role of Source Inductance in Rectifier circuits (*Ls*), 188
Thermal Management and Heat sinks, 385
Three Phase-Single Phase Cyclo converters, 303, 308
Three Phase-Three Phase Cycloconverters, 303, 307
Three Phase bridge Inverter, 273–274
Three Phase Cycloconverters, 307, 312
Three Phase half controlled bridge rectifier with *RL* load, 180
Three Phase half-wave controlled rectifier, 178–180
Three Phase half-wave controlled rectifier with *R* Load, 178
Three phase induction fed drive, 408
Three Pulse Converter, 178
Thyristor turn on and turn off characteristics, 85
Thyristors, 76, 80, 87, 91, 99–100, 102, 108, 123, 152
Time Ratio Control, 208
To delete a row or a column, 20
Torque and Speed Control, 393
Total Harmonic distortion (THD), 33, 265
Total resistance, 43
Triac, 76, 107–108, 290–291, 301–302
Trigonometric Functions, 11
Turn on methods of SCR, 93
Two transistor model of thyristor, 92
Type A Chopper, 216
Type B Chopper, 217
Type C chopper (regenerative chopper), 220
Type D Chopper, 222, 224
Type E Chopper, 222, 224
Types of other Circuit Simulators, 52

UJT Firing Circuit, 116, 118
Uncontrolled rectifiers, 146–147, 396, 409
Uninterruptible Power Supply (UPS), 372–373, 407

V-I characterisitcs of SCR, 78
VAR Compensators, 380–382
Variable Frequency Control or Frequency Modulation, 208
Variable Speed Operation, 393, 404
Variables and functions in memory, 28

Vector and Matrix operations, 12
Vector operations, 12, 17
VisSim, 52, 53
Voltage Commutated Chopper, 231–234
Voltage derating, 92
Voltage Source Inverters, 257, 260
VSI Induction FED Drive, 407

Why is Power Electronics important, 61
Workspace, 3, 4, 21, 27–28, 31

Zener Diode, 49, 73, 75, 120, 123
Zero Current Switching, 280
Zero Voltage Switching, 280
ZVS and ZCS –PWM Converters, 280